TREATING INDUSTRIAL WASTE INTERFERENCES AT PUBLICLY-OWNED TREATMENT WORKS

TREATING INDUSTRIAL WASTE INTERFERENCES AT PUBLICLY-OWNED TREATMENT WORKS

by

Edward D. Wetzel
Scott B. Murphy

James M. Montgomery, Consulting Engineers, Inc.
Pasadena, California

NOYES DATA CORPORATION
Park Ridge, New Jersey, U.S.A.

Library of Congress Catalog Card Number: 90-24376
ISBN: 0-8155-1282-1
ISSN: 0090-516X
Printed in the United States

Published in the United States of America by
Noyes Data Corporation
Mill Road, Park Ridge, New Jersey 07656

10 9 8 7 6 5 4 3 2 1

Library of Congress Cataloging-in-Publication Data

Wetzel, Edward D.
Treating industrial waste interferences at publicly-owned treatment works / by Edward D. Wetzel, Scott B. Murphy.
p. cm. -- (Pollution technology review, ISSN 0090-516X ; no. 203)
Includes bibliographical references (p.) and index.
ISBN 0-8155-1282-1 :
1. Factory and trade waste--Purification. 2. Sewage disposal plants. I. Murphy, Scott B. II. Title. III. Title: Publicly -owned treatment works. IV. Series.
TD897.5.W47 1991
363.72'8--dc20 90-24376
CIP

Foreword

The discharge of heavy metals, toxic organics and variable strength conventional pollutants from industrial (and other nondomestic) sources can have negative impacts on the operation of publicly-owned treatment works (POTWs). Such industrial discharges can result in an interference at POTWs, recently redefined by the U.S. Environmental Protection Agency (EPA) to mean ". . . the causation of a POTW's noncompliance with its permit or inability to lawfully use or dispose of its sludge."

The EPA-funded studies in this book consider the sources of and contaminants causing interference, the impact on the POTW, and the mitigation techniques available for interference prevention. The intent is not to assess blame, but rather to identify cause and effect relationships and to seek techniques and provide guidance for the mitigation of upsets created by the introduction of abnormal contaminants into POTWs.

Part I of the book is a guidance manual which provides POTW operators with guidance on dealing with interferences caused by pollutants from nondomestic sources. In addition, some guidance is provided on distinguishing interferences caused by nondomestic discharges from problems resulting from poor operation and maintenance of the POTW. Part I is divided into three major sections, which correspond to the order in which a POTW should address interference. These sections are: Detecting Interference, Source Identification, and Mitigation.

The section on detecting interference is intended to help identify the types of interferences and substances which are known to cause problems. The way in which interference occurs in both the sewer collection system and the treatment plant is also discussed, along with analytical tests and monitoring that can be conducted by POTW operators. The second major section deals with the identification of the industrial sources of the interference-causing substances. Sources can be separated into chronic dischargers of industrial pollutants, isolated spill events, and hauled wastes. Identification techniques range from simple sensory observations to the use of sophisticated monitoring equipment for tracing problems at the POTW back to a source. The final section, on mitigation, discusses ways in which municipalities can cope with interference problems. In-plant control, source control and legal and enforcement remedies are addressed in the section. Operators should be cautioned that there are few straightforward solutions to these problems, and that often a combination of techniques will need to be employed to properly mitigate an interference. The section concludes with a discussion on planning, so that future industrial discharges will not interfere with successful POTW operations.

Part II of the book contains a review of the technical literature, a telephone survey of all state environmental agencies and identified municipalities, and data on site visits to nearly 30 POTWs that have been successful in their attempts to mitigate interference problems.

The information in the book is from the following documents:

Guidance Manual for Preventing Interference at POTWs, prepared by Edward Wetzel and Scott Murphy of James M. Montgomery, Consulting Engineers, Inc., for the U.S. Environmental Protection Agency, September 1987.

Interferences at Publicly Owned Treatment Works, prepared by Edward D. Wetzel and Scott B. Murphy of James M. Montgomery, Consulting Engineers, Inc. for the U.S. Environmental Protection Agency, September 1986.

The table of contents is organized in such a way as to serve as a subject index and provides easy access to the information contained in the book.

Advanced composition and production methods developed by Noyes Data Corporation are employed to bring this durably bound book to you in a minimum of time. Special techniques are used to close the gap between "manuscript" and "completed book." In order to keep the price of the book to a reasonable level, it has been partially reproduced by photo-offset directly from the original reports and the the cost saving passed on to the reader. Due to this method of publishing, certain portions of the book may be less legible than desired.

NOTICE

Contents and Subject Index

PART I

GUIDANCE MANUAL

1. INTRODUCTION 3
   1.1 Background 3
   1.2 Definition of Interference 4
   1.3 Guidance Manual Objectives 6

2. DETECTING INTERFERENCE 8
   2.1 Types of Interference 8
      2.1.1 Chronic Inhibition 10
      2.1.2 Upset Conditions 10
   2.2 Interference-Causing Substances 11
      2.2.1 Conventional Pollutants 13
      2.2.2 Metals and Other Inorganics 13
      2.2.3 Organic Compounds 14
   2.3 Sewer Collection System 14
   2.4 Plant Operations 16
      2.4.1 Observation 16
      2.4.2 Instrumentation 17
      2.4.3 Analytical Results 17
   2.5 Wastewater Monitoring 18
      2.5.1 POTW Influent 18
      2.5.2 Other POTW Locations 19
      2.5.3 Inhibitory Effects Testing 20

3. SOURCE INDENTIFICATION 31
   3.1 Chronic Discharges 32

3.1.1 Routine Monitoring 33
3.1.2 Tracking Program 34
3.2 Isolated Spills and Unauthorized Discharges 35
3.2.1 Hauled Wastes 37
3.3 Rapid Screening Techniques 38
3.4 Summary 39

4. MITIGATION 45
4.1 Treatment Plant Control 45
4.1.1 Biological Process Control 45
4.1.2 Biological Augmentation 47
4.1.3 Chemical Addition 47
Chlorine 47
Nutrients 48
pH Adjustment 48
Coagulants 48
Activated Carbon 48
4.1.4 Operations Modification 48
Activated Sludge Alternatives 48
Staged Treatment 49
Excess Biomass 49
4.1.5 Physical Modification 49
Flow Equalization 49
Instrumentation/Control 50
Special Treatment Operations 50
Pure Oxygen Activated Sludge 50
Oxygen Transfer 50
4.1.6 Summary 50
4.2 Pretreatment and Source Control 51
4.2.1 Local Limits 51
4.2.2 Accidental Spill Prevention 51
4.2.3 Pretreatment Facilities 52
4.2.4 Regulation of Waste Haulers 52
4.2.5 Planning for Future Sources 53
4.3 Legal and Enforcement Remedies 54
4.3.1 Penalties 55
4.3.2 Orders and Compliance Schedules 56
4.3.3 Litigation 56
4.3.4 Sewer Disconnection or Permit Revocation 56

REFERENCES 63

APPENDIX A: CASE STUDIES 65
Back River Wastewater Treatment Plant (Baltimore, MD) 68
Patapsco Wastewater Treatment Plant (Baltimore, MD) 72
Bayshore Regional Sewerage Authority (Union Beach, NJ) 76
East Side Sewage Treatment Plant (Oswego, NY) 80
Hamilton Township Wastewater Treatment Plant (Trenton, NJ) 84
Horse Creek Pollution Control Facility (North Augusta, SC) 88
Maiden Creek Wastewater Treatment Plant (Blandon, PA) 92
Metro-West Point Treatment Plant (Seattle, WA) 96
Neuse River Wastewater Treatment Plant (Raleigh, NC) 100
Newark Wastewater Treatment Plant (Newark, OH) 104
North Shore Sanitary District Gurnee Plant (Gurnee, IL) 108
Passaic Valley Wastewater Treatment Plant (Newark, NJ) 112
Sioux City Waste Treatment Plant (SCWIP) (Sioux City, IA) 116
Tolleson Wastewater Treatment Plant (Tolleson, AZ) 120

APPENDIX B: INTERFERING SUBSTANCES 124

PART II
LITERATURE REVIEW, TELEPHONE SURVEY, AND CASE STUDIES

1. INTRODUCTION 129
Background 129
Scope of Work 130

2. CONCLUSIONS 132
Summary 132
Research Needs 133
Treatment Processes 133
Plant Operations 134
Toxicity Testing 134
Management Aspects 135

3. LITERATURE REVIEW 136
General Wastewater 138
Process Comparisons 140
Toxicity and Inhibition 141
Primary Treatment 142
Secondary Biological Treatment 144
Suspended Growth Processes 145
Activated Sludge 145
Conventional Pollutants 146
Phenols and Cyanides 148

Metals 150
Other Organics 154
Miscellaneous Industries 156
Powdered Activated Carbon/Activated Sludge Process 159
Lagoons 159
Anaerobic Reactors 160
Fixed Film Processes 161
Trickling Filters 161
Rotating Biological Contactors 162
Conventional Pollutants 162
Metals 163
Pretreatment 163
Aerated Submerged Filters 164
Anaerobic Filters 164
Advanced Treatment 165
Biological Nutrient Removal 166
Nitrification 166
Metals 166
Miscellaneous Industries 167
Process Comparisons 168
Phosphorus Removal 169
Physical/Chemical Processes 169
Organic Pollutants 169
Inorganic Pollutants 171
Miscellaneous Processes 172
Sludge Processing 173
Sludge Treatment 173
Digestion 173
Metals 173
Organics 174
Dewatering and Thickening 175
Disposal 175
Summary 176
Acclimation 176
Staged Treatment 176
Operations 177
Influent Wastewater Characteristics 177

4. TELEPHONE SURVEY 179
Procedure 179
General Survey Results 180
Regional Results 182
Northeast 182

Connecticut, Delaware, Maine, Massachusetts,
New Hampshire and Vermont 182
New Jersey 182
New York 182
Pennsylvania 182
Southeast 184
Alabama and Mississippi 184
Florida 184
Georgia, North Carolina, South Carolina 184
Kentucky and Tennessee 184
Maryland 185
Virginia and West Virginia 185
Midwest 185
Illinois 185
Indiana 185
Iowa and Nebraska 185
Kansas 185
Michigan 186
Missouri 186
Ohio 186
Wisconsin 186
Southwest 186
Arizona 186
Arkansas 187
California 187
Colorado, Nevada, New Mexico, Oklahoma and Utah 187
Hawaii 187
Louisiana 187
Texas 187
Northwest 187
Alaska, Idaho and Montana 187
Minnesota 188
North Dakota, South Dakota and Wyoming 188
Oregon 188
Washington 188

5. CASE STUDIES 189
Selection Process 189
Site Visit Logistics 200
Case Study Reports 206

6. RESULTS 207
Monitoring 207

POTW Influent 207
Other POTW Locations 208
Operations Monitoring 208
Industrial Discharges 209
Analytical Testing 210
Sewer Use Ordinances 210
Interference 212
Identification of Interference 212
Types of Interference 213
Interfering Compounds 214
Toxicity Testing 220
Mitigation 221
Treatment Plant Control 221
Plant Operations Modifications 222
Biological Process Control 222
Treatment Step Operations Modifications 223
Treatment Plant Modifications 224
Pretreatment and Source Control 225
Industrial Waste Discharge Agreement Development 226
Industrial Pretreatment Works 227
Industrial Waste Agreement Monitoring and Enforcement 227
Legal and Enforcement Remedies 228
Fines 228
Litigation 229
Sewer Disconnection 229

REFERENCES 230

APPENDIX A: INTERFERING SUBSTANCES 253

APPENDIX B: PROJECT FORMS 256

APPENDIX C: CASE STUDY REPORTS 267
Bayshore Regional Sewerage Authority (Union Beach, NJ) 268
Hamilton Township Wastewater Treatment Plant (Trenton, NJ) 272
Passaic Valley Wastewater Treatment Plant (Newark, NJ) 276
Binghamton-Johnson City Joint Sewage Treatment Plant (Binghamton, NY) 280
City of Canandaigua Wastewater Treatment Plant (Canandaigua, NY) 284
East Side Sewage Treatment Plant (Oswego, NY) 288

Hatfield Township Advanced Treatment Facility (Colmar, PA) 292
Maiden Creek Wastewater Treatment Plant (Blandon, PA) 296
Rocky Creek Water Pollution Control Plant (Macon, GA) 300
City of Baltimore, Maryland 304
Back River Wastewater Treatment Plant (Baltimore, MD) 305
Patapsco Wastewater Treatment Plant (Baltimore, MD) 308
City of Raeford Wastewater Treatment Plant (Raeford, NC) 312
Neuse River Wastewater Treatment Plant (Raleigh, NC) 316
Horse Creek Pollution Control Facility (North Augusta, SC) 320
North Shore Sanitary District Gurnee Plant (Gurnee, IL) 324
Sanitary District of Rockford Sewage Treatment Plant (Rockford, IL) 328
Lake Mills Wastewater Treatment Plant (Lake Mills, IA) 332
Marshalltown Water Pollution Control Plant (Marshalltown, IA) 336
Sioux City Waste Treatment Plant (SCWTP) (Sioux City, IA) 340
Newark Wastewater Treatment Plant (NWTP) (Newark, OH) 344
91st Avenue Wastewater Treatment Plant (Phoenix, AZ) 348
Tolleson Wastewater Treatment Plant (Tolleson, AZ) 352
Victor Valley Regional Water Reclamation Plant (VVRWRP) (Victorville, CA) 356
Duck Creek Sewage Treatment Plant, Paw Paw Sewage Treatment Plant (Denison, TX) 360
Paris Wastewater Treatment Plant (Paris, TX) 366
Newberg Wastewater Treatment Plant (Newberg, OR) 370
Metro-West Point Treatment Plant (Seattle, WA) 374
Post Oak Wastewater Treatment Plant (Sherman, TX) 380

CONVERSION FACTORS 381

Part I

Guidance Manual

The information in Part I is from *Guidance Manual for Preventing Interference at POTWs,* prepared by Edward Wetzel and Scott Murphy of James M. Montgomery, Consulting Engineers, Inc., for the U.S. Environmental Protection Agency, September 1987.

Note to Users of the Guidance Manual: The case studies contained in Appendix A, and referred to throughout the text were conducted over a period from December, 1985 to March, 1986. The information contained in each discussion was current at that time, but since then the status of some of the activities at the case study sites is likely to have changed.

Acknowledgments

This document was prepared by James M. Montgomery, Consulting Engineers, Inc. Mr. John Grantham was the Project Manager, with Dr. Edward Wetzel and Mr. Scott Murphy acting as Project Engineer and Assistant Project Engineer, respectively. Messrs. Wetzel and Murphy are the principal authors of this manual. Treatment plant site visits conducted as part of the case study effort were made by Messrs. David Harrison, Paul Skager and Roger Stephenson and Ms. Sheila McShane, in addition to the authors. The contributions of a number of other members of the Pasadena office support staff to the production of this manual are gratefully acknowledged.

This manual was prepared under the technical direction of Ms. LeAnne Hammer and Mr. Gregory McBrien of the Permits Division, Office of Water Enforcement and Permits, and Dr. Sidney Hannah of the Water Engineering Research Laboratory (WERL). Additional guidance and assistance were provided by the other members of the Pretreatment Support Group of WERL, consisting of Messrs. James Kreissl, Dolloff Bishop, Richard Dobbs, Kenneth Dostal and Henry Tabak. Peer review and comments were also provided by Science Applications International Corporation (SAIC). Special thanks are also offered to Mr. Guy Aydlett of the Hampton Roads Sanitation District and to Mr. Paul (Kip) Keenan of the City of Baltimore Pollution Control Section.

1. Introduction

1.1 BACKGROUND

Sections 307(b)(1) and (c) of the Clean Water Act (CWA) direct the U.S. Environmental Protection Agency (EPA) to establish pretreatment standards "to prevent the discharge of any pollutant through treatment works... which are publicly owned, which pollutant interferes with, passes through, or is otherwise incompatible with such works." These sections address the problems created by discharges of pollutants from nondomestic sources to municipal sewage treatment works. Specifically addressed are discharges of pollutants which either interfere with the operation or performance of the works or pass through the works to navigable waters untreated or inadequately treated. Pretreatment standards are intended to prevent these problems from occurring by requiring nondomestic users of publicly owned treatment works (POTWs) to pretreat their wastes before discharging them to the POTW. In 1977, Congress amended Section 402(b)(8) of the CWA to require POTWs to help regulate their industrial users (IUs) by establishing local programs to ensure that industrial users comply with pretreatment standards.

In establishing the national pretreatment program to achieve these goals, the EPA adopted a broad-based regulatory approach that implements the statutory prohibitions against pass-through and interference at two basic levels. The first is through the promulgation of national categorical standards which apply to certain industrial users within selected categories of industries that commonly discharge toxic pollutants. Categorical standards establish numerical, technology-based discharge limits derived primarily to control the discharge of toxic pollutants which could interfere with or pass through POTWs. The EPA has promulgated categorical standards for many major and minor industry categories (See 40 CFR Parts 400-469). The EPA will be evaluating these industries and other industries for the control of additional toxic pollutants.

Implementation of the categorical standards will not remedy all the interference and pass-through problems that may arise at a POTW. The potential for many pass-through or interference problems depends not only on the nature of the discharge but also on local conditions (e.g., the type of treatment process used by the POTW, local water quality, the POTW's chosen method for handling sludge), and thus needs to be addressed on a case-by-case basis. Such problems can result from discharges by categorical industries of pollutants not covered by a categorical standard or from nondomestic sources not regulated by the categorical standards. In addition, since categorical standards are established industry-wide, they cannot consider site-specific conditions, and therefore, may not be adequate to prevent all pass-through and interference even for the regulated pollutants. The second level of EPA's regulatory approach, contained in the General Pretreatment Regulations (40 CFR Part 403), addresses these areas of concern. First, Section 403.5(b) establishes specific prohibitions which apply to all nondomestic users and are designed to guard against common types of pollutant discharges that may result in interference and pass-through (e.g., no discharge of flammable, explosive, or corrosive pollutants). Second, Section 403.5(a) establishes a general prohibition against pass-through and interference which serves as a back-up standard to address localized problems that occur. In addition, POTWs with total design flow greater than 5 mgd and which receive pollutants which pass through or interfere with POTW operation or are otherwise subject to pretreatment standards must establish formal pretreatment programs which must be approved by the EPA or a designated State agency. POTWs with design flow less than 5 mgd may also be required to develop pretreatment programs if circumstances warrant in order to prevent pass-through or interference. As part of their programs, POTWs must develop and enforce specific local limits to prevent pass-through and interference. POTWs not required to develop pretreatment programs may still be required to develop local limits if they experience pass-through or interference that is likely to recur (Section 403.5(c)).

The need for guidance on preventing interference was identified by the Pretreatment Implementation Review Task Force (PIRT). PIRT was established on February 3, 1984 by the EPA Administrator. The task force was composed of 17 representatives from POTWs, States, industry, environmental groups and EPA Regions. The charge given to PIRT was to make recommendations to EPA concerning the problems faced by POTWs, States, and industry in implementing the national pretreatment program. In its Final Report to the Administrator (U.S. EPA, 1985b), one of the specific problems identified by PIRT was the difficulty experienced by POTWs in the recognition, tracking, and mitigation of interferences caused by industrial discharges. PIRT's recommendation was for EPA to provide guidance to municipalities regarding such interference problems. This report is EPA's response to that recommendation.

1.2 DEFINITION OF INTERFERENCE

The U.S. EPA recently promulgated revised definitions for the terms "pass through" and "interference" (52 Federal Register 1586, January 14, 1987). As defined in 40 CFR, Part 403.3(i):

(i) The term "Interference" means a discharge which, alone or in conjunction with a discharge or discharges from other sources, both:

(1) Inhibits or disrupts the POTW, its treatment processes or operations, or its sludge processes, use or disposal; and

(2) Therefore is a cause of a violation of any requirement of the POTW's NPDES permit (including an increase in the magnitude or duration of a violation) or of the prevention of sewage sludge use or disposal in compliance with the following statutory provisions and regulations or permits issued thereunder (or more stringent State or local regulations): Section 405 of the Clean Water Act, the Solid Waste Disposal Act (SWDA) (including

Title II, more commonly referred to as the Resource Conservation and Recovery Act (RCRA), and including State regulations contained in any state sludge management plan prepared pursuant to Subtitle D of the SWDA), the Clean Air Act, the Toxic Substances Control Act, and the Marine Protection, Research and Sanctuaries Act.

In the same rulemaking that established the new definitions (52 Federal Register 1586, January 14, 1987), EPA amended the General Pretreatment Regulations to establish affirmative defenses to liability on the part of an industrial user for violating the general prohibitions or certain of the specific prohibitions. These defenses address situations where an industrial user did not know or have reason to know that its discharge would cause interference. The reader is referred to the Federal Register citation above for additional information and perspective, as provided in the preamble to the regulation.

The interference prohibition addresses situations where an industrial user's discharge, either alone or in conjunction with other discharges, disrupts the POTW or its sludge practices, and the disruption is a cause of a permit violation or prevents the POTW from lawfully using its chosen sludge use or disposal method. In contrast, the pass-through prohibition addresses situations where an IU's discharge exits the POTW to waters of the United States in quantities or concentrations which, alone or in conjunction with other discharges, cause a permit violation. Pass-through is not necessarily related to an inhibition or disruption of the POTW processes, but instead is related to a pollutant discharge which is not susceptible to adequate treatment by the POTW.

An industrial user whose discharge is found to cause pass-through or interference is legally liable for violating the general prohibitions, and may be subject to enforcement action. However, as discussed in the Federal Register preamble to the new definitions of pass-through and interference, an industrial user's discharge is considered to be interference or pass-through only if the discharge is a cause of the POTW's noncompliance. If a malfunction or improper operation by the POTW, rather than an industrial user's discharge, causes the POTW's noncompliance with its NPDES permit or sludge requirements, interference and/or pass-through are not occurring. The EPA intends the definitions to be interpreted and implemented in a manner consistent with the Congressional intent that pretreatment technology not be required as a substitute for adequate operation and maintenance of the POTW. Thus, if the POTW's improper operation alone prevents it from meeting the effluent limitation in its NPDES permit, the POTW must correct its operational problem.

The interference definition does not directly address situations in which a discharge causes problems other than NPDES permit violations or impairment of sludge use or disposal. For example, POTW worker health and safety problems or unacceptable air emissions could result from IU discharges. The EPA is currently considering whether and how to address these problems more specifically through guidance and future regulations, if appropriate, and by encouraging POTWs to address these concerns in their local ordinances. This manual addresses such concerns to only a limited degree, focusing mainly on interference as defined in the above-mentioned regulation.

1.3 GUIDANCE MANUAL OBJECTIVES

The purpose of this manual is to provide POTW operators with guidance on dealing with interferences caused by pollutants from nondomestic sources. In addition, some guidance is provided on distinquishing interferences caused by nondomestic discharges from problems resulting from poor operation and maintenance of the POTW. This manual is divided into three major sections, which correspond to the order in which a POTW should address interference. These sections are:

- Detecting Interference
- Source Identification
- Mitigation

The section on detecting interference is intended to help identify the types of interferences and substances which are known to cause problems. The way in which interference occurs in both the sewer collection system and the treatment plant is also discussed, along with analytical tests and monitoring that can be conducted by POTW operators.

The second major section deals with the identification of the industrial sources of the interference-causing substances. Sources can be separated into chronic dischargers of industrial pollutants, isolated spill events, and hauled wastes. Identification techniques range from simple sensory observations to the use of sophisticated monitoring equipment for tracing problems at the POTW back to a source.

The final section on mitigation discusses ways in which municipalities can cope with interference problems. In-plant control, source control and legal and enforcement remedies are addressed in the section. Operators should be cautioned that there are few straightforward solutions to these problems, and that often a combination of techniques will need to be employed to properly mitigate an interference. The section concludes with a discussion on planning, so that future industrial discharges will not interfere with successful POTW operations.

There are two appendices included in the back of this manual. Appendix A contains case studies of 14 POTWs, visited as part of this project, that have previously experienced interferences but have mitigated the problems over the past few years. These cases are referred to throughout the manual wherever a certain case study illustrates a particular problem or solution that is discussed. While the case studies represent some of the types of interference problems experienced by POTWs, they should not necessarily be viewed as examples of pretreatment programs which are ideally implemented or fully endorsed by the EPA. It is hoped that the case studies will be useful to people who are experiencing problems similar to those described. A summary of the case studies including the names and phone numbers of individuals who can be contacted for further information is provided on Table A-1. Appendix B is a list compiled from the published literature and actual treatment plant studies that includes many of the inorganic and organic substances now recognized as having the potential to cause interference.

2. Detecting Interference

2.1 TYPES OF INTERFERENCE

Interference can be broken down into two basic types: (1) interference with the POTW's ability to meet its NPDES permit; and (2) interference with its ability to utilize its chosen sludge disposal method. With either type, several sources may contribute to the interference. For example, contamination of sludge with unacceptable levels of metals may be due to the cumulative contributions from several industries. Domestic sewage background concentrations can also be a significant source of some metals. Unless the interference is caused solely by domestic sources or inadequate operation and maintenance of the POTW, each nondomestic source of the interfering pollutant should be identified and its role in causing the interference assessed. The individual sources must then be controlled as necessary to allow the POTW to meet its NPDES permit and utilize its chosen sludge disposal method.

Industrial users' discharges can cause the first type of interference, involving a permit violation, by several means. These include, but are not limited to:

- physically disrupting the flow of wastewater through the POTW's system
- chemically, physically, or thermally inhibiting the treatment processes
- hydraulically overloading the plant so that proper settlement does not occur or wastes are retained for too short a time to receive adequate treatment before discharge.

The pollutants discharged by the industrial user that cause the POTW to violate its permit may be the same as, or different from, the pollutants discharged in violation of the permit. For example, an industrial user discharging excessive BOD that causes a disruption of the biological treatment process and results in the POTW exceeding its BOD discharge permit limit may be causing an interference. Likewise, the same industrial user discharging a toxic pollutant that inhibits the POTW's performance and results in effluent BOD permit violations is also causing an interference. It should be noted that in the example of an excessive BOD discharge causing a BOD permit violation, the problem could be pass-through rather than interference. For example, a heavy discharge of relatively non-degradable organic matter might pass through the plant without causing an upset. The distinction between pass-through and interference must be made depending of the individual circumstances of such cases.

The second type of interference, impairment of sludge use or disposal, results when the POTW's sludge no longer meets applicable requirements for its chosen use or disposal alternative. Thus, if the POTW has elected to apply the sludge to land but industrial discharges prevent the lawful implementation of this method, interference occurs. Detection of this type of interference is generally performed by sludge monitoring coupled with monitoring of industrial users.

As mentioned in Chapter 1, any type of interference is a violation of the general prohibition (40 CFR Part 403.5(a)). Some interferences are also violations of the specific prohibitions (40 CFR Part 403.5(b)). The specific prohibitions bar discharges which:

1. create a fire or explosion hazard;
2. are corrosive to POTW structures;
3. obstruct wastewater flow resulting in interference;
4. release pollutants (including BOD) at rates or concentrations which will cause interference; or
5. increase the influent wastewater temperature above 40°C, or inhibit biological activity due to heat, resulting in interference.

The problems referred to by the specific prohibitions do not always result in interference (ie., permit violations), yet they are detrimental to POTW operations. In fact, many local sewer use ordinances contain additional or more stringent local prohibitions, such as a prohibition against discharges which release toxic vapors endangering POTW worker health and safety. The EPA strongly supports and encourages such local prohibitions.

Another way of looking at types of interference is to classify them by the location of impact: either the collection system or the treatment plant. Collection system problems (corrosion of sewer mains, explosions in sewers, etc.) are generally easier to relate to industrial or commercial discharges, while interference at the treatment plant requires detailed analysis to ensure that it is not the result of poor operation and maintenance practices or from domestic sources. Since it is often the most difficult to detect and trace to industrial sources, this chapter emphasizes treatment process interference. The chapter looks at both chronic inhibition and upset conditions. The ability of a particular waste discharge to cause inhibition or upset is considered in terms of three factors:

- industrial discharge practices
- acclimation of POTW treatment processes to the specific pollutants
- impacts on the POTW

The next two subsections will discuss chronic inhibition and upset conditions in more detail. Collection system problems are discussed in Section 2.3.

2.1.1 Chronic Inhibition

Chronic inhibition refers to a more or less consistent pattern of impairment of the functioning of the biomass in a biological treatment process caused by influent pollutant concentrations that are above tolerable levels. Inhibition is usually defined by a decrease in oxygen uptake rate or a decrease in COD/BOD removal. If the inhibition leads to a permit violation, it then is classified as interference. This type of interference results from either a continuous or semi-continuous discharge of an industrial pollutant to the POTW. Chronic inhibition may also result from the total effect of several industries discharging a variety of inhibitory pollutants. Industrial sources of chronic problems tend to be by-products of production activities such as chemical derivatives, rinse waters and contact cooling water.

The effects of an inhibitory pollutant on plant biomass vary depending on how frequently and at what level the pollutant is discharged. The more consistently a pollutant is fed to the biological treatment process, the more chance the biomass has to develop a "resistance" to the pollutant. If a pollutant is fed at a fairly even rate and concentration, the biomass will generally eventually become accustomed to or "acclimate" to the polluant, and BOD removal efficiency will no longer suffer. For this reason, a plant may experience operational problems unless there has been sufficient time for the biomass to become acclimated. In addition, discharges of toxics at high enough concentrations can cause inhibition even in acclimated systems.

Although it does not always result in a POTW violating its NPDES permit limits, chronic inhibition can increase the overall expense and difficulty of operating a treatment plant in compliance with NPDES permit limits. For example, a plant may have to be operated at an increased MCRT or require additional aeration capacity to counteract the negative effects of inhibition. Depending on the circumstances, this may involve significantly increased operating costs for recirculating sludge at a higher rate or providing more aeration. It may also take away any reserve capacity that the plant might otherwise have had for future growth. Therefore, POTW's experiencing chronic inhibition should take steps to mitigate it even when there is no immediate threat of an NPDES permit violation.

2.1.2 Upset Conditions

The results of 29 case studies performed in conjunction with the development of this manual showed that most interference problems are caused by intermittent discharges of high-strength conventional wastes which overload a POTW's organic capacity, causing plant upset. The term "upset" is used in this manual to refer to an exceptional incident which creates a temporary non-compliance with permit limits due to the impacts of the incoming waste characteristics on the treatment processes. Discharges causing upset commonly come from food processors such as bakeries, dairies, breweries, canneries, poultry farms and meat packaging plants. Examples of interferences due to high-strength conventional wastes are provided by the Bayshore Regional Sewerage

Authority (New Jersey), and Hamilton Township (New Jersey) case studies in Appendix A. In some cases, upsets have occurred even when the total industrial contribution was significantly less than 5 percent of the total plant flow. It is frequently the intermittent discharge of concentrated wastes which leads to the upset.

Similarly, in cases of plant upsets due to the discharge of toxic pollutants, it is usually the intermittent discharges of toxics which produce the most drastic effects. These types of discharges may result from:

- process tank contents disposal
- cleanup operations
- industrial spills
- waste hauler discharges
- midnight dumping (illegal waste hauling)

Biological populations are typically not acclimated to either the specific compounds or concentration levels observed in such discharges. The impacts on biological processes can therefore be severe and rapid, often requiring long recovery periods. Such interferences commonly affect the effluent quality rather than the stabilized and dewatered sludge characteristics, although the loss of an anaerobic digester due to slug loads of heavy metals is not unusual. A slug load is defined as any pollutant in a discharge at a flow rate and/or pollutant concentration which will cause interference at the POTW.

It is important that POTWs monitor the occurrence of upset conditions caused by industrial waste discharges. In some cases, the problems may recur in a cyclical pattern, such as once-per-week or once-per-month. Recognition of the pattern coupled with contaminant identification will go a long way toward discovering the source of the problem. Intermittent discharges of this type tend to produce similar impacts on the POTW from incident to incident. Changes in dissolved oxygen (DO) levels, mixed liquor suspended solids (MLSS), sludge volume index (SVI), reactor temperature, etc. are all indications of process changes potentially resulting from industrial wastes.

The impacts of interference-causing substances are not restricted to biological systems within a treatment facility. Interference problems can also surface in physical treatment systems (clarifiers, thickeners, filters, etc.) or in the sewer collection system. Municipalities should make every effort to mitigate discharges that threaten any treatment process as well as the integrity of the collection system, not only to avoid interference, but for the protection of worker health and safety as well.

2.2 INTERFERENCE-CAUSING SUBSTANCES

POTW interference can be caused by a wide variety of chemical, biological and physical factors. Chemical factors such as the types and concentrations of industrial wastewater constituents which cause interference are highly variable.

The studies reported in the literature discussing chemical interference/inhibition range from research done in the laboratory to studies of actual treatment plant operations. There has been a substantial body of work published and many researchers have devoted a great deal of effort to these types of studies. Previous reviews (U.S. EPA, 1979; Geating, 1981; U.S. EPA, 1981a; Russell, et al., 1983; U.S. EPA, 1986a) have presented ranges of concentrations for a variety of pollutants which inhibit biological processes. The reader should refer to these documents for a more thorough presentation of pollutant treatability and process inhibition. As an aid to POTW operators, this manual compiles available information on the types and concentrations of pollutants and compounds which inhibit some biological treatment systems.

There are various ways of measuring inhibition and the fact that different researchers use different methods results in a range of published "inhibiting concentrations", even for nearly identical study conditions. The two most typical methods of determining activated sludge inhibition are by measuring 1) decreases in COD or BOD removal or 2) decreases in oxygen utilization rates. Threshold inhibition levels as measured by these two methods are usually defined differently by individual researchers, but are most typically set at the 10-50 percent range. Anaerobic treatment inhibition is typically defined as increased volatile acid levels or decreased methane generation, but once again the threshold levels are variously defined. Nitrification inhibition is specified as a decrease in the degree of ammonia conversion.

The most important conditions that affect biological inhibition are:

- the nature and strength of the inhibiting agent
- biomass characteristics
- pH
- temperature
- synergism
- antagonism
- acclimation

For most studies, biomass characteristics are not described in the literature, except as related to whether or not the biomass was acclimated. The diverse biomass population is likely to be very different from one reported study to the next. Characteristics such as sludge age or food to microorganism (F/M) ratio will have a significant impact on the inhibitory concentration levels of pollutants. Actual test conditions, including temperature and pH, vary dramatically from study to study, with the result being that inconsistent values are reported. Wastewater pH plays a particularly important role in metal-caused inhibition. The pH affects the solubility of metal ions, and it is primarily the soluble metal that is toxic to microorganisms. Synergism, or the increase in the inhibitory effect of one substance by the presence of another, is most important when considering combinations of metals. Toxic organics do not exhibit this effect as often as metals. On the other hand, some compounds are antagonistic towards each other, decreasing the inhibitory effect of either compound alone. Examples are chelating agents, such as EDTA, which are antagonistic toward metal ions and reduce their toxic effects.

Substances which cause interference/inhibition problems can be divided into three groups:

- conventional pollutants
- metals and other inorganics
- organic compounds

Each of these categories is considered separately in the subsections to follow.

2.2.1 Conventional Pollutants

The term "conventional pollutants" as used in this manual includes BOD, suspended solids, pH, and oil and grease. Since BOD and suspended solids (SS) form the usual basis of secondary treatment plant design, interference/inhibition problems result from exceeding the peak mass loadings specified by the design. Such "shock loadings" (slug loadings) of conventional pollutants are a common cause of permit violations resulting from oxygen transfer limitations, insufficient biodegradation and solids carryover. Oil and grease are normal constituents of domestic wastewater that if present in elevated concentrations can interfere with normal waste treatment by preventing bacteriological floc from properly settling and disrupt mechanical equipment operation. The pH of a wastewater can also cause interference if it is too high or too low, or is widely fluctuating.

2.2.2 Metals and Other Inorganics

More research efforts have been directed toward the impacts of heavy metals on biological treatment than any other classification of contaminant found in wastewater. A large percentage of the insoluble metals and metal salts that enter a POTW settle out during primary clarification. Consequently, a significant impact of metals is in rendering sludges unacceptable for a variety of disposal options, notably landspreading for agricultural purposes.

The soluble fractions of the metals can upset the secondary treatment processes. Table 2-1 presents ranges of metal and other inorganic pollutant concentrations inhibiting biological processes. Important factors affecting these ranges of values are pH, solubility, and the definition of inhibition used by the researchers reporting the results. The wide range of concentrations presented results from apparently contradictory data published in the literature. The values presented in Table 2-1 represent the range of reported threshold inhibition concentrations. Acclimation is an important issue, which in many studies was either not reported or was not known. However, it would be reasonable to expect the lower end of a range to correspond to threshold levels inhibiting an unacclimated system while the upper end of the range would correspond to threshold levels inhibiting an acclimated system. The primary references, for this table are U.S. EPA (1981a), Russell, et al., (1983) and Geating (1981). U.S. EPA (1986a) provides a more complete reference list.

The inhibition levels presented in Table 2-1 are for the dissolved metal unless otherwise indicated. The dissolved form is the most toxic. However, POTWs should control the total metal entering the plant because particulate metal or metal compounds may exert some toxicity or may later be resolubilized. The dissolved metals present in the secondary treatment process are derived both from dissolved metals in the plant influent and from desorption of metals from sludges that are recycled to secondary treatment. A large percentage of the toxic metals present in the aeration basins at some treatment plants has been found to be contributed by recycled solids handling sidestreams. Such contributions can cause a continued toxic effect long after the source has been controlled.

2.2.3 Organic Compounds

Considerable interest exists among the EPA and public health officials concerning the fate and effects of toxic organic compounds in POTWs. Organic substances which enter municipal facilities are either removed in the biological treatment processes, inhibit biological degradation, or pass through the plant. The principal removal mechanisms are:

- volatilization
- biodegradation
- adsorption to biological flocs and settling

The amount of information available on the impacts of organic contaminants is small compared with the metals, due in large part to the number of compounds of interest and also to the sophisticated analytical equipment required to measure these organics. Table 2-2 presents ranges of concentrations for toxic organic compounds which inhibit biological systems.

The classification scheme used in Table 2-2 involved grouping compounds of similar structure and characteristics which might tend to inhibit biological processes at similar concentrations. The reader is cautioned, however, that chemicals with similar structure do not always possess similar inhibition characteristics. For a more detailed summary of what is known about inhibition by individual organics, see Russell, et al. (1983), U.S. EPA (1981a), Geating (1981), and U.S. EPA, (1977b).

It is important to note that the categories in Table 2-2 are very broad and the concentration ranges presented are simply typical values for some compounds and should not be interpreted as defining an inhibition range for all compounds within the classifications. Appendix B lists the compounds that fall into these broad classifications.

2.3 SEWER COLLECTION SYSTEM

Industrial and waste hauler discharges can be very detrimental to the condition of the sewer collection system. The types of substances responsible for such problems are generally the same pollutants addressed by the specific prohibitions. Table 2-3 defines four categories of these substances with examples given for each.

Corrosivity problems can be identified by observing the deterioration of the pipe material or measuring the pH of the wastewater at several locations within the collection system. Corrosion rates generally increase significantly below pH 5 and above pH 12.5. The best approach is to maintain a program of regular sewer inspection coupled with the use of recording pH meters located at strategic interceptor locations in the sewer system. A proper inspection program should include the detection of unusual colors or odors by trained personnel.

Detecting substances which may result in ignitability or reactivity problems is a more complicated task. Instrumentation is available to detect explosive conditions, lack of oxygen and the presence of hydrogen sulfide. Such equipment is typically used for worker safety prior to entering confined, below grade areas such as manholes and sewer interceptors. To install and maintain sensitive instruments of this type, along with the recording devices needed for proper monitoring, would be very expensive if placed in numerous locations. A more practical approach is to survey the industries discharging to a POTW as a means of identifying potential dischargers of these substances (see Section 3). Once likely industrial candidates are identified, portable detection instrumentation can be used to spot check the sewer environment or permanent equipment can be installed in a few, selected locations.

Baltimore, Maryland and Passaic Valley, New Jersey are examples of locations where sewer collection system problems have been identified and addressed (see Appendix A). Discharges of volatile organics such as ethyl benzene, xylene, toluene and tetrachloroethylene (PCE) into the Baltimore collection system have resulted in pump station and other building evacuations in the past. By successfully tracing these problems to the source, the City has reduced the occurrence of such incidents dramatically. Passaic Valley experienced both sewer clogging problems from a pulp and paper mill and high concentrations of flammables from a number of industrial sources. Lower explosion limits (LEL) were established and industries identified as being dischargers of flammables were required to install LEL detection equipment in their effluent piping.

Another example of the use of LELs is by the Los Angeles County Sanitation Districts, where the wastewater ordinance requires all significant potential dischargers of flammable substances to install, operate and maintain an adequate combustible gas monitoring system. The system provides early detection so that preventive measures can be taken. Systems must be installed in a fixed location and continuously operated, incorporating an indicator, as well as an automatic continuous recorder, adjustable two-stage alarm system, calibration for methane detection, and a means for diverting flow from the sewer to a holding vessel when the combustible gas level is 20% of the LEL or greater. Industrial users, primarily petroleum refineries, storage and transfer facilities, and chemical manufacturing plants, must submit engineering drawings of their proposed systems for review and approval by the Districts prior to construction.

2.4 PLANT OPERATIONS

There are numerous tools available to the plant operator to monitor the condition and performance of the facility. Suspended growth biological treatment systems generally provide more operational control (such as sludge wasting and recycle, aeration tank D.O., process modifications), and therefore monitoring opportunities, than do fixed film systems. However, all POTWs have processes that can be easily checked on a daily basis which can signal the onset of an interference problem. Making use of the available tools may be the difference between total process failure and catching the problem before it fully develops.

The operational tools available fall into the following categories:

- observation
- instrumentation
- analytical results

2.4.1 Observation

Some of the most powerful tools in the operation of a treatment facility are the senses of sight, sound, touch and smell. Maintenance personnel are typically trained to listen for the improper operation of a blower or to feel for signs of an overheated bearing. Similarly, plant operators should be trained to observe changes in the appearance or smell of unit processes which might be indicative of a problem. A major thrust of the Hamilton Township, New Jersey, interference identification program was to require that operators spend a minimum number of hours each work shift "walking the grounds" (see Appendix A). Such a requirement can (and did) result in the identification of late night spill events that might otherwise go unnoticed until morning, when it may be too late for biological processes to recover.

Examples of what operators should notice as they work around a POTW are the:

- surface appearance of clarifiers
- amount and color of foam in aeration tanks
- presence of nuisance organisms, insects or odors near fixed film systems
- common odors at each plant location
- sludge and recycle flow appearance at each processing step

The EPA has provided troubleshooting and process control guidance to operators in previously published manuals (U.S. EPA, 1977a; U.S. EPA, 1978). These documents assist POTWs in troubleshooting process performance problems, and provide numerous tables to help the operator identify problems through visual inspection. Many problems uncovered through plant observation do not result from industrial discharges, but rather from equipment malfunction, inadequate maintenance or design deficiencies. The two manuals referenced above provide assistance in distinguishing between the potential sources of such problems. Table 2-4 identifies the operational indicators of process malfunction which may be related to industrial waste discharges. The processes listed in Table 2-4 are for typical secondary treatment plants. Advanced wastewater treatment systems and sludge thickening and dewatering processes are not included in this manual.

2.4.2 Instrumentation

Instrumentation is designed into treatment facilities as an aid to the operations staff. Whether located at the central control panel or at the piece of equipment being monitored, digital and dial gauge readouts provide instant feedback to an experienced operator concerning the conditions in the plant. Strip chart recorders maintain permanent records of the critical parameters, such as raw wastewater feed, to identify long-term trends and isolated excursions. Despite the availability of instrumentation and level of sophistication, much of the hardware may be unused or simply ignored by operators because of a perceived complexity and/or unreliability.

When monitoring instruments are incorporated into a POTW, it is important that such equipment be maintained in accordance with manufacturer's specifications and recalibrated at regular intervals. The utility of these instruments depends upon the operator's understanding of the readout. Proper training of operations personnel is therefore a critical element in using instrumentation as possible early warning signals of a pending interference problem.

The use of simple portable instruments and equipment for routine POTW site inspection can be quite useful to the operator. The use of a device to measure the depth of sludge in clarifiers may be the best way to learn that a sludge pump did not operate as expected or that unusual wastes have entered the plant. In the Tolleson, Arizona treatment plant (see Appendix A), the operators discovered that a rapidly increasing sludge depth in the primary clarifiers was indicative of upset conditions caused by high solids discharges from a meatpacking industry.

A number of commercially available instruments can be utilized by plant operators either for permanently-mounted, continuous monitoring and control or as portable devices. Table 2-5 lists the types of instruments, where they can be utilized in the treatment facility, and the parameters of interest in interference identification. The instruments listed in the table are useful for both process evaluation as discussed in this section, and for wastewater monitoring (Section 2.5). The instrumentation selected by a POTW should be a function of the wastewater characteristics of the industrial discharges as determined by the industrial survey conducted during pretreatment program development.

2.4.3. Analytical Results

The subsection on observation (2.4.1) outlined the benefits of noting the smell and appearance of unit processes during routine plant inspections. There are a number of standard analytical techniques that can be used to confirm the presence and extent of problems identified by sensory observation. Table 2-6 lists the common test procedures that can be performed on typical treatment plant processes. The testing frequency indicated is typical of well-operated facilities. The actual frequency used will be site-specific as a function of process problems, industrial discharges, and staff and equipment availability. The monitoring procedures listed on Table 2-6 are relevant to assessing overall plant performance, not just interference problems. As was the case in both Tolleson, Arizona (primary sludge depth) and Oswego, New York (activated sludge SVI), the relationship between operating parameters and specific industrial discharges is oftentimes correlated by trial and error.

Conductivity, D.O., flow and pH are the parameters that are measured most reliably by the instrumentation specified in Table 2-5. Some of the devices, such as the selective ion electrodes, are adversely affected by the wastewater environment, and are therefore not well-suited to on-line monitoring applications. TOC analyzers are expensive instruments that should be housed in controlled laboratory environments. In order to be effective in identifying slug discharges of organics, the instrument must be provided with representative wastewater samples on a regular basis.

2.5 WASTEWATER MONITORING

A critical aspect to any successful industrial waste management program is comprehensive monitoring of industrial discharges, POTW influent, effluent and sludge, and important process streams within the plant. The benefits derived by the municipalities in terms of understanding their influent wastewater characteristics and sources of specific contaminants are many. Monitoring is also performed to provide data from which to develop local limits and later to evaluate an industry's compliance with those limits.

Developing a large database of analytical results on a POTW's wastewater provides a baseline for future comparison. When industrial discharges cause a significant deviation from the baseline, noting such changes will help detect a potential interference problem and may prove useful in later identifying the source. In the Hamilton Township, New Jersey plant, the discharges from a pharmaceutical manufacturer were correlated to high POTW influent soluble BOD through an extensive analytical testing program. A pharmaceutical manufacturer was also implicated in the discharge of high ammonia levels to the Neuse River Plant in Raleigh, North Carolina. Year-round sampling of 70 metal finishing/electroplating industries and strict enforcement of local limits for metals has substantially reduced the concentrations of heavy metals (Cr, Cd, Cu, Pb, Ni, Zn) at the Metro-West Point Plant in Seattle, Washington over the past five years.

2.5.1 POTW Influent

Most municipalities now have some form of influent wastewater monitoring. The most common approach is to install an automatic sampler at the headworks of the plant. State-of-the-art sampling equipment provides the POTW with four options:

- timed sampling with collection in discrete sample containers
- flow-proportioned sampling with collection in discrete sample containers
- time-proportioned composite sampling
- flow-proportioned composite sampling

Samples collected in discrete containers provide a means of identifying diurnal fluctuations in wastewater characteristics. Such an approach can be costly if hourly samples are analyzed, but is particularly useful if "midnight dumping" of prohibited substances is suspected, since discrete samples do not mask the impact of short-term discharge concentrations by averaging over a 24-hour period. An alternative approach is to preserve and store the discrete samples, and then analyze only if problems occur at the POTW.

Composite sampling involves the collection of a fixed volume of wastewater at regular intervals into a single, large container. A typical approach is to collect 100 ml every 15 minutes for 24 hours into a 10-liter sample bottle. This is the most common method of obtaining average daily influent samples. A better approach is to proportion the sample volume consistent with the influent volumetric discharge at the time of collection. This technique requires a feedback signal from an influent flowmeter to the sampler, but results in a sample that is consistent with the mass loadings to the POTW.

Analyses performed on a POTW influent should routinely include BOD, SS and other pollutants (such as NH_3 and P) included in the NPDES permit. When evaluating the potential for inhibition caused by toxic pollutants, additional testing is necessary. The testing intervals for the toxic organics and metals are determined on a site-specific basis as a function of permit violations, pretreatment program requirements, process upsets, types of industrial discharges and budgetary constraints.

A suggested approach is for the POTW to survey its nondomestic users to find out what toxic metals and organics are reasonably expected to be present in its influent at detectable levels. The POTW should then analyze its plant influent for those pollutants. In addition to the pollutants expected to be present, it is recommended that the POTW sample for the metals and cyanide listed in Table 2-1. Among the toxic organic pollutants, standardized analytical methods are available primarily for EPA's list of "priority pollutants". These pollutants are covered by EPA methods in the 600 series. The reference for these methods is the Federal Register (40 CFR Part 136), October 26, 1984, and June 30, 1986. Most full-service commercial analytical laboratories, as well as some of the large municipal laboratories are capable of analyzing for the priority pollutants. A once per year scan for the priority pollutants is recommended. For pollutants which are detected in the influent at least once, additional sampling should be conducted to determine variability and evaluate trends.

2.5.2 Other POTW Locations

POTW effluent is generally composite sampled and analyzed in accordance with NPDES permit requirements. Operators may, however, select other process streams within a facility for intermittent monitoring. For example, sampling primary effluent allows for calculations of loadings to the secondary treatment system. The response of a biological process is more easily explained if one knows the specific wastewater feed characteristics, as opposed to assuming a primary clarifier performance based on influent characteristics.

POTW sludge monitoring is necessary to determine if the POTW is meeting applicable sludge use or disposal requirements and to detect sludge contamination. If sludge contamination is found, it will trigger the need for additional sampling of both domestic wastewater and nondomestic discharges in order to identify the source(s) of contamination.

An informative yet infrequently employed sampling method is to evaluate the strength of sidestream flows, particularly from solids processing. Recycle flows can add 50 to 100 percent of the influent solids and organics to the liquid processing trains when inefficient sludge solids capture persists. POTW design often neglects the impact of recycle streams, a problem magnified when unanticipated quantities of heavy metals and priority organics are discharged from industrial sources. While monitoring such sidestreams on a daily or weekly basis may prove impractical (and costly), periodic sampling and flow measurement permits mass balancing around solids processing units, and can provide insight into the presence of substances in the POTW effluent not necessarily present in the influent.

Recycle flows can be intermittent, or at least shift dependent, and as such are poor candidates for 24-hour composite sampling. Grab sampling is done by extracting a representative sample of sufficient quantity to perform the necessary analytical tests. Some procedures, such as the extraction methods for oil and grease, require grab sampling to prevent deposition of the material on the container over the 24-hour composite period.

2.5.3. Inhibitory Effects Testing

Testing which measures the inhibitory effects of industrial discharges might prove useful in evaluating the impacts of those discharges on the POTW. One of the simplest methods of detecting inhibition due to an industrial discharge is to add incremental volumes of the waste to seeded dilution water and analyze for 5-day BOD. If the wastewater is inhibitory to the POTW bacteria, higher concentrations will result in less oxygen depletion and lower BOD. If completely biodegradable, larger volumes of the industrial waste should produce proportionately higher oxygen depletion. The advantage of this technique, termed serial addition, is that the concentration at which the waste changes from biodegradable to inhibitory can be estimated by this technique. The major disadvantages are the five day waiting period and the questionable correlation between degradation in a BOD bottle as compared with a full-scale biological reactor.

Other test procedures have been developed which overcome some of the disadvantages of the BOD procedure. One such procedure is to add varying concentrations of an industrial wastewater to a BOD bottle containing an active biological culture (usually mixed liquor activated sludge) from the secondary treatment system. A DO meter equipped with a BOD probe can be used to measure the oxygen uptake rate after the sample is saturated with oxygen. If the industrial wastewater is inhibitory, increased doses will result in reduced oxygen utilization. A similar approach using respirometers allows for the use of larger reactors (up to 10-liters), continuous oxygen feed and strip-chart recording of the uptake rate with time. At the Patapsco Wastewater Treatment Plant in Baltimore, Maryland (see Appendix A) daily routine operation of a respirometer is used as a tool for measuring the inhibitory characteristics of the plant influent. The standard operating procedure for the respirometer involves the use of plant biomass and simulates the plant's biological system (Slattery, 1986).

Recent variations of the respirometry approach utilize special cultures of microorganisms, instead of the POTW bacteria, as more precise predictors of toxic effects. One manufacturer uses specially prepared and packaged bacterial cultures in conjunction with a DO meter to plot families of inhibition curves and to develop lethal concentration dosages analogous to those obtained by bioassay testing. A second technique uses photo-luminescent marine microorganisms, whose light output decreases proportionally to the level of toxic shock when fed varying concentrations of industrial wastewater. This approach has been used extensively in Baltimore, Maryland (see Appendix A) and Chattanooga, Tennessee to evaluate the toxicity of influent wastewaters to the POTW and to measure toxicity reduction through the biological treatment process. At the Patapsco plant in Baltimore, pure oxygen activated sludge treatment reduced the wastewater toxicity by up to 40 percent from the raw wastewater feed.

A summary of the methods available to measure biological inhibition is presented in Table 2-7. The table also includes some cost, testing and training time estimates, of concern to municipalities developing a program for determining inhibitory effects of industrial or waste hauler discharges. A potential problem with all of the techniques discussed is that the results do not accurately reflect the treatability of the wastewater if the biological treatment populations become acclimated to the industrial waste over a long period of exposure.

TABLE 2-1
METAL, CYANIDE AND INORGANIC COMPOUND CONCENTRATIONS INHIBITING BIOLOGICAL PROCESSES
(in mg/l)

	Biological Process			
Pollutant	Activated Sludge	Nitrification	Aerobic Fixed Film	Anaerobic Digestion
Ammonia	$\geq$480	N/A	N/A	1,500-3,000
Arsenic	0.04 - 0.4	N/A	290	0.1 - 1
Boron	0.05 - 10	N/A	N/A	2
Cadmium	0.5 - 10	5-9	5-20	0.02 - 1
Calcium	2,500	N/A	N/A	N/A
Chloride	N/A	180	N/A	20,000
Chromium (Tot.)	0.1 - 20	0.25 - 1	50	1.5 - 50
Copper	0.1 - 1	0.05 - 0.5	25 - 50	0.5 - 100
Cyanide	0.05 - 20	0.3 - 20	N/A	0.10 - 4
Iodine	10	N/A	N/A	N/A
Iron	5 - 500	N/A	N/A	5
Lead	0.1 - 10	0.5 - 1.7	N/A	50 - 250
Manganese	10	N/A	N/A	N/A
Magnesium	N/A	50	N/A	1,000
Mercury	0.1 - 5.0	2 - 12.5	N/A	1,400
Nickel	1 - 5	0.25 - 5	N/A	2 - 200
Silver	0.03 - 5	0.25	N/A	N/A
Sodium	N/A	N/A	N/A	3,500
Sulfide	>50	N/A	N/A	50 - 100
Tin	N/A	N/A	N/A	9
Vanadium	20	N/A	N/A	N/A
Zinc	0.30 - 20	0.01 - 1	N/A	1 - 10

N/A - Not Available

Sources: U.S. EPA (1981a), Russell, et al. (1983), Geating (1981) and U.S. EPA (1986a).

TABLE 2-2
ORGANIC COMPOUND CONCENTRATIONS
INHIBITING BIOLOGICAL PROCESSES
(in mg/l)

	Biological Process		
Compound Type	**Activated Sludge**	**Nitrification**	**Anaerobic Digestion**
Agricultural Chemical			
Common Pesticides	0.05 - 0.10	N/A	N/A
Lindane	5 - 10	N/A	N/A
Aromatics	5 - 150	N/A	100 - 870
Chlorinated Benzenes	0.1 - 5	N/A	0.1 - 1
Halogenated Aliphatics	150 - 250	< 0.1 - 18	0.1 - 100
Nitrogen Compounds	1 - 500	0.1 - 100	5 - 500
Oxygenated Compounds	120 - 500	N/A	20 - 1,000
Alcohols	1,000	N/A	N/A
Acids	N/A	N/A	10
Phenol	90 - 1,000	1 - 10	100 - 200
Chlorophenols	5 - 100	N/A	0.2 - 100
Nitrophenols	50 - 200	150	100
Methylphenols	N/A	5 - 50	N/A
Phthalates	>10	N/A	N/A
Polynuclear Aromatic Hydrocarbons	500 - 2,500	N/A	N/A

N/A - Not available

Principal Sources: Russell, et al. (1983), U.S. EPA (1977b), U.S. EPA (1981a), Geating (1981), U.S. EPA (1986a)

TABLE 2-3
WASTE CHARACTERISTICS PERTAINING TO HAZARDS IN COLLECTION SYSTEMS
(From Busch, 1986)

Term	Description	Examples
Ignitability	Pose a fire hazard	- Gasoline - Industrial solvents
Corrosivity	Corrode standard construction materials	- Acids - Caustics
Reactivity (Explosivity)	- Spontaneous reaction with air or water - Pose explosion hazard - Generate toxics	- Calcium carbide - Cyanides - Sulfides - Industrial solvents - Petroleum hydrocarbons
Fume Toxicity	- Build up of toxic fumes - Pose a hazard to human health	- Metals - Pesticides - Industrial solvents (benzene, toluene)

TABLE 2-4
INTERFERENCE IDENTIFICATION THROUGH PLANT OBSERVATION

Process	Observation	Possible Cause
Primary Clarification	Black/odorous wastewater	Inadequate pretreatment of waste
	Scum overflow	Inadequate pretreatment of waste
	Low solids content of sludge	Hydraulic overloads
Activated Sludge		
- Aeration Tank	Excessive air rates to maintain D.O.	Organic overloads
	Low density sludge	Acid waste, low influent pH, nutrients
	White, billowy foam	Toxics (metal, bacteriocides)
	Dark-brown sudsy-foam, black mixed liquor	Organic overloads, anaerobic conditions
- Clarifier	Pin floc in overflow	Toxic shock load
	Ash-like material floating on surface	High grease content of mixed liquor
	Straggler floc (< 1/4" dia) in supernatant	Changing organic loads
	Cloudy supernatant, poor settleability	Increased organic loads, toxic shock loads, low nutrients
	Sludge rising throughout tank	Low D.O. or low pH (<6.5) in aeration tank

Source: U.S. EPA (1977a), U.S. EPA (1978)

TABLE 2-4 (Continued)
INTEFERENCE IDENTIFICATION THROUGH PLANT OBSERVATION

Process	Observation	Possible Cause
	Localized rising sludge	Organic overloads in aeration tank
	Sludge clumps and bubbles rising to surface	Septic conditions
Trickling Filters		
- Filters	Surface ponding	Organic overloads, excessive biological growth
	Odors	Organic overloads, anaerobic conditions
	Slime colors	Metals, toxic shock
- Clarifier	Increased effluent suspended solids	Excessive sloughing due to pH or toxic shock loads
Lagoons	Poor effluent quality	Organic overloads, toxic shock
	Odors	Low D.O., sulfides
Rotating Biological Contactors (RBC)	Poor effluent quality	Organic overloads, poor pH conditions
	Excessive sloughing	Toxic shock, pH fluctuations
	White, stringy biomass	First stage organic overloads, sulfides
	Odors	Septic influent, sulfides

TABLE 2-4 (Continued)
INTEFERENCE IDENTIFICATION THROUGH PLANT OBSERVATION

Process	Observation	Possible Cause
Anaerobic Digestion	Rotten egg odor	Organic overloads, sulfides
	Rancid butter odor	Toxic shock (metals, ammonia)
	Poor supernatant quality	Organic overloads, toxic shock
	Foam in primary supernatant	Organic overloads
	Sludge temperature drop	Hydraulic overloads
	Scum blanket too thick	High grease content
Aerobic Digestion	Excessive foaming	Organic overloads
	Odor	Low D.O., organic overloads

TABLE 2-5
INSTRUMENTATION OF PLANT PROCESSES AND WASTE STREAMS

Instrument	Locations	Parameters Measured
Conductivity Meter	Industrial discharge Primary effluent Final effluent	Metals, Dissolved solids
Density Meter gamma radiation ultrasonic	Aeration basins Clarifier underflow Conditioned sludge Anaerobic digesters	MLSS Solids concentration
D.O. Meter membrane electrode	Aeration basins RBC stages Final effluent	Dissolved oxygen,
Flow Meter flume, venturi, magnetic, weir	Raw wastewater Sidestream flows Return/waste sludge Chemical feed Final effluent	Flow rate
Gas Analyzers	Collection system Confined spaces Aeration basin off-gas	CO, CO_2, CH_4, H_2S Oxygen transfer
Oxidation-Reduction Potential (ORP) Meters	Industrial discharge Primary effluent	Metal forms
pH Meter	Industrial discharge Collection system Raw wastewater Aeration basins Anaerobic digester Final effluent	Acids, Bases
Selective Ion Electrodes	Industrial discharge Primary effluent Final effluent	Cl^-, CN^-, Cu^+, Cu^{++}, F^-, NH_3, NO_3^-, Pb^{++}, $S^=$
Total Organic Carbon (TOC) Analyzer	Industrial discharge Raw wastewater Primary effluent Final effluent	Organic slugs, oil and grease

Source: James M. Montgomery, Consulting Engineers, Inc.

TABLE 2-6
ANALYTICAL MONITORING OF PLANT PROCESSES

Process	Parameters	Testing Frequency
Clarification	Dissolved oxygen	Daily
	Sludge solids content	Weekly
	Sludge blanket depth	Daily
Activated Sludge	Dissolved oxygen	Daily
	Mixed liquor suspended solids	Daily
	Oxygen uptake rates	Daily/Weekly
	Microscopic examination	Daily/Weekly
	Nutrients	Daily/Weekly
	Sludge volume index	Daily
Trickling Filters	Slime thickness	As needed
	Influent pH, temperature, H_2S	As needed
	Effluent solids content	As needed
RBCs	Dissolved oxygen (each stage)	Daily
	Soluble BOD (each stage)	Weekly
	Biomass thickness	As needed
	Shaft weight	Daily/Weekly
	Effluent solids content	As needed
Anaerobic Digestion	Temperature	Daily
	Solids content	Weekly
	Metals content	Weekly/Monthly
	Volatile acids/alkalinity	Daily
	Supernatant solids, NH_3	Weekly
	Methane content of gas	Daily

Source: James M. Montgomery, Consulting Engineers, Inc.

TABLE 2-7
METHODS FOR EVALUATING INHIBITORY EFFECTS
OF INDUSTRIAL WASTEWATERS

Method	Testing Time	Approximate Equipment Costs	Operator Training
1. BOD Serial Addition	5 days	$1,000	2-4 hrs.
2. Respirometry	30-60 min.	$1,000 - $5,000	4-8 hrs.
3. Packaged Bacteria	30-60 min.	$1,000 - $2,000	4-8 hrs.
4. Photo Luminescense	15 min.	$5,000 - $10,000	20-40 hrs.

Source: James M. Montgomery, Consulting Engineers, Inc.

3. Source Identification

There are two aspects of source identification that should be considered by POTWs when investigating interference problems:

- specific causative pollutants
- industrial source(s) of the pollutants identified

The ease with which a causative agent is identified depends upon the nature of the permit violation. For example, if the interference results from the inability to dispose of sludge, the problem nearly always results from an unacceptable concentration of a particular heavy metal. However, if the plant effluent has a BOD above the permit limit, the problem can range from a shock loading of influent BOD to an inorganic or organic pollutant that is toxic to the biological population in secondary treatment. Isolated spill events are difficult to trace to a specific pollutant unless the pollutant is detected in routine influent and effluent screening or the spill is accompanied by distinct, recognizable odors, appearance, pH or solid residues. Recurring discharges may be linked to a substance with time by process of elimination and analytical testing.

Once an interference is linked to a specific pollutant, the next step is to identify the industrial source. If the POTW has sufficiently charcterized its industrial users as part of its initial pretreatment program development, this task will be greatly simplified. As part of the development of a federally-approved pretreatment program, POTWs are required to conduct a survey of industrial users to characterize their wastes. The POTW should be familiar with each IU's industrial processes and the chemicals which are used, produced, stored, disposed, or otherwise handled on the site. The potential for intentional or accidental discharge of pollutants should be evaluated. The IU survey information should be updated at least annually. Another approach to identifying industrial sources is a tracking program that monitors the interfering pollutants at key interceptors and traces the substance back to its discharge point.

While industries are sometimes responsible for POTW permit violations, the fault can be with operation and maintenance practices at the POTW. Where plants experience chronic operational problems that cannot be linked to industrial waste discharges, the plant staff may wish to conduct a Composite Correction Program (U.S. EPA, 1984) to identify operational problems. If violations persist, then a more comprehensive search for industrial sources is justified. The CCP was developed by the U.S. EPA as a means to provide "information on methods to economically improve the performance of existing POTWs". It outlines an approach for POTW personnel to evaluate POTW operations and implement systematic improvement steps.

3.1 CHRONIC DISCHARGES

Industrial waste monitoring is the key to successfully identifying most chronic industrial waste sources. Industries should be monitored for conventional pollutants, with the testing of other compounds determined by the nature of the specific industrial waste. In the case of categorical industries, some substances of concern and pretreatment requirements are already specified by the regulations. For noncategorical industries, information such as permit applications and questionnaire responses or specific analytical testing of industry effluent should provide sufficient data to establish a monitoring program.

Sewer use ordinances and industrial waste management programs typically provide for some means of monitoring an industry's discharge to the municipal collection system. Such ordinances require measurements of both quantity and quality of the industrial or combined domestic/industrial flow. Industrial discharges are usually monitored both by industry, with regular self-reporting requirements, and by the municipality.

If it has been determined that a plant upset is being caused by industrial wastes and is not a result of other POTW deficiencies, then it is up to POTW personnel to identify the specific source of that upset. This may necessitate expansion of the POTW's monitoring program as discussed in the next subsection. POTWs experiencing interference problems tend to fall into one of three categories regarding interference:

1. A single major industry in town dominates the waste characteristics at a relatively small POTW.

2. One or two industries among several are primarily responsible for waste strength fluctuations in small to medium-sized POTWs.

3. Industrial wastewater from numerous sources controls the wastewater feed characteristics, with no single dominant industry.

The first category listed above is by far the easiest situation to deal with from an identification standpoint. By monitoring the industry's discharge, POTW influent and effluent and other relevant plant operations, the impact of the industrial waste on the POTW can be determined. The cities of Oswego, New York and Tolleson, Arizona are examples of small facilities significantly impacted by a single industry.

Category two is a more difficult interference to trace. A monitoring program may be sufficient if a large database exists covering a period of time. Unfortunately, when numerous industries must be tested on a frequent basis, the sampling and analysis costs can be high. Routine sampling for all industries with additional sampling for troublesome industries may provide a solution for some POTWs. For example, Paris, Texas set up a comprehensive short term (90 day) sampling program that industry supported financially. Through this effort, Paris was able to distinguish which industries were likely to be problems and then could adjust their long-term sampling accordingly.

The third category generally applies to larger facilities which are less likely to be susceptible to any particular industrial effluent. Baltimore, Maryland and Passaic Valley, New Jersey are examples of facilities which fit into this third category, but have experienced interference (see Appendix A). Large plants may be less likely to experience permit violations due to industrial waste, but they have frequently experienced inhibition and other operational and maintenance problems. Intermittent discharges are particularly difficult to pinpoint by POTW personnel because of large service areas.

3.1.1 Routine Monitoring

In order to have the ability to utilize POTW influent characterization to identify the source of interfering pollutants, adequate background and supporting information must be available to POTW personnel. A database obtained over several years of routine monitoring enables a POTW to develop action level criteria for key parameters. When monitoring shows that these criteria have been exceeded, it can be suspected that a spill or unauthorized discharge of industrial waste has occurred, which triggers a tracking program. Specific details of industrial monitoring programs have been outlined by EPA and others (EPA, 1983; WPCF, 1982).

Routine compliance monitoring, which is part of any local industrial waste control program, will sometimes serve to generate an adequate background database. However, POTWs which have interference problems may need to perform additional monitoring until the source of the problem can be identified. For compliance monitoring purposes, monitoring methods and frequency are generally specified by each municipality in its pretreatment program documents or sewer use ordinance and in discharge permits, contracts or orders issued to industrial users. Self-monitoring by industry with monthly checking by the municipality enables the POTW and the industrial users to share the expenses of monitoring. Such an approach is most successful when:

- key manholes or representative sampling points are available
- sampling procedures are clearly outlined and followed
- a qualified laboratory performs the analytical testing
- rigorous reporting requirements are established for the industries
- spot checking by the municipality is performed on a frequent yet random basis

The alternative to self-monitoring is for a municipality to perform all sampling and analytical services on a once-per-month or once-per-quarter basis, depending on the significance of the specific industry to the POTW. Under this scenario, split samples should be made available to the industry, if requested, to provide them with the opportunity to verify the test results from which their compliance status and user fees will be determined. Many municipalities prefer not to place major reliance on industrial self-monitoring for compliance determinations; they are able to recover the costs for their monitoring programs by assessing fees for industrial discharge permits or by directly billing the sampling costs to the industrial user.

Regardless of the approach taken, the objective of any industrial monitoring program is to obtain representative analytical results of the wastewater flow and characteristics. An industry with highly variable quality and quantity should be sampled more frequently than one with a consistent effluent quality. An appropriate sampling schedule or discharge schedule for batch processes should be determined for the industry.

If industrial wastes have been well characterized and adequately monitored, then the identification of an interfering or potentially interfering pollutant source will be facilitated. As an example, if a POTW suspects a change in their influent wastewater characteristics by observing a change in one or more operational parameters, this triggers influent sampling. The interfering pollutant and concentration are determined through analytical testing, which is then compared with the information from the monitoring database to identify industries that discharge (or have the potential to discharge) the problem pollutant. In some cases, especially large sewer systems, it is not easily determined which of many industrial contributors is responsible for a particular pollutant that is causing an interference. However, several large POTWs including Baltimore, Maryland and Hampton Roads, Virginia have experienced success after setting up their monitoring programs. It has even been suggested that the mere fact that they set up a program motivated some industries into cleaning up, rather than risking the consequences. Those large POTWs that have put effort into their monitoring program have been successful.

3.1.2 Tracking Program

A tracking program is a procedure developed for locating the source(s) of a pollutant or impact which has been identified at a POTW. Depending on the size of the POTW, the sewer system and the type and number of industrial users, this procedure may be very simple or rather complex. A small system with only a few industrial contributors will probably not require anything more than a procedure for comparing POTW influent sample characteristics with industrial monitoring results. On the other hand, large systems may require sophisticated programs involving computer analysis.

The City of Baltimore has a computer program that attempts to trace contaminants back to the source, knowing the necessary background data (see Appendix A). Batch printouts, called the "Daily Average Mass Discharge Reports," provide monthly listings of companies grouped by sewer service area and chemicals used, stored, and/or discharged. If a chemical compound (such as a solvent) can be identified by the tracking team, or later by means of sample analysis, a search of the Data Management System's batch printouts can identify possible industrial sources.

Rapid toxicity testing procedures may become valuable tools for identification of interference sources as they gain acceptance by municipalities. A toxic impact can be traced upstream through a collection system very rapidly when the test procedure takes less than 30 minutes. Such a system has been used at Baltimore's Patapsco Plant to identify influent toxicity problems. This approach to interference tracing is most useful if the troublesome industry discharges toxicants. Municipalities must continue to rely on more conventional monitoring practices for upsets resulting from non-toxic contamination.

One of the most comprehensive tracking programs is maintained by the Hampton Roads Sanitation District (HRSD) in the Tidewater area of Southeastern Virginia. The HRSD operates nine treatment plants handling 130 million gallons per day generated over a service area covering 1,700 square miles. Industrial wastes from 300 sources are dominated by military installations, with other significant discharges from manufacturing and food processing. Industrial discharges are categorized according to which of the following methods of tracking is employed:

- sensory observations
- measurements with field equipment
- sampling and analysis

For the first two types of tracking methods, the HRSD has personnel on stand-by duty supplied with radio equipped vehicles and extensive field sampling and lab equipment capable of qualitative, as well as quantitative, analyses. Tracking begins by HRSD personnel checking pump stations and sewer lines in a downstream to upstream fashion until the source is isolated. Along the way samples are collected, labeled and preserved as evidence.

For the third type of tracking method, automatic sampling equipment is set up at key locations throughout a service area. The samples are collected each day and analyzed. After pollutant concentration trends are determined, the samplers are moved upstream. This general procedure is continued until the source of the problem is found.

In either case, once the source(s) is located, the industry is contacted directly and actions taken appropriate to the circumstances. All costs associated with the investigation, clean-up and any other item are billed directly to the source. The HRSD has found that just by having a highly visible industrial waste investigative team, users are deterred from unauthorized discharge to the sewer system. As a result, incidences have decreased by more than 50 percent in the last eight years.

Tracking programs such as HRSD's are most successful at tracking chronic discharges. Although not as easily accomplished, isolated spills and unauthorized slug discharges of short duration can be tracked if quick, aggressive action is taken. The next section discusses isolated spills in more detail.

3.2 ISOLATED SPILLS AND UNAUTHORIZED DISCHARGES

Interference-causing materials frequently enter POTWs as spills and unauthorized discharges. The sources generally fall into one of the following categories (Busch, 1986):

- transportation accidents or leaks
- storage tank or transfer pipe leaks
- industrial discharges

- industrial accidents
- fires in warehouses and commercial operations
- waste haulers
- midnight dumpers

The focus of this manual is on industrial discharges, industrial spills, and waste hauler discharges (both legal and illegal), because these are the problems over which the POTW usually has the most control. However, POTWs may be able to control some of the other problems listed by extending the spill prevention and control plan procedures described in this manual to any business that has toxic or hazardous materials on site. The POTW would have to assess its legal authority to set up this type of comprehensive program.

The extent of the spill and illegal discharge problem in POTWs is severe. In the spring of 1985, the Association of Metropolitan Sewerage Agencies (AMSA) surveyed 107 of their member municipalities concerning hazardous waste discharges to their facilities. The respondents to the survey represent 308 POTWs, corresponding to 39 percent of the estimated total flow and 47 percent of the estimated industrial flow nationwide. The results of the survey indicated that hazardous wastes, if improperly discharged, can have serious effects on POTWs. Specifically, the survey showed:

- nearly all POTWs receive hazardous wastes

- the most commonly discharged wastes are corrosives, solvents, electroplating baths and sludges

- the most commonly reported sources of these wastes are spills, illegal discharges from industries and routine discharges from industries

- half of the respondents indicated the discharge of explosive or flammable materials (gasoline, toluene, naphthalene, benzene, xylene, jet fuel) and nearly half reported corrosion of the sewer lines due to acids and hydrogen sulfide gas

- approximately 30 percent of the respondents have experienced one or more biological treatment system upsets since 1980 resulting from the presence of metals, cyanide, diesel fuel, toluene, paint thinner or stripper, iodine, thiocyanate and pesticides

It is clear that slug discharges resulting from spills, batch releases, dumps, and illegal discharges are a common concern for many POTWs. It is the responsibility of industry to notify a POTW of a slug discharge under federal regulations (40 CFR Part 403.12(f)). The regulations describe a slug loading as any pollutant, including oxygen-demanding pollutants (BOD, etc.), released in a discharge at a flow rate and/or pollutant concentration which will cause interference at the POTW. However, POTWs do not always receive proper notification. One POTW (HRSD) has responded to slug loads by contacting its major industries in the service area immediately upon detection. This action is taken for the following reasons:

1. an IU may not be aware that it is causing a problem;
2. it brings the problem to industry management attention;
3. it provides visibility for the POTW's control program;
4. it discourages illegal discharges;
5. if the problem is later tracked to an industry, the fact that the industry was notified of the problem immediately may stengthen enforcement proceedings against an uncooperative industry; and
6. there may still be time to correct the problem.

The mitigation efforts described in Section 4 related to industrial spills focus mainly on prevention measures and in-plant corrective measures that are best implemented if proper notification is received by the POTW. The use of permanent gas detection equipment in sewer lines or treatment plant headworks is a method of detecting certain types of pollutants that does not rely on industry notification.

Tables 3-1 and 3-2 provide some examples of industrial spill incidents as documented by Busch (1986) and Attachment 2 of the AMSA survey report.

3.2.1 Hauled Wastes

Identification of a waste hauler as the source of an interference is sometimes a difficult task. Hauled wastes can be discharged to convenient manholes and the hauler gone before the waste reaches the POTW. There are examples where hazardous waste haulers have paid industries for the seclusion their facility provides during such illegal discharge events. Approaches used to help alleviate the problem include:

- periodic sampling of suspected sewer lines
- surveillance of waste haulers and suspected discharge points
- education of industries concerning the seriousness of these violations
- increased public awareness of illegal dumping
- increased enforcement

Many states have enforcement programs to assist POTWs in detecting illegal discharges. Local law enforcement officials can also be requested to assist in surveillance activities and enforcement. Video surveillance of suspected manholes or storm drains is also a possible option. Some POTWs use locking manholes to discourage illegal dumping at suspected sites.

Table 3-3 gives examples of the impacts of hauled wastes on both the collection system and treatment plant in cities identified through the AMSA survey. A number of problems indicated by the AMSA survey showed the source as "unknown", which is indicative of the problems associated with tracking hauled waste interferences. In the Louisville, Kentucky example given in Table 3-3, the waste hauler discharged the hexachloropentadiene to a manhole located within a tobacco warehouse (Busch, 1986).

Interference can also occur when hauled wastes are discharged legally to treatment plants. POTWs that accept discharges of hauled waste should establish control procedures to ensure that the wastes are compatible with treatment processes. Procedures for regulating waste haulers are discussed in Section 4. Identification of a waste hauler as the source of interference can be facilitated by employing such measures as:

- restricting hauled waste disposal to designated, monitored sites in the collection system or at the treatment plant
- permitting waste haulers
- requiring submission of a tracking form that documents the origin, transportation, and disposal of the waste
- sampling hauler loads (samples only analyzed if there is a plant impact)
- permitting, sampling, and inspecting the waste generator

Submission of a tracking form, called a waste manifest, is already a federal requirement when haulers are discharging hazardous wastes. The Resource Conservation and Recovery Act (RCRA) places requirements on hazardous wastes received by truck, rail, or dedicated pipeline into POTWs. It is important that POTW operators become aware of these RCRA requirements and the need to coordinate their local procedures for accepting hazardous wastes with State and EPA personnel. To provide information and guidance on the RCRA hazardous waste requirements and their implications for POTWs, EPA has published a manual titled RCRA Information on Hazardous Wastes for Publicly Owned Treatment Works (EPA, 1985c).

3.3 RAPID SCREENING TECHNIQUES

Once an interference is suspected, a number of rapid chemical tests, available from chemical supply houses, can provide preliminary indication of the presence of substances thought to be producing the interference. These tests help determine in seconds the need for more thorough quantitative analysis and tracking. In addition, these screening tests are also useful when evaluating the loads of waste haulers at dumping stations (Section 4.2.4).

1. Metals — Chemical test strips utilizing color change indicators may be used to detect the presence and concentration of specific metals.

2. Solvents — Gas detection tubes, sensitive to gases and vapors, can indicate the presence and concentration of solvents, but may not be reliable for determining the specific solvent type due to chemical interferences among similar-type solvents. A portable hand pump draws in a calibrated amount of air through the detector tube, and the amount of color change indicates the concentration.

3.4 SUMMARY

Source identification is the key aspect of any industrial waste management program. Identifying the source(s) of interference-causing substances being discharged from a variety of industries is not an easy task and must be approached with an aggressive, well conceived program if it is to be successful.

There is no simple step-by-step procedure to follow to efficiently identify the source of every interference problem. However, a rational approach to the problem can be employed for some interferences which can minimize the effort required. Figure 3-1 is a flow chart that suggests a possible approach to dealing with permit violations or upsets. It basically outlines steps to be taken at the treatment plant to identify possible pollutants causing problems. Figure 3-2 is a flow chart developed by the HRSD that outlines the steps they take in the event that the treatment plant is upset or unusual influent is detected. It must be recognized that each POTW presents a unique management and operations structure to go along with process variations. Therefore, it is important to realize that Figures 3-1 and 3-2 are only examples, and not necessarily applicable to all POTWs. The most important aspects of any source identification or tracking program are well thought out procedures coupled with an aggressive approach to enforcement.

TABLE 3-1

INDUSTRIAL SPILLS OF HAZARDOUS MATERIALS: IMPACT ON SEWER COLLECTION SYSTEM

City	Industry	Pollutants	Impact
Akron, OH	Rubber Mfg.	Naphtha, Acetone, Isopropyl Alcohol	Explosion
Bayville, NJ	Pharmaceutical	Sulfides from high BOD	Corrosion
Bergen, NJ County	Water Treatment	High and low pH	Corrosion
Bloomington, IN	Grain Processing	Hexane	Explosion
Dayton, OH	Electroplating Food Processor	Acids	Corrosion
Forth Worth, TX	Gasoline Station	Gasoline	Explosion
Hillborough, FL	Battery Salvaging	Acids	Corrosion
Jacksonville, FL	Organic Chemicals	Solvents	Corrosion, Odors
Los Angeles, CA County	Petroleum Refining	Sulfides	Corrosion
St. Paul, MN	Metal Finishing	Acids	Corrosion
Toledo, OH	Adhesives	Glue	Plugged Sewers
WSSC, MD	Photofinishing	Sodium Bisulfite, low pH	Corrosion

Sources: Busch (1986), AMSA

TABLE 3-2

INDUSTRIAL SPILLS OF HAZARDOUS MATERIALS: IMPACT ON TREATMENT PLANT

City	Industry	Pollutants	Impact
Boise, ID	Electroplating	Cu, Ni, Zn	Reduced treatment efficiency
Camas, WA	Pulp Mill	Chlorine	Biological upset (2 days)
Camden, NJ County	Dye Mfg.	Aniline	Biological upset, sludge contamination
Dallas, TX	Organic Chemicals	Xylene, Toluene	Fouled carbon scrubbers
Depue, IL	Fertilizer Mfg.	Sulfuric Acid	Biological process wiped out

Sources: Busch (1986), AMSA

TABLE 3-3

IMPACTS OF WASTE HAULER DISCHARGES ON POTWs

City	Pollutants	Impact
Central Contra Costa, CA	Solvents	Biological process wiped out
Louisville, KY	Hexachloropentadiene	Treatment plant out of operation for 3 months
Rockford, IL	Electroplating sludge Cd, Cr, Pb, Zn, CN^-	Hydrogen cyanide gas production potential
San Diego, CA	Gasoline	Sewer explosion

Source: U.S. EPA (1986a)

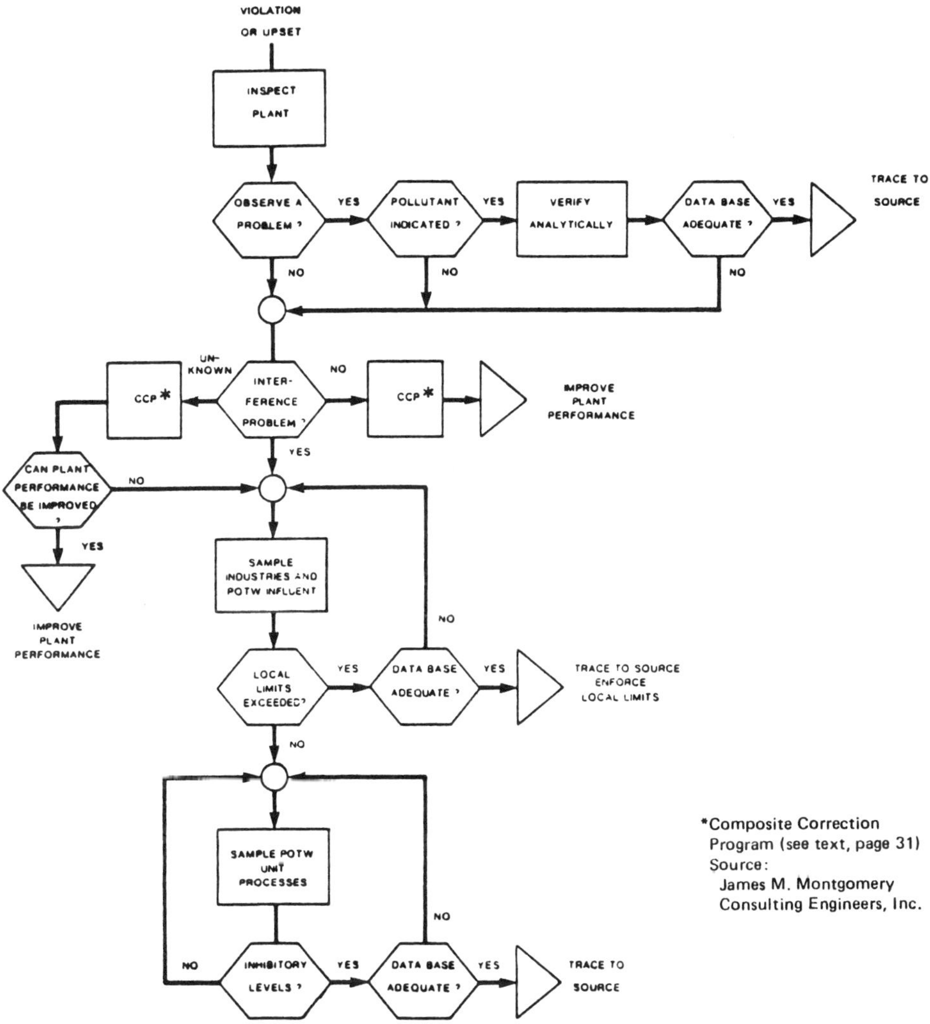

*Composite Correction Program (see text, page 31)
Source:
James M. Montgomery Consulting Engineers, Inc.

FIGURE 3-1
TREATMENT PLANT UPSET IDENTIFICATION PROCEDURES

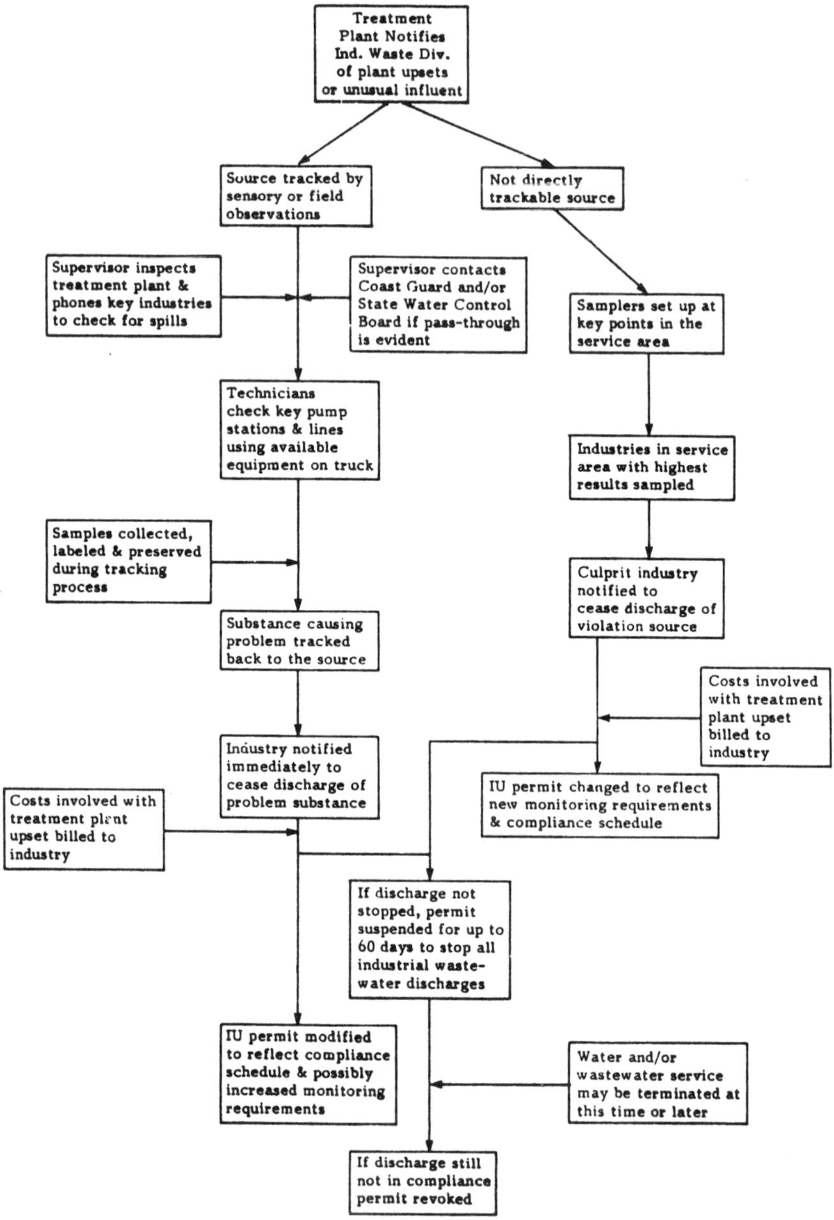

FIGURE 3-2
HRSD SOURCE TRACKING PROCEDURE

4. Mitigation

Mitigating an interference is the goal following the detection of an interference problem. Whether source identification precedes mitigation depends on the success of the POTW's tracking program, its knowledge of its IUs, and the other issues discussed in Section 3. In certain cases, interim measures to address interference can be taken without initially defining the interfering pollutant substance or source, although this information can be very helpful. However, even if an isolated interference event can be handled by process modification at the treatment plant, the source of the interfering discharge should be identified and controlled. Interference mitigation by pretreatment and source control or legal and enforcement remedies obviously requires information about the discharger(s) causing the problem, but results in a more reliable solution.

The success of any effort to mitigate interference is dependent to a great extent on the characteristics of the pollutants causing the interference, the characteristics of the treatment plant (capacity, capacity utilization, biological process, etc.) and the type(s) of mitigation attempted. It is important to emphasize that mitigation of an interference problem is generally not a straightforward process. Each POTW possesses unique characteristics that exclude generalized solutions or approaches so that a combination of techniques is often necessary to realize satisfactory results.

4.1 TREATMENT PLANT CONTROL

The effects of industrial pollutants on a typical POTW can be eliminated or minimized through a number of measures initiated at the treatment plant, often in combination. They can be generally categorized as:

- biological process control
- biological augmentation
- chemical additions
- operations modifications
- physical modifications

The list above is generally in order of increasing implementation difficulty, e.g., biological process control generally requires only minor changes in plant operation while physical modifications can include costly capital improvements.

4.1.1 Biological Process Control

Biological process control is generally limited to activated sludge systems, although some modifications to fixed film processes (e.g., trickling filters, rotating biological contactors) might be considered as a form of biological process control. An activated sludge system is generally monitored or controlled by utilizing one or more of three process parameters: mean cell residence time (MCRT), mixed liquor suspended solids (MLSS) and food to microorganism ratio (F/M). The biomass and its characteristics are controlled by varying these interrelated process parameters. The following changes to these parameters have been observed to mitigate the effects of industrial pollutants on an activated sludge system:

1. **Increase the Mean Cell Residence Time.** Increasing the MCRT (sludge age) has been shown to have the effect of reducing the inhibitory effects of all forms of toxic industrial contaminants. By increasing the MCRT at the first sign of a possible toxic upset, (by decreasing the solids wasting rate) the inhibitory effect of any toxicant will generally be less than if no action is taken.

2. **Increase the Mixed Liquor Suspended Solids.** High mixed liquor suspended solids (MLSS) concentrations have been shown to offset some of the effects of industrial pollutants. A high MLSS provides the best conditions for biosorption and acclimation to a toxic substrate. Increasing the sludge return rate to the aeration basin at the first indication of toxic upset, while at the same time diverting and storing any remaining toxic influent away from the aeration basins, will lessen the impact of a short term upset and cause quicker biomass acclimation to a long term problem.

3. **Decrease the Food-to-Microorganism Ratio.** This parameter is directly related to both the MCRT and the MLSS. It has been observed that decreasing the F/M causes improved biodegradation of toxic comtaminants, and expedites biomass acclimation.

Table 4.1 summarizes these process control steps.

The process control steps described apply to both activated sludge systems treating for carbonaceous removal and nitrifying systems. Generally, the steps described are beneficial for treating any type of interfering pollutant, whether it be a metal, toxic organic or high-strength conventional pollutant.

For a fixed film process, control of the biomass characteristics is not as easily accomplished. However, varying the amount and point of recirculation in a trickling filter can modify the inhibitory effect of industrial pollutants. Recirculating secondary clarifier effluent is a means of achieving the greatest dilution effect, which may be desirable for high-strength organic waste or toxics. Should excessive biomass sloughing be a problem due to toxic pollutants, returning uncontaminated secondary clarifier underflow may help in maintaining a proper biomass population.

4.1.2 Biological Augmentation

Biological augmentation is a method by which selected microorganisms are added to an existing biological population in an attempt to improve some characteristic of the biological system. Conclusive evidence is lacking, but biological augmentation of secondary treatment systems has been reported to improve some industrial pollutant treatment by promoting the specific microorganism populations that successfully degrade particular pollutants. Other enhancements reported include reduced sludge production and increased COD removal rates (Grubbs, 1986). The addition of selected microorganisms to an aeration basin is relatively inexpensive and in the worst case will have no effect on treatment. The EPA, through ongoing experiments at the Wastewater Engineering Research Laboratory in Cincinnati, Ohio, is presently studying the subject in greater depth. Maiden Creek, Pennsylvania employed biological augmentation to improve treatment, however the results of these efforts were clouded due to other modifications made at the same time (see Appendix A). Rotating biological contactors plants have used selected bacteria under substrate-limiting conditions as a control on biomass growth, but with limited success.

After a treatment upset has occurred, biological augmentation by reseeding with viable microorganisms is a useful step in getting a plant up and running quickly. Having commercially packaged microorganisms available and in supply at a treatment facility may help in speeding such a recovery if reseeding from another treatment facility is difficult.

4.1.3 Chemical Addition

The addition of chemicals or nutrients to the wastewater stream in existing treatment steps has been shown in many instances to mitigate the effects of some industrial pollutants. The following are examples of chemicals or additives that have been shown to improve industrial wastestream treatability or biological process stability:

- chlorine
- nutrients
- lime or caustic
- organic polymers
- inorganic coagulants
- powdered activated carbon

Table 4-2 lists these chemicals and additives, the reasons for their use and the resulting effects. The reader is cautioned that the generalizations in the table do not apply to all situations. Some exceptions are pointed out in the text.

Chlorine. Chlorine has been shown to be successful in controlling bulking activated sludge caused by industrial pollutants from such industries as textiles, breweries and wood and paper products. Points of chlorine addition vary, but best results generally occur when chlorine is added to the aeration basin effluent or return activated sludge (RAS). The Horse Creek Plant in North Augusta, South Carolina and the East Side Plant in Oswego, New York are examples of facilities which have successfully employed chlorination to control bulking sludge (see Appendix A).

Nutrients. Phosphorus addition and, to a lesser extent, sulfur and nitrogen addition, occasionally improve biological treatment and sludge settleability of industrial wastewater with high carbonaceous content. In general, better treatment and settleability is attributed to correcting a nutrient deficient condition resulting from a high industrial/domestic wastewater ratio.

pH Adjustment. Lime and caustic are sometimes successful at mitigating the effects of some heavy metals on activated sludge systems. Addition of either before primary treatment has the effect of raising the pH which generally improves precipitation of heavy metals in primary clarifiers. There are exceptions to this generalization, however. For example, it makes a difference whether the pH is being raised from 2 to 6 or from 7 to 11. In this latter case, iron and chromium will go into solution rather than precipitate. Optimum pH ranges exist for metal insolubilities, but these ranges are affected by many factors and are therefore system dependent. Lime can also be used for pH adjustment of an acidic wastewater prior to aeration to provide a more conducive environment for biodegradation.

Coagulants. Polymers and inorganic coagulants such as alum and ferric chloride are introduced to POTW wastestreams in part to help mitigate the effects of industrial pollutants. Added prior to primary treatment, the coagulants improve primary sedimentation and may increase the removal of toxic pollutants before they reach the aeration basins. Added after the aeration basins, the coagulant aids can assist in controlling bulking sludge and reducing effluent suspended solids. Jar testing is an important part of any chemical addition program as the best means of determining optimum dosages. The North Shore Sanitation District in Gurnee, Illinois has utilized coagulants successfully for mitigating the effects of interference (see Appendix A). It should be noted that chemical coagulants affect the characteristics of the sludge and could alter ultimate disposal methods. If added after secondary treatment, they could increase the toxicity of the recycle sludge. Therefore, their use should be carefully evaluated and contamination potential should be investigated.

Activated Carbon. The addition of powdered activated carbon (PAC) to an activated sludge unit has been successful at reducing the inhibitory effect of toxic organic chemicals. By providing adsorption sites, the organic pollutants not biodegraded are removed by the activated carbon. The activated carbon also improves sludge settleability by providing dense floc nuclei. A patented process (PACT, licensed and sold by Zimpro, Inc) exists employing this treatment concept at full scale. However, even a slug additon of PAC to an aeration basin known to contain toxics can significantly reduce the effects of the toxics on the biomass.

4.1.4 Operations Modification

Activated Sludge Alternatives. A further means of mitigating the effects of industrial pollutants on POTWs is through modifying the operation of existing treatment steps. Activated sludge systems are often designed to operate in several different "modes" (e.g., step aeration, contact stabilization, etc.) by providing the appropriate physical layout. Some modes of operation have been shown to be more successful than others at mitigating the effects of industrial contaminants, particularly those dosed in highly variable concentrations. It has been shown at the laboratory and plant-scale that extended aeration and step aeration (step feed) are generally more resistant to upset than complete mix and conventional activated sludge (see East Side Plant, Oswego, New York, Appendix A). It appears that complete mix generally provides more consistent treatment, particularly under shock loading conditions, than conventional plug flow treatment. The contact stabilization mode is generally less successful at treating industrial pollutants than other modes, particularly when the organic matter is predominantly soluble and waste strength fluctuations are common.

Staged Treatment. A successful means of mitigating the effects of industrial contaminants on any biological treatment process is through the use of staged treatment. Many treatment systems have realized improved conventional and industrial pollutant removal when switching from parallel treatment to series treatment. For example, two aeration basins operating in series are generally more successful at mitigating the effects of industrial contaminants than the same two basins operating in parallel. The same principles have been observed to apply equally to fixed film processes and fixed film/suspended growth combinations.

Excess Biomass. A typical response of a fixed film process to some industrial waste stressing is excess biomass growth, resulting in clogged media and reduced treatment efficiency. Should this be a problem, treatment is generally improved if the biomass population (thickness) can be reduced. By increasing or altering shearing forces, biomass sloughing increases. This can be accomplished by altering the direction of flow through RBCs and submerged fixed film basins, or by increasing or altering the aeration pattern (if any) in the basins. A second means of inducing increased biomass sloughing is through chemical addition, but this approach is potentially harmful to the biomass and should only be attempted under the guidance of professionals skilled in the use of such chemicals.

4.1.5 Physical Modification

The most permanent type of industrial pollutant mitigation effort that can be undertaken at the POTW itself comes in the form of physical addition to or modification of the treatment system. Successful modification of treatment plants for industrial waste effects mitigation have included the addition of new plant facilities such as flow equalization and physical/chemical treatment steps, the addition of facilities for adding chemicals (as previously discussed) to existing treatment processes, and the modification of existing biological systems (i.e. converting to oxygen activated sludge or replacing rock trickling filter media with plastic media).

Flow Equalization. Adding flow equalization prior to biological treatment units has the effect of dampening any slug or diurnal loads of noncompatible or high-strength industrial contaminants entering a treatment plant. Pollutants that intermittently enter a POTW in inhibitory concentrations can be diluted by flow equalization to noninhibitory levels and thus, not adversely impact the biological system. Maiden Creek, Pennsylvania provides a dramatic example of the effects of non-equalized industrial flows (see Figure A-4). In this particular case, hydraulic shocks were accompanied by organic shocks that resulted in solids carryover, reduced BOD removal efficiency and sometimes total biological process failure.

Instrumentation/Control. Some POTWs use a variation of the flow equalization principle with success, especially when toxic metal pollutants are involved. pH and conductivity of the influent wastewater is measured and recorded continuously in the influent. When the pH drops or conductivity rises drastically, possibly indicating an increased heavy metal level, the influent flow is diverted to a holding basin until such time that the pH and conductivity in the influent return to normal. At that time, the diverted wastewater can be bled back to the influent wastestream in a manner such that metal concentrations are diluted and do not inhibit the biological system. This type of technique may become more useful in the future as continuously recording specific ion electrodes are developed for more pollutants.

An example of similar control steps is Chicago Heights, IL. Officials there were alerted to a pesticide spill that entered the sewer system. Operators were able to isolate the incoming spill to some parallel primary clarifiers, activated sludge and aerobic digester tanks where the toxic materials were subsequently treated chemically and biologically (Busch, 1986). Passaic Valley, New Jersey and Newark, Ohio employ similar procedures when necessary (see Appendix A).

Special Treatment Operations. Other treatment steps that might be added depend on the interfering industrial pollutants. The addition of flotation/skimming tanks is beneficial for removing pollutants like oils, greases or other water-immiscible compounds. Separate settling basins may be beneficial in some cases for chemical treatment to precipitate metals or cause coagulation of unsettleable solids.

Pure Oxygen Activated Sludge. Another type of treatment plant modification that has experienced some mitigation success is the replacement of an existing air activated sludge unit with oxygen activated sludge. Pure oxygen activated sludge has been reported by U.S. EPA (1981c) to be a more biologically stable process with improved sludge settleability over conventional air facilities when responding to toxic or high-strength organic loadings. However, a disadvantage with a covered oxygen system is that volatile organics can build up to potentially explosive levels inside the covers. Baltimore, Maryland and Passaic Valley, New Jersey have both experienced problems of this type.

Oxygen Transfer. Increasing the efficiency of oxygen transfer in aeration basins will help mitigate the effects of high-strength conventional pollutants. Retrofitting existing coarse bubble or turbine aeration units with fine bubble units may provide additional treatment capacity for a high-strength waste (see Newark, Ohio and Maiden Creek, Pennsylvania in Appendix A). However, maintaining oxygen levels above 2-3 mg/l has not been shown to consistently result in a better treatment of conventional or organic pollutants.

4.1.6 Summary

Table 4-3 summarizes the available measures that may be employed at a treatment plant to mitigate interference effects.

4.2 PRETREATMENT AND SOURCE CONTROL

Pretreatment and source control of interfering industrial pollutants is the most direct and efficient way of mitigating the effects of industrial pollutants because the cause of the interference never reaches the POTW. This reasoning was the impetus for the General Pretreatment Regulations which specify the guidelines under which municipalities must develop pretreatment programs. It is not the intent of this guidance manual to discuss pretreatment guidelines, complete program development or details of industrial treatment processes. Rather, this discussion is intended to document elements important to bringing about pollutant source control, whether as part of a municipal/industrial cooperative agreement or a fully approved pretreatment program.

4.2.1 Local Limits

Setting local industrial discharge limits is one of the best and most direct ways of mitigating any industrial interference that may exist at a POTW. Federal Categorical Pretreatment Standards must be applied by POTWs with federally-approved pretreatment programs, but this does not guarantee interference prevention because of the uniqueness of POTWs and the waste they treat. In addition, noncategorical industries are not regulated by such federal standards. Setting rational, technically-based local limits in a fair and equitable manner is a sound approach to preventing interference. The General Pretreatment Regulations (40 CFR Part 403.5(c)) require POTWs with federally-required pretreatment programs and other POTWs which experience pass-through or interference to establish local limits. Details on the development of local discharge limits are contained in the "Guidance Manual for POTW Pretreatment Program Development" (U.S. EPA, 1983). In addition, a computer program/model for helping municipalities develop local limits has been developed (U.S. EPA, 1985a) and is available from the EPA Office of Water Enforcement and Permits.

4.2.2 Accidental Spill Prevention

It is in the best interests of any municipality to consider developing an accidental spill prevention program (ASPP). The purpose of an ASPP is to provide "...a set of procedures and a regulatory structure that will minimize the chance that accidental spills of toxic materials will damage a municipality's collection system or treatment plant" (U.S. EPA, 1986b). The principal elements of an effective municipal ASPP are:

- identification of potential sources and types of spill materials
- adequate regulatory control
- POTW review of industrial user spill prevention programs
- complete emergency response procedures
- documentation of the development strategy

Spill materials would include all sources and types identified for industrial pretreatment, but would also include apparently insignificant users who have the potential for spillage into floor drains connected to a POTW. Facilities such as chemical warehouses, radiator shops, etc., which are supposedly "dry" or usually recycle all harmful wastes, could have an accident that would impact a POTW.

A POTW should require industrial users to develop their own in-house ASPP and the program should be reviewed for thoroughness and effectiveness by the POTW. Industrial user ASPPs, as well as the overall ASPP should include complete emergency response procedures by all involved parties. These procedures must be outlined in enough detail to be effective and all the appropriate personnel must be adequately familiar with the necessary emergency steps.

Finally, the development of the ASPP must be well documented so that as time passes and modifications become necessary, a written record of the program development will be available for consultation. This record should prevent needless rethinking of old ideas.

An active spill prevention program with a high degree of visibility can have a positive impact on reducing unauthorized discharges of industrial wastes. Figure 4-1 outlines the fundamental procedures in the development of an ASPP.

4.2.3 Pretreatment Facilities

There exists a wide variety of treatment processes applicable to industrial pretreatment, depending on the wastestream pollutants, the volume of the wastestream and the extent to which the waste must be treated. The application of specific treatment streams is not addressed by this document. However, many typical municipal treatment processes can be applied to some industrial wastestreams. There are many other types of treatment processes, usually physical/chemical, applicable to pretreatment applications.

In many cases where industries have been required to pretreat wastes, it has been found that wastewater flow equalization, pH neutralization or conservation and recycle/reuse have been all that are necessary to meet discharge limits and eliminate interferences. Process modifications or wastestream recovery processes (such as for metals) have in some cases ended up saving industries money in addition to reducing pollutant loads. These aspects of pretreatment should be emphasized in discussions with industries. The Horse Creek facility in North Augusta, South Carolina, experienced significant operational improvement from relatively small industrial operation changes (see Appendix A). Modifications such as discharging sump water from the surface rather than the drain and equalizing pumping schedules, so as to minimize hydraulic peaks were typical of successful adjustments.

4.2.4 Regulation of Waste Haulers

POTWs that accept discharges of hauled waste should establish procedures to control the wastes so as to ensure that they are compatible with the treatment process. A waste hauler permit or "manifest" system is an effective method of regulation. Use of such a system to document the origin, transportation, and disposal of the waste, along with a source control program (permitting, sampling and inspecting the generator) and predischarge sampling, will provide a high degree of control over incoming wastes. Figure 4-2 presents an overview of the procedures of a waste hauler permit system.

POTWs may choose to restrict the discharge of hauled waste either to a designated point in the collection system or to the plant itself. These restrictions may be implemented through a permit or license. Larger POTWs that can handle the slug load from a hauler, may grant access to the headworks.

In other cases, where storage or equalization capacity is available, hauled waste may be discharged to equalization or holding tanks, where it can be characterized prior to introduction to the system. Sioux City, Iowa has developed a successful method to regulate the impact of waste hauler discharges (see Appendix A). A large holding receptacle is utilized for all wastes and the contents are metered to the treatment plant in controlled dosages, so as to prevent any upsets from high strength waste.

If hauled waste discharges are restricted to a single site, the POTW can easily inspect and sample the waste, verify tracking records, supervise the discharge of the waste, and prohibit the discharge of wastes that would be incompatible with the POTW. Such supervision will also discourage illegal discharges. Monitoring of a collection system discharge point is more difficult than monitoring a headworks discharge point. However, dilution of the waste is achieved when discharged at a remote location in the collection system.

Waste generators may be regulated by permits specifying conditions such as self-monitoring requirements, local limitations, categorical standards, specific prohibitions, etc., which must be met before allowing discharge. Procedures to control generators of hauled wastes should be similar to those employed for generators of fixed discharges, since both are covered by the General Pretreatment Regulations (40 CFR Part 403). The POTW may inspect the generator's facility and sample the wastes for pollutants of concern to the POTW, as well as determine if any other wastes have the potential for being mixed with the wastes that are to be hauled. Based on inspection results, the POTW may sample for those pollutants which should be limited before discharge is allowed.

If the POTW monitors the waste prior to discharge, one sample of the waste may be analyzed for a single indicator parameter and a second sample preserved in case any problems occur after introduction to the POTW. While this might subject the POTW to unknown pollutants, it would save the cost of analyzing each load extensively. In the case of manifest discrepancies, or where the sample failed the indicator parameter test, or where an interference resulted, more comprehensive testing could occur.

A waste hauler permitting and monitoring program should serve as a deterrent to haulers against discharging illegal or harmful wastes. If deterrence alone is unsuccessful, such a program could trigger enforcement action such as fines, refusal of wastes, permit revocation, or assignment of liability for damages.

4.2.5 Planning for Future Sources

To prevent the likelihood of future interferences developing, POTW officials must plan for future sources of industrial pollutants. Future pollutant loadings should be considered from two sources: new industries, and new pollutant streams of existing industries. Planning for future sources is particularly important as it relates to local limits development. Future pollutant sources and quantities must be considered in setting local limits, so that a treatment facility is able to handle increased pollutant loadings adequately.

4.3 LEGAL AND ENFORCEMENT REMEDIES

Interference is costly to POTWs in terms of worker safety, physical plant integrity, effectiveness of operation, and liability for NPDES permit violations. Interference is also a violation of a federal prohibition applicable directly to industrial users. POTWs are required to establish local limits as necessary to prevent interference and to take appropriate enforcement actions against violators.

In order to prevent and quickly remedy interference, the POTW must be ready to exercise its authority to take effective enforcement and legal actions. These actions should be clearly defined and readily understood by all parties involved. The range of enforcement mechanisms available to the POTW will depend on the legal authorities given to it by the municipality, county, and state. Wastewater treatment personnel who have not had extensive experience with enforcement and legal proceedings in the past should consult with the POTW's attorney, city solicitor, or comparable city official to determine what options are available. POTWs which have federally-approved or state-approved pretreatment programs should consult their program submission documents regarding legal authority and enforcement procedures.

EPA has recently distributed a comprehensive guidance document for POTWs titled Pretreatment Compliance Monitoring and Enforcement (PCME) Guidance (EPA, 1986c). It provides detailed discussions on compliance monitoring, establishing enforcement priorities, and conducting enforcement actions. The PCME guidance should be examined by POTW personnel for the development or review of their enforcement response procedures. POTWs are encouraged to develop an enforcement response guide containing procedures which will define, in a nonsubjective way, the type of enforcement response that can be expected for a particular kind or level of violation.

Enforcement actions for POTW interference or industrial discharge noncompliance are typically spelled out in the local sewer use ordinance, permits or contracts with industrial users, or an approved pretreatment program. In addition, the enforcement procedures can be described in the POTW's enforcement response guide or their NPDES permit. It is important that the enforcement options be strong enough to provide a real deterrent to the regulated industries. This requires that adequate manpower and documentation exist to pursue enforcement actions. Documentation will primarily consist of industrial waste monitoring as discussed elsewhere in this manual. An in-depth evaluation of all documentation concerning monitoring results, methods, and techniques, as well as quality assurance procedures should be part of the preparation for enforcement proceedings.

In interference situations in which there is imminent endangerment to human health, the environment, or the POTW, it is important that the POTW have the ability to immediately notify the discharger and bring about a halt to the discharge. POTWs with approved pretreatment programs are required to have this authority by the General Pretreatment Regulations (40 CFR 403.8 (f)(vi)(B)). In situations in which there is no threat of immediate harm, enforcement steps usually begin with noncompliance warnings, meetings and other informal actions. Should these measures prove inadequate, other more stringent measures should be taken. These commonly include:

- penalties
- orders and compliance schedules
- litigation
- sewer disconnection and permit revocation

Bayshore Regional Sewerage Authority (Union Beach, New Jersey) and METRO-Seattle, Washington are examples of POTWs which have shown aggressive enforcement efforts (see Appendix A). These POTWs have not hesitated to levy fines and take other enforcement actions after documenting the source of interference problems.

Both formal and informal actions are important parts of an effective enforcement program. Informal actions are likely to be more successful if the POTW has developed a cooperative relationship with its industrial users. Virginia's Hampton Roads Sanitation District provides a good example of the advantages of developing and maintaining a good working and monitoring relationship between an authority and the industrial user community. Once the interfering source is located, the District technicians, along with a supervisor, directly contact the industry to notify them of the problem and see to it that the discharge ceases. The District approaches the source of any interference in a cooperative manner, with ample documentation in hand. The source is normally willing to rectify the problems and agreement on administrative and other measures is reached informally, without the need to resort to legal remedies. If a clean-up is warranted, the responsible industrial user contracts for the necessary work to be done, with District personnel overseeing the operation until completion. All costs involved with the investigation, clean-up and any other District expenditures as a result of the upset are then billed to the industrial source.

4.3.1 Penalties

After compliance warnings and efforts to encourage industrial pretreatment have failed, the enforcement option most commonly initiated is the use of penalties. The amount of a penalty is generally limited through state or municipal laws. EPA's "Guidance Manual for POTW Pretreatment Program Development" (October, 1983) recommended that POTWs have the ability to assess penalties of at least $300 per day of violation to act as a sufficient deterrent. However, this limit may be inadequate for discharges which interfere with the POTW. Appropriate action may involve seeking the assistance of the state or EPA for obtaining penalties under state or federal law, which may be substantially greater (up to $100,000 per day and 6 years in jail for a repeat knowing criminal violation). Penalties may be used in conjunction with billing procedures for minor violations which may be detected during inspections or compliance review of self-monitoring data. Such penalties should appear as a separate item on a bill with the violation identified. The amount of the penalty imposed will usually depend upon the nature and severity of the interference caused or the quantity of the interfering pollutant.

Surcharges are not penalties, but rather recover the POTW's cost of treating industrial wastewaters. Payment of surcharges is not a justification for an IU to violate pretreatment standards or cause interference. POTWs should make it clear to their industrial users, as part of the IU permit or contractual agreement, that IUs may be subject to both surcharges for the additional treatment costs, as well as substantial penalties for causing interference.

4.3.2 Orders and Compliance Schedules

In order to force an industrial user to install acceptable pretreatment equipment, some POTWs may issue administrative orders to place an industrial user on an enforceable compliance schedule to meet pretreatment standards. Additionally, orders are sometimes used to require increased monitoring or installation of slug notification systems.

The Hampton Roads Sanitation District, for example, may modify an industry's discharge permit to reflect increased monitoring for a period of time to show compliance. Also, a compliance schedule from the industry is required to show what steps are taken to prevent recurrence. Depending on the severity of the problem, the District may require the industry to permanently install some type of alarm system and/or automatic shut-off.

4.3.3 Litigation

POTW-initiated litigation can be used as a further attempt to cause compliance after earlier measures have failed to bring about the desired result. In many cases, litigation is a way of obtaining an injunction against the discharger to cease the discharge or to clean it up, or to obtain a sewer disconnection or the payment of substantial penalties which go beyond routine fines. Litigation also serves to bring media attention and public pressure to bear when pressure from a sewer authority has failed. For example, the City of Canandaigua, New York obtained an out-of-court settlement dictating a compliance schedule for a user following the City's initiation of court action. The user was required to expand its pretreatment facility and the City's POTW operation was then able to meet its NPDES permit.

In some cases, litigation has been initiated to collect unpaid fines, which may amount to sizeable sums. New Jersey's Bayshore Regional Sewerage Authority found adverse publicity to have little effect on a major industrial employer, and was forced to initiate legal action in an attempt to recover $1.25 million in back surcharge payments and costs.

4.3.4 Sewer Disconnection or Permit Revocation

Sewer disconnection or permit revocation is used by many POTWs under serious circumstances such as when there is imminent endangerment to public health, the environment or the POTW, or when other methods to obtain compliance have failed.

A local ordinance can provide this authority by allowing the POTW to issue a suspension order and by requiring the discharger to immediately halt discharging upon notification. Furthermore, the ordinance can allow the POTW to sever the sewer connection if the industry does not respond.

Frequently, unless there is an immediate threat to human health, an administrative hearing of some type is held before discontinuing service. The industrial user is invited to appear before a local hearing board, presented with the facts demonstrating noncompliance, and asked to show cause why service should not be discontinued. The board then decides whether to pursue disconnection.

Recently in New Jersey, the Hamilton Township Wastewater Treatment Plant required an industrial user to install flow equalization equipment. However, deterioration in effluent quality continued, leading to termination of sewer service. In Pennsylvania, the Maiden Creek Wastewater Treatment Plant discontinued service to a user when the discharge from the facility caused a total process failure. Any future failure to comply with municipal requirements for flow equalization and monitoring, BOD reduction, and sampling will subject the user to another shut-off. The Bayshore, New Jersey Regional Sewerage Authority's policy is to notify recalcitrant industries of a violation, with subsequent discontinuation of service if noncompliance extends beyond 15 days.

At Hampton Roads in Virginia, if a problem represents an imminent hazard to the public health, safety or welfare, or to the local environment or to any portion of the sewerage system, the District may suspend a permit for a period of up to 60 days. Failure to immediately cease discharge of all industrial wastewater into the sewerage system may also result in termination of water and/or wastewater service. If cooperation is not received from the user, then the District may revoke the industrial user's permit.

TABLE 4-1

BIOLOGICAL PROCESS CONTROL STEPS

Operating Parameter	Control Objective	Method of Implementation	Result
Mean Cell Residence Time (MCRT) also known as Sludge Age and Solids Retention Time (SRT)	Increase	Decrease solids wasting rate	Quicker acclimation to toxic pollutants Better able to accommodate fluctuating conventional pollutant loads
Mixed Liquor Suspended Solids (MLSS)	Increase	Increase solids return rate	Better biosorption and acclimation
Food-to-Microorganism Ratio (F/M)	Decrease	Increase solids return rate	Improved toxic pollutant biodegradation and acclimation

Source: U.S. EPA, 1986a

TABLE 4-2

CHEMICAL ADDITIONS

Additive	Reason for Addition	Pollutant(s) Causing Problem	Result	Point of Addition
Chlorine	Reduce bulking	Various, particularly textile and wood products wastewater	Kills filamentous organisms	Varies, but RAS & aeration basin effluent common
Phosphorus Nitrogen and Sulfur	Improve biological treatment and reduce bulking sludge	High carbonaceous strength waste	Corrects nutrient deficient condition	Before aeration basin
Lime or Caustic	Reduce biological inhibition	Heavy Metals	Raises pH causing metals to precipitate	Before primary clarifier
Polymers or Coagulants	Improve sedimentation	Various	Removes toxics and improves sedimentation	Before primary or secondary clarifier
Powdered Activated Carbon	Reduce biological inhibition	Toxic organics	Adsorbs organic pollutants	To aeration basin

Note: The generalizations in this table do not apply in all situations. The text should be consulted.
Source: James M. Montgomery, Consulting Engineers, Inc.

TABLE 4-3

TREATMENT PLANT CONTROL MEASURES

Process	Biological Process Control	Biological Augmentation	Chemical Addition	Operations Modification	Physical Modification
Primary Clarification			•		•
Activated Sludge	•	•	•	•	•
Trickling Filters	•		•	•	•
Lagoons		•	•	•	•
Rotating Biological Contactors (RBC)		•	•	•	•

Source: James M. Montgomery, Consulting Engineers, Inc.

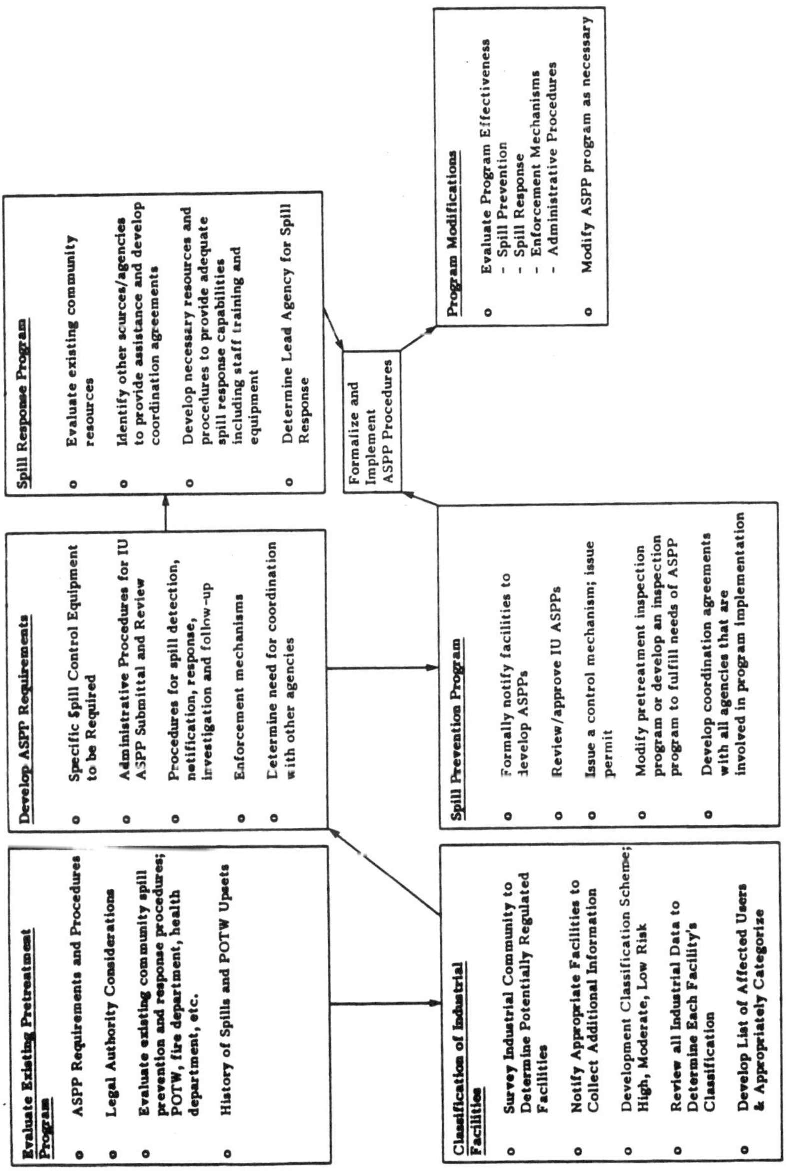

FIGURE 4-1
FUNDAMENTAL PROCEDURES FOR POTW ACCIDENTAL SPILL PREVENTION PLAN (ASPP) DEVELOPMENT
(U.S. EPA, 1986b)

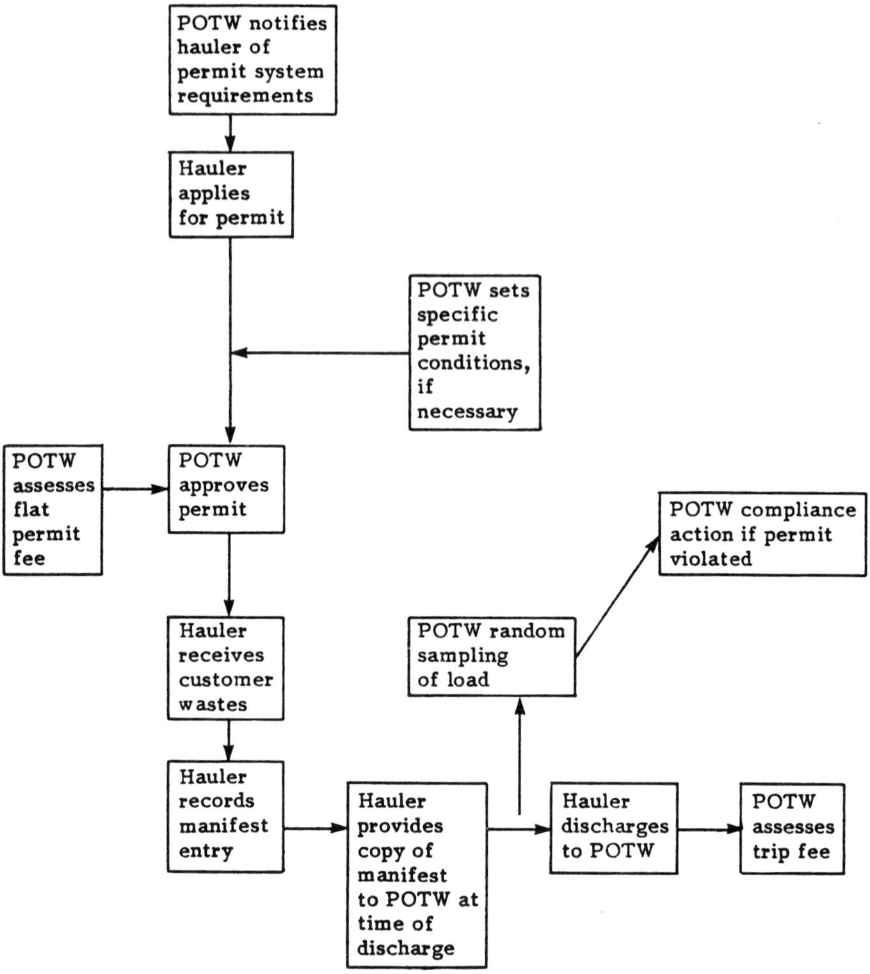

FIGURE 4-2

PROCEDURES OF A WASTE HAULER PERMIT PROGRAM
(U.S. EPA, 1985a)

References

Busch, W.H. (1986) "Protecting Your Plant From Hazardous Waste." Operations Forum, WPCF, April Issue. 11-15.

Geating, J. (1981) "Literature Study of the Biodegradability of Chemicals in Water (Vols. 1 and 2)" U.S. EPA, Cincinnati, Ohio. 241 pp.

Grubbs, R.B. (1986) "Biotechnology Proves Good Examples." Pollution Engineering, June, 1986.

Russell, L.L., Cain, C.B. and Jenkins, D.I. (1983) "Impact of Priority Pollutants on Publicly Owned Treatment Works Processes: A Literature Review." Proc. 37th Ind. Waste Conf. Ann Arbor Publishing, Ann Arbor, Michigan. 871-883.

Silva, S.J. (1981) "EPA Moving to Control Industrial Toxic Pollutants with New NPDES Permits." Civil Engr. 51:76.

Slattery, G.H. (1986) "Patapsco Wastewater Treatment Plant Standard Operating Procedure: Routine Operation of Wastewater Respirometer", Baltimore, Maryland.

U.S. EPA (1977a) "Process Control Manual for Aerobic Biological Wastewater Treatment Facilities." Prepared by Tsugita, R.A., De Coite, D.C.W. and Russell, L.L. for U.S. EPA, Washington, DC.

U.S. EPA (1977b) "Federal Guidelines - State and Local Pretreatment Programs." U.S. EPA, Municipal Construction Division MCD-43, EPA-430/9-76-017a, Washington, D.C. 3 volumes.

U.S. EPA (1978) "Field Manual for Performance Evaluation and Troubleshooting at Municipal Wastewater Treatment Facilities." EPA-430-9-78-001. Prepared by Culp, G.L. and Heim, N.F. for U.S. EPA, Washington, DC. 387 pp.

U.S. EPA (1979) "Biodegradation and Treatability of Specific Pollutants." EPA-600/9-79-034. Prepared by Barth, E. F., and Bunch, R. L. for U.S. EPA, Cincinnati, Ohio. 60 pp.

U.S. EPA (1981a) "304(g) Guidance Document: Revised Pretreatment Guidelines (Vols. I and II)." Internal Report. Prepared by JRB Associates for U.S. EPA, Cincinnati, Ohio.

U.S. EPA (1981b) "Assessment of the Impacts of Industrial Discharges on Publicly Owned Treatment Works." Report submitted to the Office of Water Enforcement, U.S. EPA, Washington, D.C. by JRB Associates.

U.S EPA (1981c) "Parallel Evaluation of Air and Oxygen-Activated Sludge." EPA-600/2-81-155. Prepared by Austin, S., Yunt, F. and Wuerdeman, D., for U.S. EPA, Cincinnati, Ohio, 43 pp.

U.S. EPA (1983) "Guidance Manual for POTW Pretreatment Program Development." U.S. EPA, Office of Water Enforcement and Permits, Washington, DC.

U.S. EPA (1984) "Improving POTW Performance Using the Composite Correction Program Approach." U.S. EPA, Center for Environmental Research EPA-625/6-84-008, Cincinnati, Ohio. 258 pp.

U.S. EPA (1985a) "PRELIM: The EPA Computer Program/Model for Developing Local Limits - User's Guide." Prepared by SAIC/JRB Associates for U.S. EPA, Office of Water Enforcement and Permits, Washington, DC.

U.S. EPA (1985b) "Pretreatment Implementation Review Task Force: Final Report to the Administrator." U.S. EPA, Washington, DC. 75 pp.

U.S. EPA (1985c) "RCRA Information on Hazardous Wastes for Publicly Owned Treatment Works." U.S. EPA, Office of Water Enforcement and Permits, Washington, D.C.

U.S. EPA (1986a) "Interferences at Publicly Owned Treatment Works (POTWs)." Submitted to U.S. EPA-WERL, Cincinnati, Ohio by James M. Montgomery, Consulting Engineers, Inc.

U.S. EPA, Region X (1986b) "Guidance Manual for the Development of an Accidental Spill Prevention Program." Prepared by Science Applications International Corp. for U.S. EPA, Region X, Seattle, Washington.

U.S. EPA (1986c) "Pretreatment Compliance Monitoring and Enforcement Guidance" U.S. EPA, Office of Water Enforcement and Permits, Washington, DC.

WPCF (1982) "Industrial Wastewater Control Program for Municipal Agencies." MOP OM-4, WPCF, Washington, DC. 166 pp.

Appendix A: Case Studies

TABLE A-1

CASE STUDY SUMMARY TABLE

Treatment Facility	Contact Person	Nature of Interference Episode	How Detected	Causative Pollutant(s)	Method of Industrial Identification	Mitigation Step
Back River, Baltimore, MD	Bob Moore Plant Manager (301) 288-6900	Restricted Sludge disposal	Lab analysis	Various metals	-	Implemented pre-treatment program
		Sewer system Explosion hazard	Visual and sensory observation	Various solvents	Monitoring	Required industry to improve in-house solvent recovery
Patapsco, Baltimore, MD	Gerald Slattery Plant Manager (301) 354-2700	Inhibition permit violations	Respirometry, operating difficulties	Insecticides, solvents, petroleum compounds	Respirometry, chemical analysis	Decrease F/M ratio, develop tracking program
Bayshore, Union Beach, NJ	David Knowles Manager (201) 739-1095	Organic overload-permit violations	Lab analysis	BOD, COD, SS	Only major industry	Enforcement of discharge limits - fines
East Side, Oswego, NY	John McGrath Lab Supervisor (315) 342-3777	Suspended solids overload- permit violation	Visual examination Lab analysis	SS, BOD	Only major industry	Treatment plant operations modifications, industrial discharge permit
Hamilton Township, Trenton, NJ	Thomas Andersen Assistant Superintendent (609) 890-3540	Upset conditions - permit violations	Lab analysis	BOD, SS, Volatile Organics	Industrial community	Industrial flow metering, disconnection
Horse Creek North Augusta, SC	Stanley Wagher Manager (803) 278-1911	Biomass inhibition bulking sludge	Visual observation Lab analysis	Alkalinity, pH SS	Industrial monitoring	Implement pre-treatment program

TABLE A-1

CASE STUDY SUMMARY TABLE (CONTINUED)

Treatment Facility	Contact Person	Nature of Interference Episode	How Detected	Causative Pollutant(s)	Method of Industrial Identification	Mitigation Step
Maiden Creek, Blandon, PA	Edward Glausman Superintendent (215) 926-4140	Treatment upset due to shock loading - permit violations	Visual observation Lab analysis	BOD, SS, NH_3	Only major industry	Treatment plant operational modifications, industrial pretreatment
METRO-West Point, Seattle, WA	Douglas Hilderbrand Industrial Waste Supervisor (206) 447-6743	N/A	Lab analysis	Various metals		Implemented pretreatment program
Neuse River, Raleigh, NC	Leon Holt Pretreatment Coordinator (919) 779-2010	Sludge contamination	Lab analysis	Various metals	Largest industry	Issue industrial discharge permit
Newark, OH	Roger Loomis Assistant Superintendant (614) 345-0549	Treatment upset due to shock loadings - permit violations	Observation Lab analysis	Phenols, ammonia Formaldehyde	Major industry, industrial monitoring	Industrial pretreatment improvements
North Shore, Gurnee, IL	Frederic Winter Director of Laboratory Services (312) 623-6060	Nitrification inhibition, process upsets	Observation Lab analysis Bioassay	Antibiotics	Industrial monitoring	Treatment plant operations modifications. Upgrade sewer ordinance
Passaic Valley, Newark, NJ	Frank D'Ascensio Manager Industrial and Pollution Control (201) 344-1800	Treatment upset - permit violation	Lab analysis	SS, Volatile organics, metals	Industrial monitoring	Implementing pretreatment program
Sioux City, IA	A.V. Flores Project Manager (712) 279-6169	Treatment upset due to shock loadings - sludge contamination	Lab analysis	Zinc	Industrial monitoring	Treatment plant operations modifications
Toileson, AZ	Sterling Dillow Manager (602) 936-3381	Treatment upset	Visual observation Lab analysis	BOD, SS	Only major industry	Industry-POTW contract

BACK RIVER WASTEWATER TREATMENT PLANT
Baltimore, Maryland

The City of Baltimore owns and operates two wastewater treatment facilities, Back River and Patapsco, with a combined flow rate of approximately 250 million gallons per day. The plants serve a combined population of nearly 1.7 million in an area which includes approximately 4,700 sources or potential sources of nondomestic wastewater. In accordance with the requirements of the General Pretreatment Regulations (40 CFR Part 403) established by the U.S. EPA, the City developed an extensive industrial waste control program requiring a significant commitment in terms of personnel, equipment, office space, and supplies.

The Back River facility is currently undergoing a major renovation to replace the 30 acres of trickling filter rock media with complete-mix activated sludge, along with significant alteration and expansion of most process units. The renovation work is in preparation for new NPDES permit limits of 10/10 (BOD and TSS) and 2 mg/l (NH_3), which will require extensive modification of the system for nitrification and multi-media filtration. Industrial flows to Back River total approximately 27 mgd, resulting in metals and solvents in the discharge.

The primary source of metals in the system is from the 12 metal plating operations identified by the industrial waste survey. If too high, the metals content in the wastewater restricts the ultimate disposal options for the digested and dewatered sludge. When local limits were calculated based on unrestricted distribution of the sludge, the limits were occasionally one-fourth of the electroplating categorical standards. A compost facility now under construction is expected to process 150 wet tons of the 450 tons produced each day, beginning in March 1987.

The benefits of pretreatment for metals removal have been demonstrated at Back River. An incinerator had been discharging 2 tons of fly ash per hour into the collection system, which was high in metal content and was responsible for 90 percent of the cadmium in the POTW influent. Other wastewater containing metals were from steel and automobile manufacturing. In each case, industrial user pretreatment facilities have come on-line during the past year, with a measureable drop in influent and sludge concentrations. A summary of the improved metal content of the sludge from 1984 to 1986 is provided on Table A-2. Based on the current metal content, the composted sludge will be acceptable for agricultural use.

The second major area of concern at the Back River plant stems from the large, batch discharges of solvents, petroleum hydrocarbons and other toxic organics. In 1985, a 2:00 am discharge of ethyl benzene, xylene and toluene resulted in the evacuation of the largest pump station and other buildings in town. The problem was traced to a paint and chemicals manufacturer, which has since improved its in-house solvent recovery system. A similar evacuation resulted from a 4,000 gallon discharge of xylene by a waste hauler, which was traced to a specific location in the collection system. Tetrachloroethylene has been discovered and traced to dry cleaning operations. While such discharges have not usually resulted in interference with the plant's ability to meet its NPDES permit limits, the health and safety issues and potential for explosion are of serious concern to the City.

TABLE A-2

AVERAGE METAL CONTENT OF BACK RIVER SLUDGE
(mg/kg dry weight basis)

Metal	Allowable[1]	1984	1986	% Reduction
Cr (total)	NA	1,491	273	82
Cu	1960	1,001	549	45
Pb	730	372	388	-4
Ni	575	266	67	75
Zn	5,130	2,747	1,522	45
Cd	48	26	17	35
Hg	12	5	3	40

[1] From Compost Contract Schedule 2, City of Baltimore, MD

An interesting aspect of Baltimore's program for preventing interference and sewer system hazards is the computer coding of the sewer collection system. By knowing the constituents of each industry's discharge, the flow rate and their location in the coded sewer system, a contaminant discovered at either Back River or Patapsco can theoretically be traced back to its potential source or sources. While such a backtracking program is of limited use for isolated discharges, it could prove beneficial in locating chronic dischargers of specific compounds.

In order to further protect the sewer system, a City Ordinance requires that the atmosphere in a manhole receiving an industrial discharge must not exceed 10% of the LEL (lower explosive limit) for any fuel. This regulation is in force by manual monitoring of the sewer manhole and has been successful in curbing intentional dumps or disposal of fuels and flammable solids.

BACK RIVER WASTEWATER TREATMENT PLANT
BALTIMORE, MARYLAND

Design Flow:	**180 mgd**
Secondary Treatment:	**Trickling Filters and Activated Sludge**

INFLUENT WASTEWATER

	Typical (Upset)
Ave. Flow, mgd	180 (270)
% Industrial	15
BOD_5, mg/l	230
SS, mg/l	190

SIGNIFICANT INDUSTRIES

Industry	Flowrate (mgd)	Problem Pollutants
Metal Plating (12)	0.18	Metals
Auto Mfr.	1.5	Cr, Cu, Ni, Zn
Paint and Chemical	N/A	Ethyl benzene, toluene, xylene
Incinerator	N/A	Cd, Hg
Waste Haulers	N/A	Solvents, petroleum hydrocarbons

PLANT LOADING

Primary Clarifiers	Typical (Upset)
Overflow Rate, gal/sf/day	730 (1,170)
Detention Time, hours	3.6
Effluent BOD_5, mg/l	180
Effluent SS, mg/l	100

Aeration Basins	Typical (Upset)
Ave. Flow, mgd	60
F/M, lbs BOD_5/lbs MLSS/day	0.4
MCRT, days	6.1
MLSS, mg/l	2,000
Detention Time, hours	3.5
Return Flow, %	30-40
D.O. Level, mg/l	2-3

Secondary Clarifiers (A.S./T.F)	Typical (Upset)
Overflow Rate, gal/sf/day	750/950
Detention Time, hours	2.5/2.1
SVI, ml/gm	95

Trickling Filters	Typical (Upset)
Ave. Flow, mgd	150 (200)
Hydraulic Loadings, ga/sf/d	120 (136)
Organic Loading, lbs BOD/1000 cf/d	20
Return Flow, %	0

PLANT PERFORMANCE

	Permit Limit	Typical (Upset)
BOD_5, mg/l	45	40 (50)
SS, mg/l	45	40 (50)

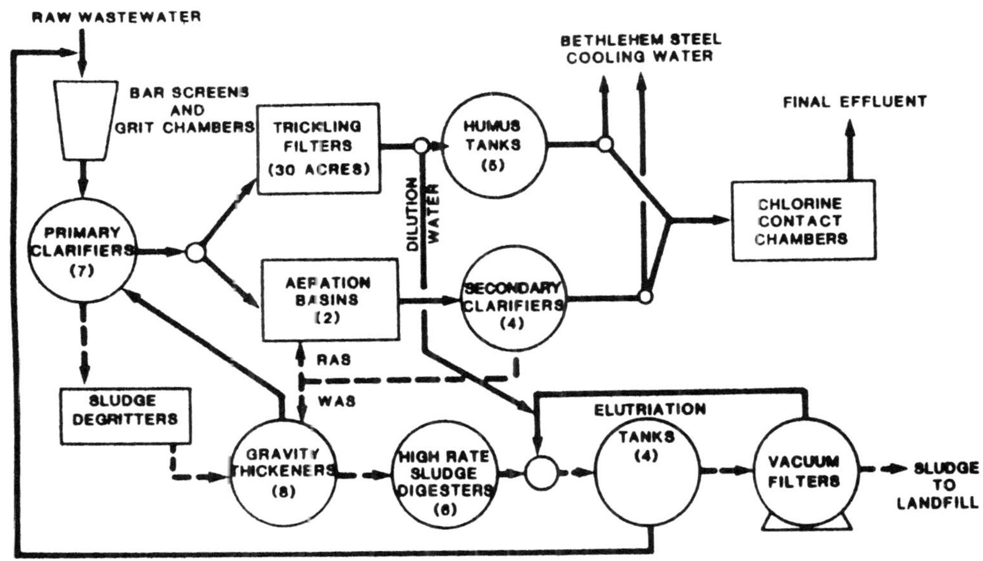

PATAPSCO WASTEWATER TREATMENT PLANT
Baltimore, Maryland

A 1981 EPA-sponsored project on biomonitoring of direct discharges rated the Patapsco plant as having the most toxic effluent of those surveyed. Ironically, the second most toxic discharge came from an agricultural chemicals manufacturer who, in 1983, ceased direct discharging and now sends their pretreated wastewater to Patapsco. The high level of toxicity has prompted the collection of much bioassay, acute toxicity and respirometer data over the past four years in order to evaluate the potential for both toxicity pass-through and toxic inhibition of the plant biomass. Despite the presence of inhibitory levels of pollutants in the influent, the plant currently meets its discharge limits for BOD and SS, indicating the ability of activated sludge to acclimate to consistent levels of many inhibitory compounds. It has, however, been necessary to operate at a reduced organic loading in order to offset the effects of the inhibition. This has reduced the wastewater treatment capacity of the plant.

The City is evaluating several measures to reduce this inhibition and thus prevent any possibility of interference. They have begun daily routine operation of a respirometer for measuring the inhibitory characteristics of the plant influent. They are also evaluating the use of respirometry as a tool for assessing the impacts of several industrial effluents on the plant.

Another concern to the City is the pass-through of toxicity. Acute influent and effluent toxicity data using a Beckman Microtox unit have been collected since November 1980. Some of the results of these analyses are shown on Figure A-1. The data are on an inverse scale, with 0% indicating complete toxicity and approximately 45 percent corresponding to no toxic effect.

Figure A-1 illustrates the highly toxic nature of the plant influent and effluent until September 1982, at which time the secondary treatment system went on-line. The acclimation of the activated sludge improved the monthly average effluent toxicity from 5 percent to 40 percent by December, where it remained until secondary shutdown in February 1983. The average effluent toxicity again increased until the secondaries returned on June 15, providing clear evidence of the detoxification capability of acclimated activated sludge. Even though overall effluent toxicity has been reduced, individual daily tests continue to show substantial day-to-day variability, with significant effluent toxicity occurring in more than one-third of the tests. Therefore, the City is continuing to study ways to reduce this toxicity pass-through. In fact, the City of Baltimore is currently performing a toxicity reduction evaluation (TRE) in conjunction with the U.S. EPA.

As a means of improving both the inhibition and toxicity pass-through situations, the State of Maryland included the following in a consent order issued to the City in 1984:

- install on-line toxicity monitoring of the plant influent
- develop a toxics emergency response plan
- enlarge the scope of the City sewer ordinance to include specifics on toxicity and flammability for industrial effluents.

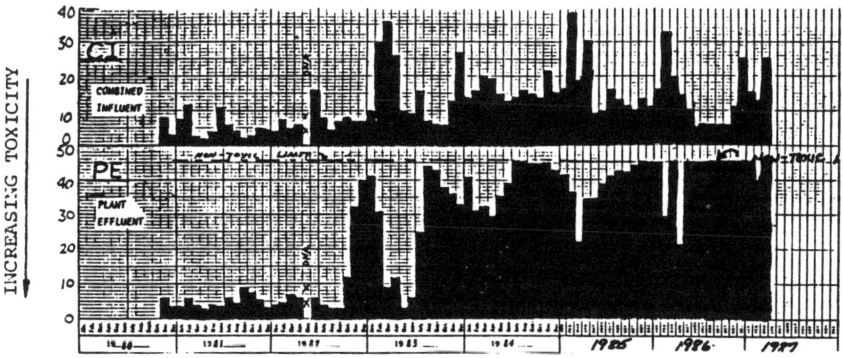

FIGURE A-1
MONTHLY ACUTE TOXICITY
(Courtesy G.H. Slattery, City of Baltimore)

In spite of high influent toxicity, the plant is not currently experiencing interference with its ability to meet its NPDES permit limits. With a mean cell residence time varying between 10 and 15 days, the plant produces reasonably stable operation and good plant performance on removals of conventional pollutants. Although compliance with the NPDES permit has been achieved for BOD and SS at Patapsco, the plant flow is well below the 70 mgd design capacity. Toxic inhibition of the activated sludge bacteria is still present despite the improvement since 1983. Evidence of this inhibition is provided by the plant actual operating F/M of 0.3, which is significantly less than the design value of 0.5, and also was verified by respirometry tests on the plant influent.

The attached data sheet indicates that Patapsco's current noncompliance has resulted from discharging excess phosphorus and an effluent pH below 6.5. The phosphorus problem is being dealt with by installing anaerobic/oxic (A/O) technology in the oxygenation basins as a means of biological phosphorus removal. The low pH is inherent in oxygen activated sludge systems, typically producing an effluent in excess of 250 mg/l of CO_2 and a pH of 6.2. The problem can be corrected with either chemical adjustment or post-aeration of the wastewater.

PATAPSCO WASTEWATER TREATMENT PLANT
BALTIMORE, MARYLAND

Design Flow:	**70 mgd**
Secondary Treatment:	**Activated Sludge (Pure Oxygen)**

INFLUENT WASTEWATER

	Typical (Upset)
Ave. Flow, mgd	42
% Industrial	30
BOD_5, mg/l	265 (320)
SS, mg/l	325 (470)
TOX, %	15

SIGNIFICANT INDUSTRIES

Industry	Flowrate (mgd)	Problem Pollutants
Chemicals	1.0	Insecticides, Volatiles, phenols, metals
Metal Finishing	0.13	pH, solvents, metals

PLANT LOADING

Primary Clarifiers	Typical (Upset)
Overflow Rate, gal/sf/day	1,150
Detention Time, hours	1.5
Effluent BOD_5, mg/l	190
Effluent SS, mg/l	80

Aeration Basins	Typical (Upset)
F/M, lbs BOD_5/lbs MLSS/day	0.3
MCRT, days	10-15
MLSS, mg/l	5,000
Detention Time, hours	2
Return Flow, %	30
D.O. Level, mg/l	2-4

Secondary Clarifiers	Typical (Upset)
Overflow Rate, gal/sf/day	450
Detention Time, hours	6.3
SVI, ml/gm	50-75

PLANT PERFORMANCE

	Permit Limit	Typical (Upset)
BOD_5, mg/l	30	18 (40)
SS, mg/l	30	25 (40)
Total-P, mg/l	2.0	3.5
pH	6.5-8.5	6-6.5
TOX, %		40

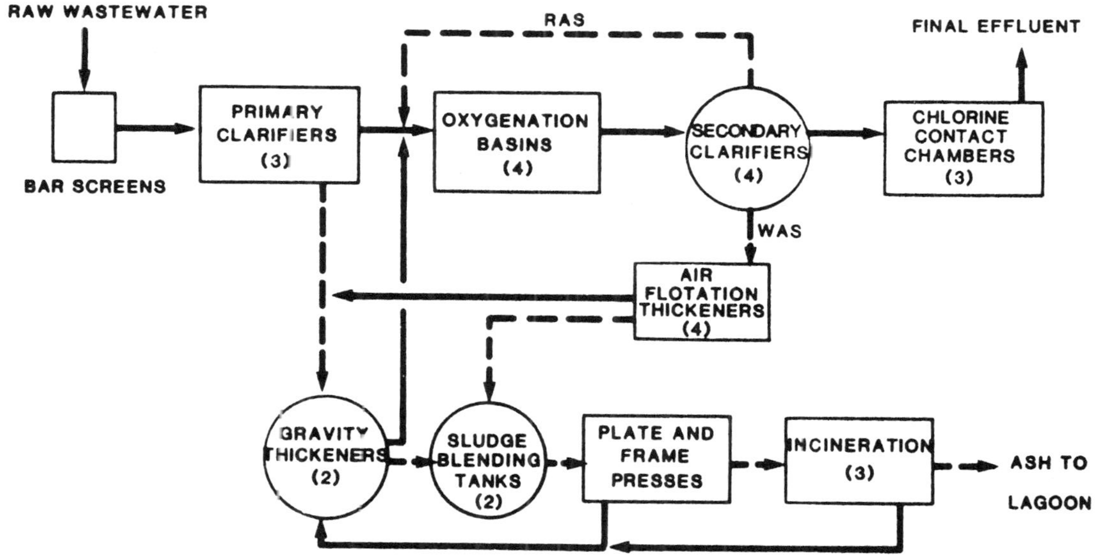

BAYSHORE REGIONAL SEWERAGE AUTHORITY
Union Beach, New Jersey

The Bayshore Regional Sewerage Authority (BRSA) operates an activated sludge treatment facility whose performance is largely dictated by a single industrial waste discharger. Three manufacturers of flavors and fragrances (one of whom is a perfume retailer) represent the total industrial wastewater flow of 325,000 gpd, or less than 5 percent of the POTW total. All three industries discharge high concentrations of conventional pollutants and routinely violate the maximum allowable monthly concentration limits for BOD (500), COD (1500) and TSS (500) as specified in their industrial waste permits. Two of the three manufacturers contribute less than 0.5 percent of the POTW flow, hence their impact is minimal. However, one building of the largest industry produces in excess of 200,000 gpd of wastewater with the following characteristics (in mg/l):

	1984			October 1985		
Parameter	**Ave.**	**Monthly High**	**Monthly Low**	**Ave.**	**Daily High**	**Daily Low**
BOD	1004	2054	245	2624	5250	522
COD	3238	4998	1440	7084	11380	2520
TSS	776	1835	94	1113	1698	672

The large variation in wastewater quality indicates that a two-stage primary pretreatment system located at the industry is not sufficient to meet the fluctuating demands of their process wastes.

The potential impact of such an industrial discharge is evident when analyzing Figure A-2. The bar graph represents the percentage of total BOD being contributed by the industry on a daily basis. The upper plot on the line graph corresponds to the mass BOD loading, with the industry's contribution plotted beneath. This graph clearly demonstrates that the effluent from this single industry has increased the BRSA plant loading above the design limit of 15,000 pounds of BOD per day. This has interfered with the plant's ability to meet its permit limit for BOD.

The BRSA has been particularly aggressive in their dealings with the industry in question. It has taken a two-pronged approach:

- notification of violation with a subsequent discontinuation of service if noncompliance persists after 15 days, and
- legal action to recover $1.25 million in back surcharge payments and costs.

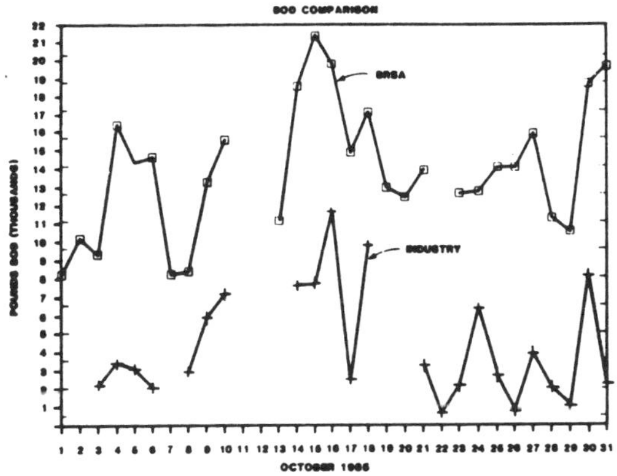

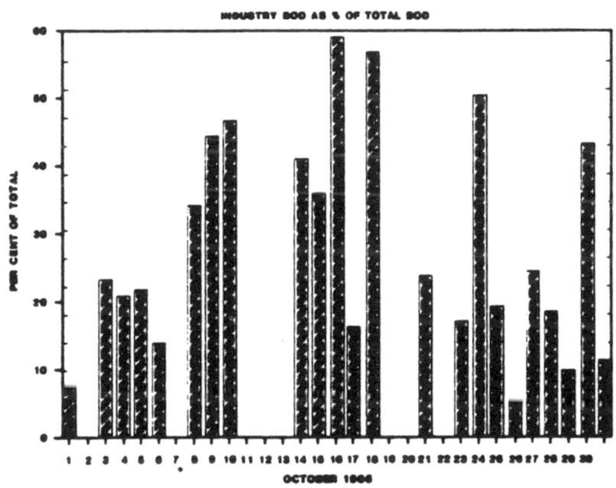

Figure A-2
Impact of Industrial Waste Discharge on POTW Loadings
October 1985

In addition to the BRSA actions, the County Prosecutor's office made a surprise visit to the industry in question, in which records were confiscated and samples collected for analysis. The result was a $5 million fine levied by the State of New Jersey in June of 1986, coupled with new NJPDES permit POTW limits. As a direct consequence of the state and local acions, the industry's wastewater BOD and SS have each been consistently below 100 mg/l since July, 1986. To date, $300,000 of the back payments have been received by the BRSA, with some litigation still pending.

BAYSHORE REGIONAL SEWERAGE AUTHORITY
Union Beach, New Jersey

Design Flow:	**8.0**	Location:	**Eastern shore**
Secondary Treatment:	**Activated Sludge (Modified Contact Stabilization)**	Population Served:	**80,000**

INFLUENT WASTEWATER

	Typical (Upset)
Ave. Flow, mgd	6.6
% Industrial	5
BOD_5, mg/l	220 (380)
SS, mg/l	250 (400)

SIGNIFICANT INDUSTRIES

Industry	Flowrate (1000 gpd)	Problem Pollutants
Flavors & Fragrances (3 industries)	325	BOD, TSS, COD

PLANT LOADING

Primary Clarifiers	Typical (Upset)
Overflow Rate, gal/sf/day	825
Detention Time, hours	1.75
Effluent BOD_5, mg/l	150 (250)
Effluent SS, mg/l	100 (200)

Aeration Basins	Typical (Upset)
F/M, lbs BOD_5/lbs MLSS/day	0.65 (1.25)
MCRT, days	8-10
MLSS, mg/l	2000-2500
Return Flow, %	25
Detention Time, hours	
Contact	3
Reaeration	12

Secondary Clarifiers	Typical (Upset)
Overflow Rate, gal/sf/day	540
Detention Time, hours	3.35
SVI, ml/gm	125 (500)

PLANT PERFORMANCE

	Permit Limit	Typical (Upset)
BOD_5, mg/l	30	35 (400)
SS, mg/l	30	27 (80)
D.O. mg/l	5	2-5

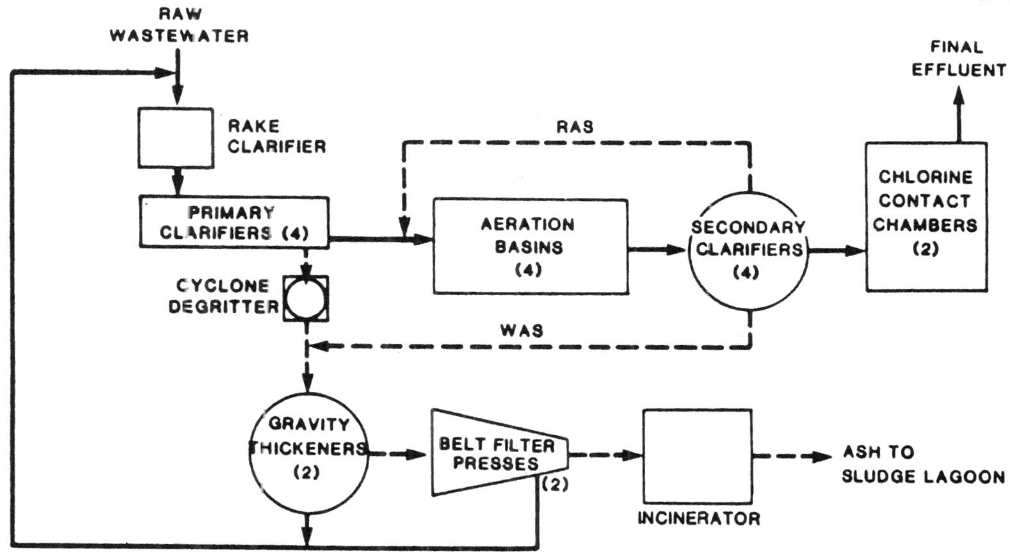

EAST SIDE SEWAGE TREATMENT PLANT
Oswego, New York

The City of Oswego, East Side Treatment Plant has experienced significant non-compliance problems associated with the loss of solids from their secondary clarifiers. Half of the plant's hydraulic flow is from a paper mill which is the only major industry in the city. From 1981 to 1983, the noncompliance problems at the plant were attributed to severe hydraulic and organic load peaks from the paper mill as well as operational difficulties such as frequent breakdowns of the return sludge pump drives. It is not known whether filamentous growth in the sludge occurred at that time. In 1983 the paper mill began reducing the hydraulic and organic peaks to the plant. Solids losses from the secondary clarifier still remained a problem. During 1984, the plant frequently exceeded their NPDES discharge suspended solids by five times the limit and the BOD by three times the limit. During that period, the plant still occasionally received hydraulic peaks from the paper mill which were twice the average rate for two to three hour periods, but a substantial cause of the problem was identified as poor sludge settleability due to filamentous growth. The frequent washout of biosolids from the secondary clarifiers resulted in a low mean cell residence time and the generation of a young sludge that did not settle well. In the spring of 1985, the belt drives on the return sludge pumps which had frequently been out of service were replaced with electronic variable speed drives. This improvement allowed the plant operators to maintain better control of the solids inventory in the aeration tanks. Plant performance was still poor, however, because of sludge bulking.

Several measures have been taken at the plant in an attempt to alleviate the sludge bulking problem. The measures that were taken include:

- switching from plug flow feed to a step feed in the aeration tanks in order to achieve better dissolved oxygen distribution;
- increasing the sludge return rate and mean cell residence time to improve settleability; and
- chlorination of the return sludge for the destruction of filamentous growth in the sludge.

The step feed operation has resulted in better dissolved oxygen distribution but did not significantly improve sludge settleability. The second two mitigation efforts were ongoing at the time of writing. A chlorination dosage of 6 lb Cl_2/1000 lb solids had been applied to the return sludge. Microscopic examination of the sludge indicated that the filaments had shrunk and the SVI level had dropped to the range of 60-80. The plant operators intend to chlorinate whenever the SVI increases to 150. It has not been determined if these mitigation measures can result in plant performance that will consistently meet the permit discharge limits.

The paper mill periodically discharges slugs of waste containing high suspended solids to the treatment plant. At these times, the sludge in the primary tanks takes on a gelatinous quality which makes sludge removal difficult. High periodic input of clay filler materials from the paper mill has resulted in poor sludge incineration with associated high fuel usage.

The City of Oswego is presently preparing an industrial discharge permit for the paper mill. The permit will restrict the monthly and daily average BOD and suspended solids levels in the influent from the paper mill as well as restrict the daily maximum hydraulic peak allowed. Under the permit provisions the paper mill will be required to submit listings of the chemicals used in their processes. The paper mill is presently investigating the possible relationship of the chemicals used in their manufacturing processes to the occurrence of filamentous growth in the activated sludge process.

EAST SIDE SEWAGE TREATMENT PLANT
OSWEGO, NEW YORK

Design Flow:	**3 mgd**	Location:	**Northern New York**
Secondary Treatment:	**Plug or Step Feed Activated Sludge**	Population Served:	**10,000**

INFLUENT WASTEWATER

	Typical (Upset)
Ave. Flow, mgd	2.5
% Industrial	50

	Municipal	Paper Mill
BOD_5, mg/l	100	300
SS, mg/l	120	450 (1000)

SIGNIFICANT INDUSTRIES

Industry	Flowrate (1000 gpd)	Problem Pollutants
Paper Mill	1,200	SS, BOD

PLANT LOADING

Primary Clarifiers

	Typical (Upset)
Overflow Rate, gal/sf/day	600
Detention Time, hours	2

	Municipal	Paper Mill
Effluent BOD_5, mg/l	70	120
Effluent SS, mg/l	40	100

Aeration Basins

	Typical (Upset)
F/M, lbs BOD_5/lbs MLSS/day	0.2
MCRT, days	7 (3)
MLSS, mg/l	2,000 (300)
Detention Time, hours	1
Return Flow, %	25 - 45
D.O. Level, mg/l	2 - 4

Secondary Clarifiers

	Typical (Upset)
Overflow Rate, gal/sf/day	800
Detention Time, hours	2
SVI, ml/gm	100 (1000)

PLANT PERFORMANCE

	Permit Limit Summer	Permit Limit Remainder of Year	Typical (Upset)
BOD_5, mg/l	30	45	20 (120)
SS, mg/l	30	70	25 (300)

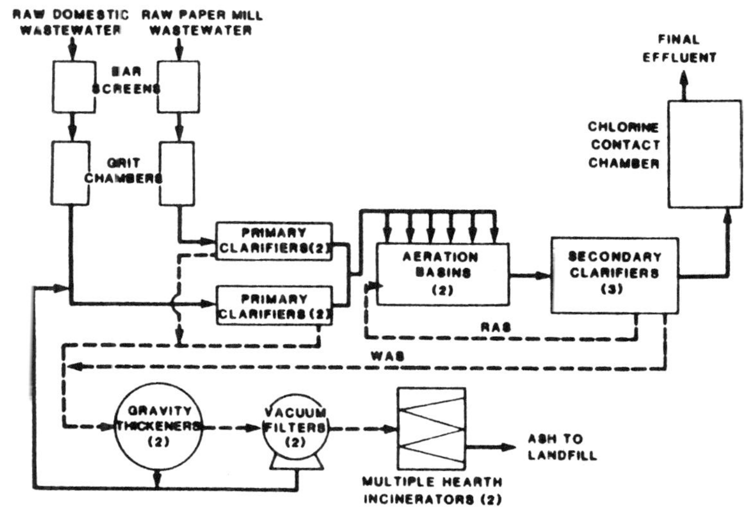

HAMILTON TOWNSHIP WASTEWATER TREATMENT PLANT
Trenton, New Jersey

The Hamilton Township Wastewater Treatment Plant (HTWTP) is an unusual facility in that plant upgrades over the past 30 years have been constructed as parallel flow processes rather than as replacements for older, outdated technology. Although this results in a complicated plant schematic (see below), parallel flow paths do provide operational flexibility and an opportunity to study the impact of a combined industrial/domestic wastewater on different fixed-film biological treatment processes. The HTWTP has had a difficult time meeting its permit limit for BOD over the past few years, and is currently under a Consent Order and Agreement and Compliance Schedule from the State Department of Environmental Protection.

Despite being at just over 50 percent of the plant's hydraulic capacity, Hamilton Township has experienced organic overloads, resulting in at least partial failure of 15 of the 48 RBC units. With the advent of an Industrial Waste Monitoring Program as part of a Sewers and Sewage Disposal Ordinance, the reasons for such overloading became apparent. Although the industrial waste program is still in its infancy, observations and analytical data have identified a pharmaceuticals manufacturer as a significant and potentially harmful discharger to the POTW.

Dating back to the summer of 1984, high concentrations of volatile organics were being discharged to the POTW on a once or twice-per-week basis. A monitoring program at the HTWTP uncovered an increase in influent BOD from 150 to 350-500 mg/l and high atmospheric levels of organic constituents with this discharge pattern. The specific industry was identified when a high influent pH reading led Hamilton Township personnel to the pharmaceuticals manufacturer in March, 1985. Sampling conducted at that time detected significant levels of ethyl benzene, toluene and xylene in the industry's effluent. These findings precipitated an extensive testing program by the Township, with an independent engineering study conducted by the industry. The results indicated a correlation between the pharmaceutical discharges and high influent soluble BOD at the POTW. Analyses conducted on the industry's flow streams resulted in the following calculated average effluent concentrations:

Parameter	**Concentration (mg/l)**
Arsenic	2.6
Phenols	25.7
Total Toxic Volatile Organics (TTVO)	1.3
BOD	21,800
TSS	557
TDS	65,800

Based on an average flow of 15,000 gpd, these wastewater characteristics should not be harmful to an 8.5 mgd facility if discharged on a steady basis. It is the intermittent discharge of this wastewater which has contributed to the overloading of the biological population of the POTW.

During a three week shutdown of the industry in July of 1985, the HTWTP recovered to the point of meeting their permit limits. Consequently, the Township only permitted the industry access to the sewer system after the installation of metering pumps to equalize flows. This requirement initially improved POTW performance during the fall of 1985, but a gradual deterioration in effluent quality (indicating possible toxicity effects) lead the Township to terminate service to the industry in late-November.

While the most recent action is being challenged, the industry is constructing an anaerobic pretreatment facility on site to reduce its loading to the POTW.

A number of operations and personnel changes have been instituted at the HTWTP to help mitigate the impact of the industrial discharges. These changes include:

- installation of aeration equipment in the influent channels to the RBCs to increase the first stage DO to 2-3 mg/l;
- extensive use of sludge depth measurement and visual monitoring to augment reliance on control room instrumentation;
- performance of bioassay testing by an independent contractor to assess toxicity effects;
- purchase of a toxicity tester to be used in calculation of local limits; and
- hiring of four more people plus the purchase of a vehicle for an extensive industrial sampling program.

HAMILTON TOWNSHIP WASTEWATER TREATMENT PLANT
Trenton, New Jersey

Design Flow:	16 mgd	Location:	Central Western Border
Secondary Treatment:	Trickling Filter and RBC	Population Served:	87,000

INFLUENT WASTEWATER

	Typical (Upset)
Ave. Flow, mgd	8.5
% Industrial	10 (est)
BOD_5, mg/l	240 (500)
SS, mg/l	160 (400)

SIGNIFICANT INDUSTRIES

Industry	Flowrate (1000 gpd)	Problem Pollutants
Pharmaceutical	15	BOD, phenol, ethyl benzene, toluene, xylene
Electroplaters (2)	160	Cd, Cr, Zn, Ni

PLANT LOADING

Primary Clarifiers	Typical (Upset)
Overflow Rate, gal/sf/day	830, 260, 320
Detention Time, hours	1.8, 4.8, 5.6

Trickling Filters	Typical (Upset)
Plant Flow (mgd)	2.5, 1.0
Hydraulic Loading. gal/sf/day	100, 210
Organic Loading, lbs BOD/1,000 cf/day	15, 16 (30)
Return Flow, %	20,100

Secondary Clarifiers	Typical (Upset)
Overflow Rate, gal/sf/day	520, 260, 265
Detention Time, hours	2.8, 4.8, 6.8

RBCs	Typical (Upset)
Plant Flow (mgd)	5.0
First Stage Organic Loading, lbs BOD/1,000 sf/day	
- Total	5.3 (10.8)
- Soluble	3.5 (6.7)

PLANT PERFORMANCE

	Permit Limit	Typical (Upset)
BOD_5, mg/l	30	45 (100
SS, mg/l	30	20 (50)
NH_3, mg/l (Effective 6/86)	10	20 (30)

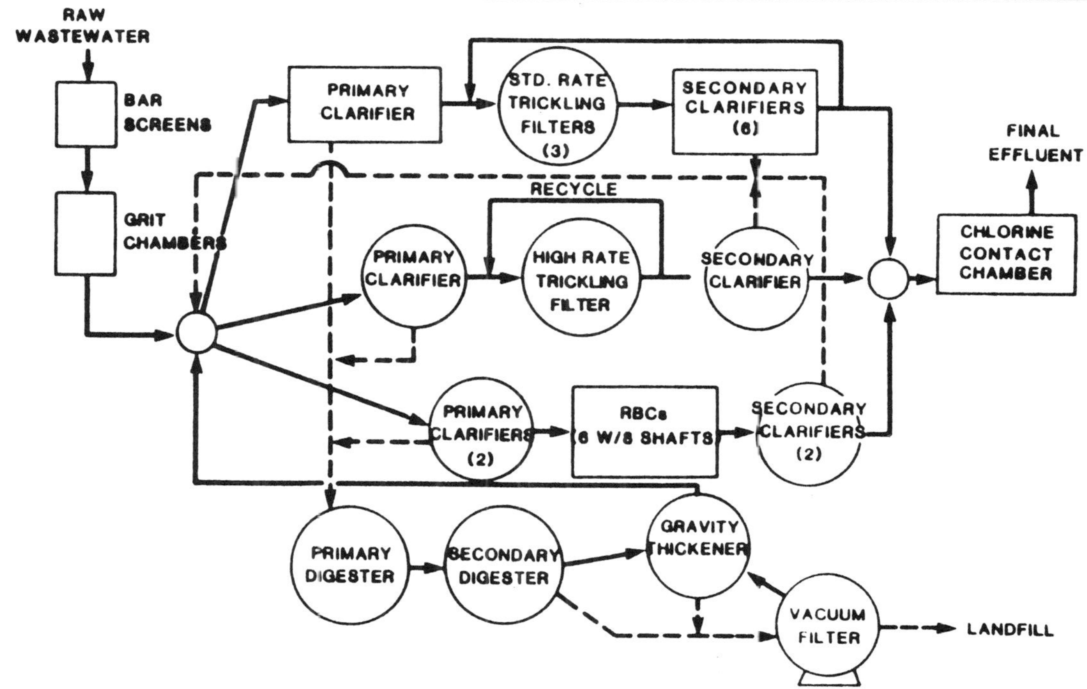

HORSE CREEK POLLUTION CONTROL FACILITY
North Augusta, South Carolina

The Horse Creek Pollution Control Facility (HCPCF) is a regional plant, operated by the Aiken County Public Service Authority (ACPSA), treating a predominantly industrial wastewater. Ninety five percent of the industrial wasteload is contributed by several large textile mills and is characterized by high COD, BOD, alkalinity and pH. Combined domestic/industrial influent wastewater pH and alkalinity fluctuations caused inhibition of the biomass, poorly settling sludge and effluent suspended solids permit violations. Since implementing a pretreatment program and issuing industrial wastewater discharge permits, the treatability of the industrial waste has improved, the result being that HCPCF has been free of NPDES permit violations for over eight months.

Local textile processes include grading operations, finishing processes utilizing dyes, and specialized textile chemical manufacturing. The textile wastewater is highly caustic with alkalinity as high as 2400 mg/l, and pH exceeding 12.5. Prior to pretreatment the combined industrial/domestic influent to the HCPCF had the following characteristics:

pH	>11
BOD	360 mg/l
COD	910 mg/l
Alkalinity	1100 mg/l
TSS	210 mg/l

Other distinguishing characteristics of the influent wastewater included the light, non flocculant nature of the suspended solids and a dark blue/black color, typical of textile wastewater from washing and dying operations.

Prior to the summer of 1985, the textile industries employed a limited type of pretreatment and flow equalization. This limited pretreatment and flow equalization resulted in plant influent pH fluctuations of 2 to 2.5 units and alkalinity fluctuations of up to 600 mg/l in a given day. These fluctuations caused some inhibition of the biomass, but because the hydraulic detention time in the aeration basins was in excess of 3.5 days, effluent BOD was within the permit limit of 33 mg/l. These pH and alkalinity fluctuations had their most detrimental effect on biomass settling characteristics and solids carryover in the secondary clarifier often resulted, lasting for 24-36 hours. During these episodes, filamentous organisms were occasionally observed in the biomass. The solids carryover problem worsened in the winter months when wastewater temperatures were lower, but chlorination of the return activated sludge, the influent to the secondary clarifier and the contents of the aeration basin was somewhat successful at improving settleability. Despite this, the HCPCF still experienced interference with its ability to meet suspended solids limits in 15 of the 19 months prior to September, 1985.

The State of South Carolina mandated that the ACPSA implement and enforce a pretreatment program in the spring of 1984. The ACPSA responded by developing such a program and issuing draft industrial wastewater discharge permits. Final State approval came in May, 1985. As presently written, the industrial wastewater discharge permits are not restrictive, allowing BOD, COD and alkalinity levels as high as 600 mg/l, 1300 mg/l and 1500 mg/l, respectively. However, the permits have caused the textile industries to make small, but meaningful alterations to their wastewater discharge practices, resulting in average plant influent pH levels dropping from 11-12 to 10 and alkalinity from 1100 mg/l to 700 mg/l. More importantly, maximum daily influent pH fluctuations have been reduced to 0.5 units or less. Figure A-3 shows the magnitude of pH fluctuations both before and after the implementation of pretreatment. Simple modifications at textile facilities to process operations and waste pumping schedules were typical of the changes that were necessary to realize the described results. Because of the more stable wastewater discharge, the HCPCF has realized more consistent plant operation and has not violated its NPDES permit in over eight months.

Some of the textile dischargers do not currently meet the pH and alkalinity limits of their industrial wastewater discharge permits and are under a compliance schedule to do so. The facilities are installing pretreatment works for caustic recovery that should significantly lower pH and alkalinity levels. The HCPCF is also presently studying the addition of floating mixing units to augment the turbine surface aerators in the aeration basins. To date, evidence indicates that a more consistent secondary clarifier solids feed is achieved which improves the quality of the secondary effluent.

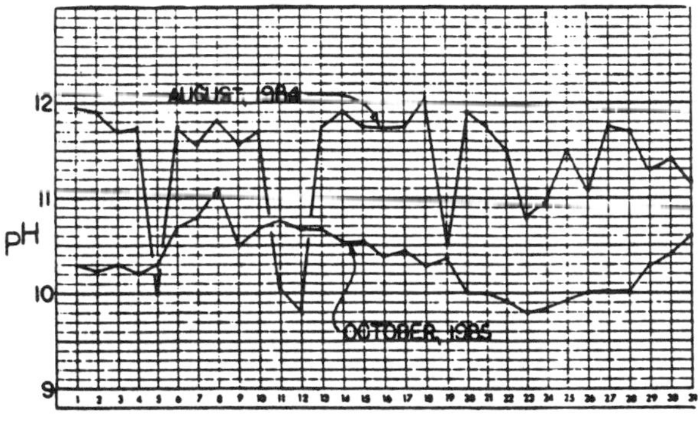

FIGURE A-3
HORSE CREEK POLLUTION CONTROL FACILITY INFLUENT pH

HORSE CREEK POLLUTION CONTROL FACILITY
Aiken County, South Carolina

Design Flow:	**20 mgd**	Location:	**West-central South Carolina**
Secondary Treatment:	**Extended Aeration Activated Sludge**	Population Served:	**70,000**

INFLUENT WASTEWATER

	Typical (Upset)
Ave. Flow, mgd	10.4
% Industrial	80
BOD_5, mg/l	360
SS, mg/l	210
COD, mg/l	910
Alkalinity, mg/l	1100 (1600)
pH	10-11 (12.5)

SIGNIFICANT INDUSTRIES

Industry	Flowrate (1000 gpd)	Problem Pollutants
Textile	8,400	COD, Alkalinity, pH
Textile chemicals	300	COD, pH

PLANT LOADING

Primary Clarifiers	Typical (Upset)
Overflow Rate, gal/sf/day	300
Detention Time, hours	4.4

Secondary Clarifiers	Typical (Upset)
Overflow Rate, gal/sf/day	195
Detention Time, hours	9.1

Aeration Basins	Typical (Upset)
F/M, lbs BOD_5/lbs MLSS/day	0.05-0.10
MCRT, days	50-90
MLSS, mg/l	3800-4500
Detention Time, hours	92
Return Flow, %	40-60
D.O. Level, mg/l	1-3 (4)

PLANT PERFORMANCE

	Permit Limit	Typical (Upset)
BOD_5, mg/l	33	15
SS, mg/l	57	40 (85)
COD, mg/l	-	175
pH	9	9 (10)

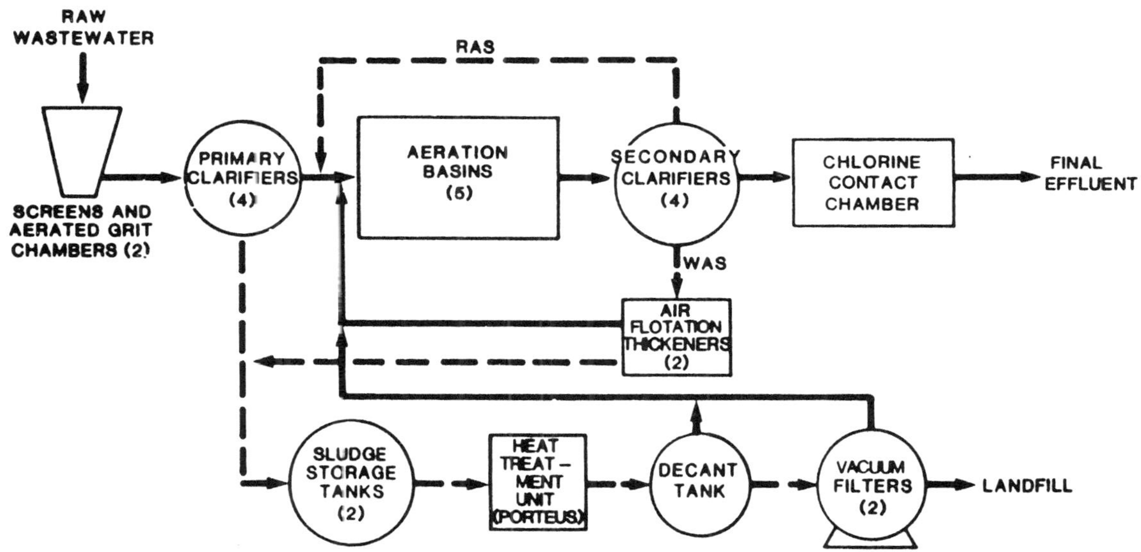

MAIDEN CREEK WASTEWATER TREATMENT PLANT
Blandon, Pennsylvania

The Maiden Creek Wastewater Treatment Plant (MCWTP) went on-line in December, 1981 as a secondary treatment facility designed to remove both carbonaceous and nitrogenous BOD. The plant uses a patented aerated submerged fixed film biological treatment system, where flat asbestos plates hanging vertically in the settled wastewater provide a growth surface for the bacteria. Each of three contact basins contains 320 plates with 200 sq. ft. of surface area. Oxygen is provided by fine bubble aeration through ceramic diffusers.

During the first six months of operation following an initial acclimation period, the MCWTP experienced gradual flow increases from 0.1 to 0.15 mgd while consistently meeting their permit limits. In August of 1981, a local mushroom processor began batch discharging high BOD wastewater to the POTW at flows sometimes exceeding 100 gpm. The hydraulic and organic shock loadings resulted in nitrifier washouts, solids carryover, reduced BOD removal efficiency and at times total biological process failure. Although the industry was not measuring their wastewater flow rates at that time, they were the only significant non-domestic contributor. After factoring out any potential infiltration/inflow from stormwater flows, the discharge pattern from the industry was obvious from an inspection of the weekly flow recordings at the POTW. Figure A-4 illustrates the dramatic effect of the industrial discharges on the MCWTP influent.

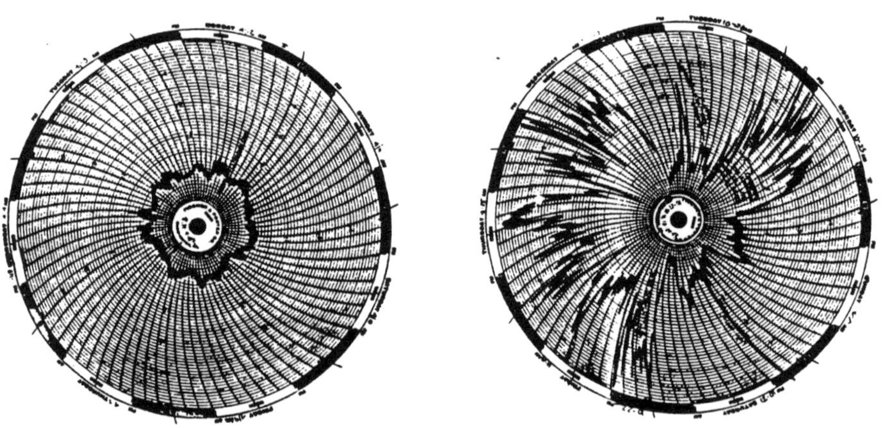

April, 1982 October, 1982

FIGURE A-4
WASTEWATER DISCHARGE AT INFLUENT METERING STATION (MGD)

As a result of significant time and effort on the part of Maiden Creek Township Municipal Authority two years ago, the food processor installed a physical-chemical treatment system which included surge control tanks and aeration. The system did reduce the solids load and partially mitigated the flow spike problem, although the surge tanks were not capable of providing complete equalization. Unfortunately, the great percentage of their organic waste is soluble, so the pretreatment facility is ineffective in reducing the BOD loading to the POTW. Additionally, wastewater production far exceeds the 50,000 gpd limit imposed by their permit, so occasional flow spikes are still evident. The industry has requested nearly ten times the current flow limit, necessitating the design of a full secondary system to reduce their waste strength to domestic levels. Such a system, including a 650,000 gallon aerated equalization basin, is scheduled to go on-line in mid-1986. In the interim, the municipality has required that the industry:

- control flow surges;
- meter and record their flows continuously;
- reduce the BOD in the effluent by in-house methods; and
- composite sample their discharge on a regular basis.

Failure to comply with the abovementioned program will result in a shut off by the POTW, a measure used previously in February, 1985 when the industry's wastewater was responsible for total process failure at the plant.

A number of operational changes were instituted in May of 1985 to help combat the high organic loads in the contact basins. These changes included:

- increasing the aeration by using all blowers at the plant, resulting in an increase in the first stage D.O. from 2 mg/l to 5 mg/l;

- addition of selective strains of bacteria to increase the rate of BOD removal;

- recycling the plant effluent to the head of the plant to dilute the incoming wastewater; and

- reducing the allowable flow from the food processor and closely monitoring their adherence to the limits.

Since these changes were implemented concurrently, it is impossible to isolate the individual impacts of each operations change. However, the collective result was a substantially improved compliance record. There have also been no flow spikes at the POTW since mid-December, 1985, indicating better flow control on the part of the food processor.

MAIDEN CREEK WASTEWATER TREATMENT PLANT
BLANDON, PENNSYLVANIA

Design Flow:	**0.45 mgd**	Location:	**Southeastern Pennsylvania**
Secondary Treatment:	**Aerated Submerged Fixed Film (Contact Aeration)**	Population Served:	**2,000**

INFLUENT WASTEWATER

	Typical (Upset)
Ave. Flow, mgd	0.25
% Industrial	20 (60)
BOD_5, mg/l	350 (900)
SS, mg/l	200
NH_3, mg/l	60

SIGNIFICANT INDUSTRIES

Industry	Flowrate (1000 gpd)	Problem Pollutants
Food Processor	50	BOD, Flow surges
Dental Office	negl.	Hg

PLANT LOADING

Primary Clarifiers	Typical (Upset)
Overflow Rate, gal/sf/day	350 (1,000)
Detention Time, hours	3.75 (1.25)
Effluent BOD_5, mg/l	260
Effluent SS, mg/l	100

Secondary Clarifiers	Typical (Upset)
Overflow Rate, gal/sf/day	450 (1,300)
Detention Time, hours	2.8 (1.0)

Contact Basins	Typical (Upset)
Organic Loading (lbs BOD_5/1000 sf/day)	
Total Plant	2.8
First Stage	8.4
Detention Time, hours	12
D.O. Level, mg/l	5 - 10

PLANT PERFORMANCE

	Permit Limit	Typical (Upset)
BOD_5, mg/l	30	15 (400)
SS, mg/l	30	10 (50)
NH_3, mg/l	10/20	1 (60)

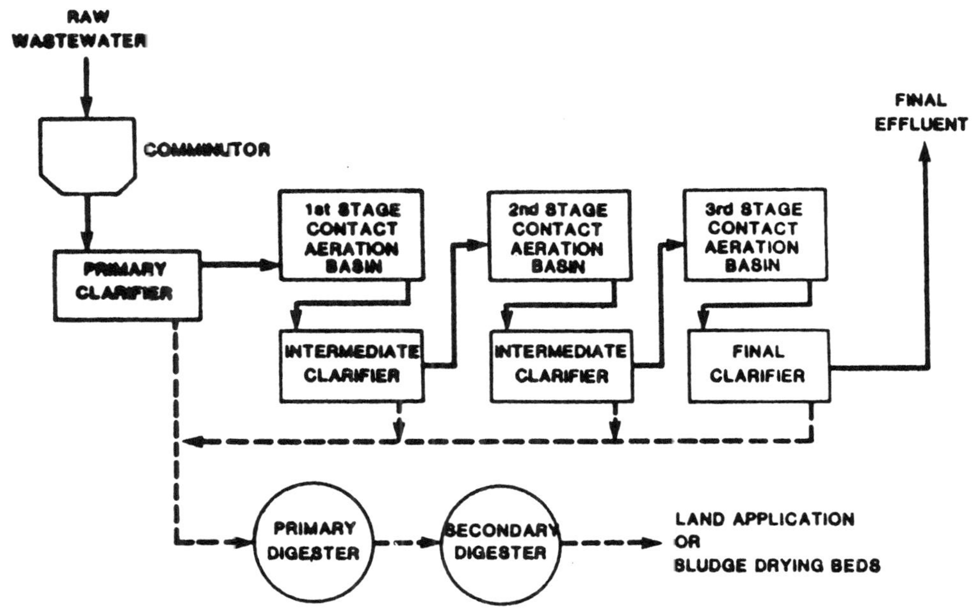

METRO-WEST POINT TREATMENT PLANT
Seattle, Washington

The Municipality of Metropolitan Seattle (METRO) has had an operational industrial pretreatment program since 1969. With minor modifications, the program was EPA-approved in 1981 as one of the first in the nation. Successful reductions in influent wastewater and primary sludge heavy metal concentrations during the last five years can, to a great extent, be attributed to implementation and enforcement of pretreatment standards. As an outcome of this, self-monitoring by industrial dischargers augmented with year-round spot monitoring by Metro's Industrial Waste Section has reduced the incidences of toxic upsets in the anaerobic digesters of the West Point Treatment Plant.

The Metro-West Point Treatment Plant provides primary treatment and sludge digestion for an average daily wastewater flow of 132 mgd, 4.7 percent originating from industrial sources. Approximately 70 metal finishing/electroplating industries discharge to the sewer system in addition to a variety of other categorical and non-categorical industries. Records of periodic digester upsets go back as early as 1967, but their occurrences have become less frequent since 1980, coinciding with substantial overall reductions in heavy metal concentrations. Past upsets directly linked to toxic metals (generally chromium) caused increased volatile acid concentrations, increased carbon dioxide content of the gas produced, and reduced gas production. An October, 1980 chromium spill to the West Point facility caused a typical upset and resulted in the plant influent chromium concentration jumping 10 fold to greater than 2 mg/l. Primary sludge concentrations of chromium reached 710 mg/l, resulting in a 30 mg/l increase in digester concentrations above their normal 16-17 mg/l level. Metro practices sludge application to forest lands. Application rates had to be decreased during upsets, although no interference occurred.

Figure A-5 below typifies the reduction in metals realized during the 1981-1985 time period. Plant influent chromium levels dropped approximately 55 percent while the digested sludge concentrations were reduced by more than 40 percent. The magnitude of these decreases are typical of other heavy metals as well, averaging 41 percent for chromium, cadmium, copper, lead, nickel and zinc combined (see the accompanying data sheet). The primary reason for the reduction of cadmium and chromium concentrations is improved industrial pretreatment. In addition to pretreatment, a less corrosive city water supply has also resulted in lower background metal concentrations for the other metals, especially for copper. The city recently began chemically conditioning its water in an attempt to extend conduit life.

Success of the Metro Industrial Pretreatment Program can be attributed to a number of important factors including:

- development of stringent local limits for industrial discharges;
- year-round industrial waste sampling programs supported financially by industry; and

- follow-up procedures to industrial waste spills, taking enforcement action and levying fines when necessary.

Metro has recently implemented the following steps to improve their pretreatment program:

- information exchange with industries through the use of quarterly newsletters and personal communication, and
- increasing public awareness of industrial discharge violators by publishing the names of violating companies in local papers along with a statement of Metro's enforcement policy.

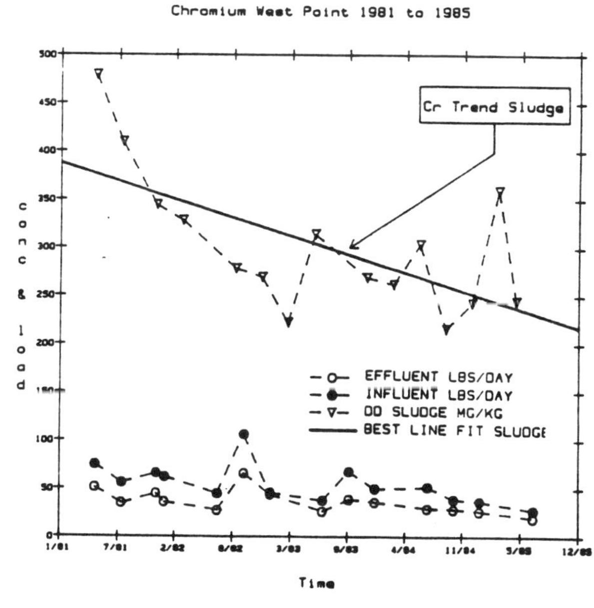

FIGURE A-5
WEST POINT CHROMIUM CONCENTRATIONS

WEST POINT TREATMENT PLANT
SEATTLE, WASHINGTON

Design Flow:	**125 mgd**	Location:	**West-Central Washington**
Primary Treatment		Population Served:	**500,000**

INFLUENT WASTEWATER

	Typical (Upset)
Ave. Flow, mgd	132
% Industrial	5
BOD_5, mg/l	160
SS, mg/l	260
Cr, mg/l	0.05 (2.0)

SIGNIFICANT INDUSTRIES

Industry	Flowrate (mgd)	Problem Pollutants
Metal finishing and electroplating	1.1	Cd, Cr, Cu, Ni, Zn

PLANT LOADING

Primary Clarifiers

	Typical (Upset)
Overflow Rate, gal/sf/day	1080
Detention Time, hours	1.58
Effluent BOD_5, mg/l	75-110
Effluent SS, mg/l	60-90

Digested Sludge Metal Concentrations

	1981 Level	1985 Level
Cadmium, mg/kg	45	28
Chromium, mg/kg	480	250
Copper, mg/kg	1300	700
Nickel, mg/kg	160	120
Lead, mg/kg	800	400

PLANT PERFORMANCE

	Permit Limit		Typical (Upset)	
	Summer	Winter	Summer	Winter
BOD_5, mg/l	135	85	110	75
SS, mg/l	125	65	90	60
Cr, mg/L	0.07	0.07	<0.05	(0.15)

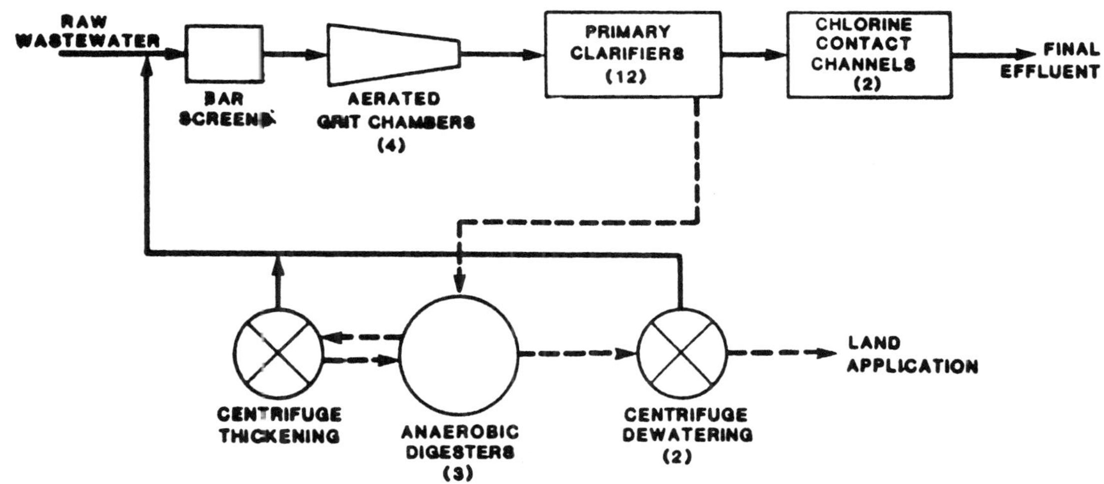

NEUSE RIVER WASTEWATER TREATMENT PLANT
Raleigh, North Carolina

The Raleigh case study illustrates the need for continuous survey and monitoring even after the implementation of an industrial waste program in any dynamic population center. In 1976, the 30 mgd Neuse River Wastewater Treatment Plant (NRWTP) went on-line to replace the overloaded 16 mgd Walnut Creek plant. In the early 1960's, influent BODs exceeded 300 mg/l at Walnut Creek, with the effluent ranging from 35 to 55 mg/l. These effluent levels violated the Walnut Creek Plant permit, established by the state in order to protect the quality of the Neuse River, which was used as the raw water source for the City of Smithfield located downstream of Raleigh. Industries were encouraged to conserve and recycle wastes, resulting in a 250 mg/l influent BOD by the mid-1960's. The City's first Sewer Use Ordinance was enacted in 1972, with continual modification to comply with changes in the federal regulations. The net effect is a current influent BOD consistently below 200 mg/l, despite an industrial flow volume representing 25 percent of the plant flow.

The only significant industrial discharger of metals to the Walnut Creek plant was a large electroplater whose occasional plating bath dumps were not prohibited by a sewer use ordinance during the 1950's. Digester upsets (decreased gas production) and high sludge metals content were traced to this particular industry. Since dried sludge was being made available to the community for landscaping purposes at the time, concern for the metals levels prompted adoption of a proposed ordinance which directed the industry to construct a physical-chemical pretreatment facility.

Two other metals-related industries have been responsible for high sludge metals since the construction of the NRWTP. In the current facility, wet sludge is land applied to farmland adjacent to the POTW, hence metal content is critical. In each case (an electroplater and a printed circuit board manufacturer), the industries were discharging levels of Cr, Ni, Zn, Pb and Cu sometimes in excess of 1,000 mg/l, with highly variable effluent pH, and were uncooperative in dealing with the City of Raleigh. Fining the former industry $ 1,000, and threatening the latter with the same, provided sufficient incentive to install pretreatment.

In the early 1980's, a producer of amino acids for pharmaceuticals in Raleigh discharged slug loads totaling 1,000 lbs of NH_3 to the POTW each day. Fortunately, an activated sludge system had been constructed for their facility for BOD reduction, which possessed sufficient capacity to nitrify their wastewater to an ammonia concentration of 50 mg/l. On one occasion, the NH_3 levels became toxic to the IU's activated sludge pretreatment, resulting in a gradual loss of nitrification at the POTW. Rapid identification of the NH_3 discharge by City personnel preserved the POTW nitrifier population, which was subsequently used to re-seed the industry's activated sludge with a viable nitrifier population for a speedy recovery. The rapid response prevented the monthly effluent NH_3 levels from exceeding the permit limit, despite high daily concentrations following the incident.

A dairy product manufacturer who cleans the stainless steel tanker trucks on-site had previously discharged these wastes directly to the sewer. Average BODs of 10,000 mg/l, with occasional values in the 30,000 to 40,000 mg/l range were typical, often resulting in effluent BODs in excess of the 6 mg/l (12 in winter) allowed for the POTW. Working with the North Carolina State University, a vacuum recovery system was developed and a market identified for the collected whey waste. The effluent BOD now averages 2,000 mg/l, still resulting in a high surcharge payment, but no permit violations at the POTW.

An unusual case at the NRWTP was the discovery of high zinc levels (1,000 mg/l) in the discharge from an office building with no manufacturing component. Through discussions with maintenance personnel, the City of Raleigh discovered that the contaminated discharges corresponded to floor stripping activities in the building. They learned that a Zn-based floor wax had been used, and stripping an entire office building over the course of a week discharged enough Zn to the POTW to significantly raise the level in their sludge. The elevated zinc level threatened to interfere with the POTW's ability to dispose of its sludge.

The Raleigh plant is currently under construction to increase the hydraulic capacity from 30 to 40 mgd, with an additional expansion to 60 mgd planned for the near future (the schematic shown on the next page is for the 40 mgd facility). The rapid growth of this community will continue to bring with it a variety of challenging new industrial wastewaters with, in some cases, unpredictable impacts on the POTW.

NEUSE RIVER WASTEWATER TREATMENT PLANT
RALEIGH, NORTH CAROLINA

Design Flow:	**40 mgd**	Location:	**Central North Carolina**
Secondary Treatment:	**Activated Sludge (Extended Aeration)**	Population Served:	**195,000**

INFLUENT WASTEWATER

	Typical (Upset)
Ave. Flow, mgd	25
% Industrial	25
BOD_5, mg/l	165 (350)
SS, mg/l	170 (500)

SIGNIFICANT INDUSTRIES

Industry	Flowrate (1000 gpd)	Problem Pollutants
Electroplaters, Metal Finishers (5)	750	Cd, Cr, Cu, Ni, Pb, Zn, Cn^-, Fe, pH
Pharmaceutical	400	NH_3
Dairy	110	BOD
Snack Foods	100	BOD

PLANT LOADING

Primary Clarifiers	Typical (Upset)
Overflow Rate, gal/sf/day	650
Detention Time, hours	3.0

Secondary Clarifiers	Typical (Upset)
Overflow Rate, gal/sf/day	680
Detention Time, Hours	3.2
SVI, ml/gm	150-200 (250)
Effluent BOD_5, mg/l	5
Effluent SS, mg/l	15
Effluent NH_3, mg/l	3

Aeration Basins	Typical (Upset)
F/M, lbs BOD_5/lbs MLSS/day	0.08-0.10
MCRT, days	12-20
MLSS, mg/l	2500
Detention Time, hours	15
Return Flow, %	50
D.O. Level, mg/l	2

Multi-Media Filters	Typical (Upset)
Hydraulic loading, gpm/sf	2.5

PLANT PERFORMANCE

	Permit Limit		Typical (Upset)
	Summer	Winter	
BOD_5, mg/l	6	12	3 (15)
SS, mg/l	30	30	4
NH_3, mg/l	3	6	1.5 (8)

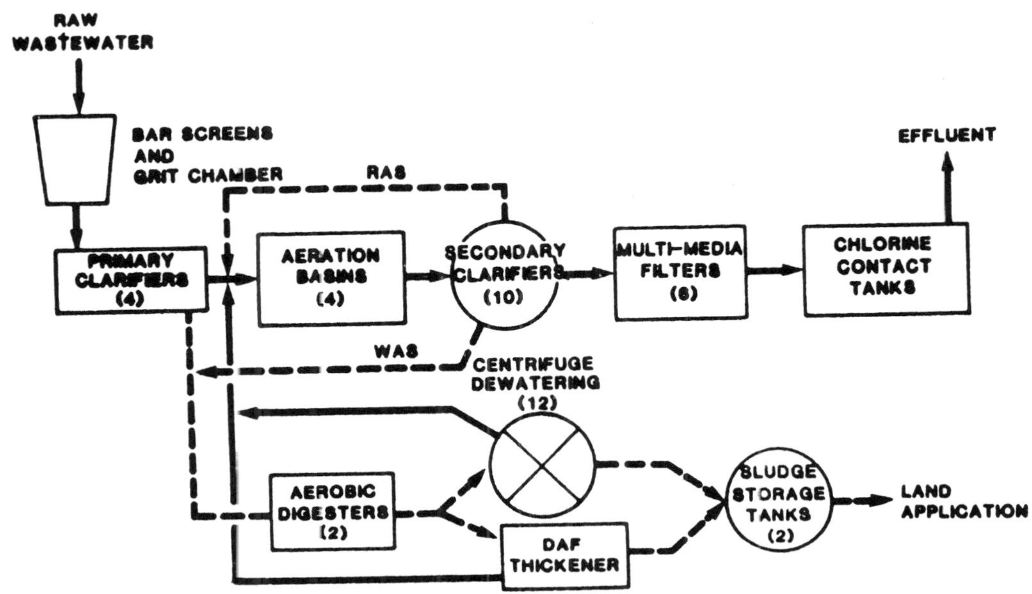

NEWARK WASTEWATER TREATMENT PLANT
Newark, Ohio

The Newark Wastewater Treatment Plant (NWTP) had been in substantial non-compliance with its 1981 NPDES permit from the beginning of 1983 until the middle of 1984. This consistent violation had resulted primarily from increased waste loads on the POTW from industrial sources. Between 1979 and 1984, the percentage of industrial wastewater increased from 12 to 22 percent by volume, with influent BOD increasing from 220 to 330 mg/l, while suspended solids increased from 200 to 350 mg/l. To complicate the non-compliance problem, four separate ammonia discharge episodes occurred from August to October, 1983 which resulted in both the loss of activated sludge viability (interference) and the pass-through of the NH_3 and the subsequent killing of 80,000 fish in the Licking River. The fish kill precipitated the submission of Verified Complaints to the Ohio EPA on August 6, 1984 by the Black Hand Gorge Preservation Association, against the City of Newark and the NWTP. Following an investigation, the Ohio EPA issued Director's Final Findings and Orders, specifying a compliance schedule and interim discharge limits for the POTW until a planned facility upgrade is completed by July 1988.

There are two significant industrial contributors to the NWTP who were also issued Director's Final Findings and Orders in May, 1985. A fiberglas insulation manufacturer had been discharging high concentrations of phenol (2-5 mg/l) and NH_3 (up to 500 mg/l), with occasional spills of formaldehyde into the collection system. The activated sludge bacteria were acclimated to the phenol in the wastewater, but were susceptible to interference from shock loadings of the NH_3 and formaldehyde. Fortunately, the industry was responsive to the problems of the NWTP, and instituted a corrective program to:

- conserve and recycle plant flows, which have reduced their discharge by 60 percent (from 1.22 to 0.45 mgd) over the past two years;

- construct an aerated equalization basin to air-strip phenol and distribute diurnal fluctuations; and

- construct a pretreatment facility for their landfill leachate.

The POTW is still subject to occasionally high NH_3 loads from the industry, which is currently the only identifiable cause of isolated interference problems in the plant. The municipality and industry continue to work cooperatively to resolve this problem through the implementation of a spill prevention and control program. Additionally, the renovated POTW will use some of the existing clarifier tankage for off-line storage in the event of future spill episodes.

The replacement of coarse bubble aerators with fine bubble equipment in mid-1984 significantly improved BOD removals and the NWTP compliance record. Nitrification, which did not occur previously, now takes place in the last two aeration basins, because of the improved carbonaceous BOD (CBOD) removal in the initial basins. The only incident of non-compliance with the interim permit in 1985 resulted from an NH_3 discharge from the fiberglass manufacturer. In this case, even though the average monthly BOD measured 29 mg/l, the carbonaceous component was less than 10 mg/l. The final permit will have a more stringent NH_3 requirement and will also designate CBOD as a permitted parameter.

A second major industry is a dairy which came on-line in 1976. Initially, the dairy stored whey waste in a silo and typically bled it into the sewer system. The discharge was high in both BOD and suspended solids (2,000 mg/l), and would occasionally be batch discharged to the POTW, resulting in a shock loading to the activated sludge and violation of the NPDES permit limits. The industry has since installed a reverse osmosis treatment system for the whey waste which has reduced the solids and organic loading to the plant.

The only categorical industry that currently discharges to NWTP is an electroplater who constructed a metals removal system in conformance with federal pretreatment regulations. In the past, dewatered sludge had been applied to corn fields adjacent to the plant property. However, when heavy metals were detected in seven of ten monitoring wells, Newark began hauling liquid sludge off-site. The planned facility upgrade will include installation of belt filter presses, so that the existing sludge (with acceptable levels of heavy metals) can once again be dewatered and more economically hauled off-site to farm land.

NEWARK WASTEWATER TREATMENT PLANT
NEWARK, OHIO

Design Flow:	**8.0 (12.0 Hydraulic) mgd**	Location:	**Central Ohio**
Secondary Treatment:	**Activated Sludge (Conventional)**	Population Served:	**41,000**

INFLUENT WASTEWATER

	Typical (Upset)
Ave. Flow, mgd	7.5
% Industrial	15
BOD_5, mg/l	305 (450)
SS, mg/l	360 (550)
NH_3, mg/l	35 (60)

SIGNIFICANT INDUSTRIES

Industry	Flowrate (1000 gpd)	Problem Pollutants
Fiberglass	450	Phenol, NH_3, Formaldehyde
Dairy	223	BOD, Phosphorus, SS
Electroplater	97	Cr, Cd, Pb, Ni, Zn, Cyanide

PLANT LOADING

Primary Clarifiers	Typical (Upset)
Overflow Rate, gal/sf/day	560
Detention Time, hours	3.2
Effluent BOD_5, mg/l	194 (280)
Effluent SS, mg/l	147 (218)

Secondary Clarifiers	Typical (Upset)
Overflow Rate, gal/sf/day	500
Detention Time, hours	3.7
SVI, ml/gm	150 (350)

Aeration Basins	Typical (Upset)
F/M, lbs BOD_5/lbs MLSS/day	0.25 (0.4)
MCRT, days	5-6
MLSS, mg/l	2,000
Detention Time, hours	6.3
Return Flow, %	50
D.O. Level, mg/l	2.0

PLANT PERFORMANCE

	Permit Limit	Typical (Upset)
BOD_5, mg/l	20	23 (60)
SS, mg/l	40	27 (95)
NH_3, mg/l (Summer)	25	15 (30)

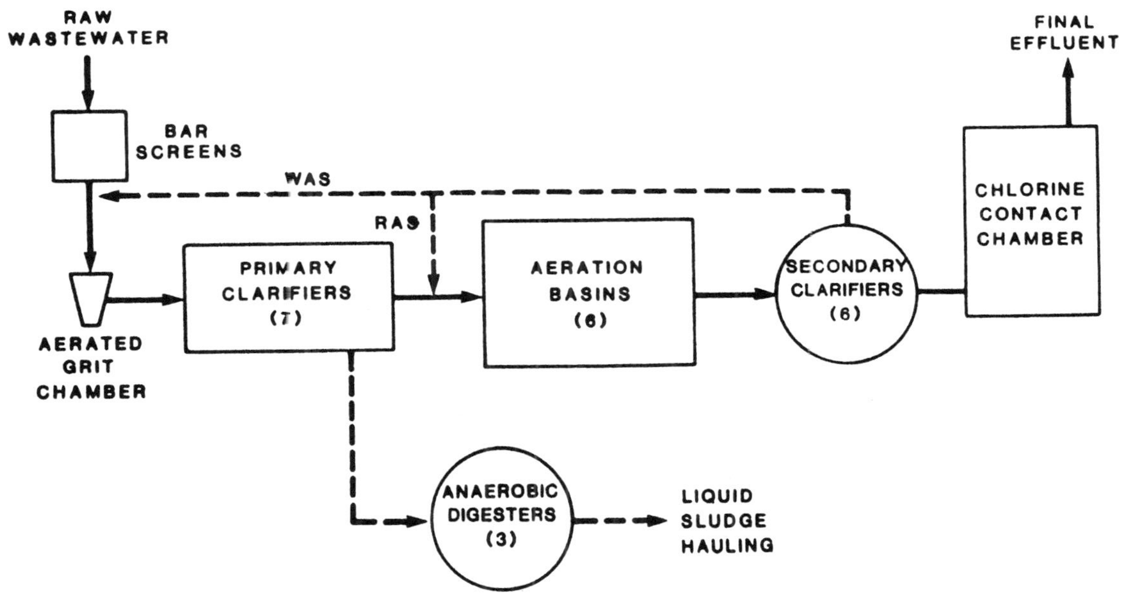

NORTH SHORE SANITARY DISTRICT GURNEE PLANT
Gurnee, Illinois

The Gurnee Plant of the North Shore Sanitary District (NSSDGP) receives an average daily wastewater flow of 12.4 mgd from a variety of sources. Those sources include a major naval installation, domestic sewage discharges, secondary effluent from the District's North Chicago Sewage Treatment Plant, and other industries which contribute 17 percent of the total flow.

Since startup in 1976, the NSSDGP has experienced periodic failures at achieving nitrification in the two-stage activated sludge system. The failures to achieve nitrification to the ammonia levels of the District's NPDES effluent limits have also, at times, been accompanied by general process upsets which have resulted in effluent SS and BOD_5 violations. One of the major industrial contributors to the Gurnee Plant, a pharmaceutical manufacturer discharging an average flow of 750,000 gpd, has similarly experienced upsets of its own activated sludge pretreatment system which have resulted in violations of the District's local sewer use ordinance. It was initially believed that the observed interferences at the NSSDGP were the result of the discharge of filamentous organisms and other solids by the manufacturer. The initiation of in-plant solids control methods (which significantly lessened the quantity of solids entering the industrial wastewater pretreatment system) and pretreatment system upgrades did not, however, eliminate interferences at the NSSDGP.

In 1980, District personnel began to suspect that the presence of a nitrification inhibiting antibiotic, erythromycin, in the pharmaceutical wastewater was the main cause of the process upsets at the NSSDGP. By 1983, test and control bench-scale activated sludge reactors were placed in operation and the effects of the pharmaceutical wastewater and erythromycin on the NSSDGP were investigated. A bioassay test for the presence of erythromycin and other nitrification inhibitors was also developed, along with a Direct Insertion Probe/Mass Spectrometric technique for confirmation. The results of the bench-scale testing indicated that the presence of soluble and/or solid constituents of the pretreated pharmaceutical wastewater inhibited nitrification and, at high levels, could completely suppress nitrification. Additionally, it was found that although erythromycin inhibited nitrification, acclimation to low concentrations of erythromycin could occur in the absence of extreme concentration fluctuations.

During January of 1984, an observed average industrial pretreatment effluent erythromycin concentration of 53 mg/l with mass loading fluctuations of greater than two orders of magnitude completely inhibited nitrification in the Gurnee Plant. The resulting BOD_5 and SS concentrations were as high as 26 mg/l and 67 mg/l, respectively. Lower concentrations of erythromycin in the absence of such strong concentration fluctuations did not interfere with the performance of the Gurnee Plant during August of 1984, with average effluent BOD_5 and SS concentrations of 11 mg/l and 8 mg/l, respectively, and effluent ammonia concentrations ranging from 0.4 mg/l to 1.5 mg/l as N. Experience at the Gurnee Plant and with the bench-scale test systems has also indicated that a lag period of two to three mean cell residence times is required before the effects of erythromycin on the activated sludge process become apparent. Erythromycin also was found to disrupt the settling of the first-stage carbonaceous organisms.

Measures undertaken by District personnel to lessen the effect of the pharmaceutical discharge on plant performance have included:

- The addition of inorganic coagulants to aid primary clarifier performance;
- the addition of polymer to the first-stage activated sludge system,
- daily bacterial (staphylococcus aureus) bioassays of industrial wastewaters for the presence of inhibiting substances; and
- the development of an ordinance governing the discharge of erythromycin to the NSSDGP.

Since passage of the ordinance in November, 1985, in which the discharge limits for erythromycin were established, the NSSDGP has substantially been in compliance with its NPDES permit and ammonia levels of 0.25 mg/l to 1 mg/l as N have been consistently achieved.

NORTH SHORE SANITARY DISTRICT GURNEE PLANT
GURNEE, ILLIONOIS

Design Flow:	**13.8 mgd**	Location:	**Northeastern Illinois**
Secondary Treatment:	**Activated Sludge (Two-Stage, Modified-Contact)**	Population Served:	**65,000**

INFLUENT WASTEWATER

	Typical (Upset)
Ave. Flow, mgd	12.4
% Industrial	37
BOD_5, mg/l	140
SS, mg/l	180
NH_3, mg/l	15

SIGNIFICANT INDUSTRIES

Industry	Flowrate (1000 gpd)	Problem Pollutants
Pharmaceutical	750	Antibiotics, SS
Electroplating	100	Cu, CN
Chemical	170	Organics
Nonferrous Metals	90	W
Military Installation	3,500	pH

PLANT LOADING

Primary Clarifiers	Typical (Upset)
Overflow Rate, gal/sf/day	695
Detention Time, hours	2.7
Effluent BOD_5, mg/l	100
Effluent SS, mg/l	100

First Stage Clarifiers	Typical (Upset)
Overflow Rate, gal/sf/day	780
Detention Time, hours	2.5

Second Stage Clarifiers	Typical (upset)
Overflow Rate, gal/sf/day	645
Detention Time, hours	3.1

First Stage Aeration Basins	Typical (Upset)
F/M, lbs BOD_5/lbs MLVSS/day	0.95
MCRT, days	7
MLSS, mg/l	3000
Detention Time, hours	4.2
Return Flow, %	25
D.O. Level, mg/l	2.5

Second Stage Aeration Basins	Typical (Upset)
F/M, lbs NH_3-N/lbs MLVSS/day	0.07
MCRT, days	13
MLSS, mg/l	3500
Detention Time, hours	5.8
Return Flow, %	50
D.O. Levels, mg/l	2.5

PLANT PERFORMANCE

	Permit Limit	Typical	(Upset)
BOD_5, mg/l	10	5	(17)
SS, mg/l	12	5	(23)
NH_3, mg/l (summer)	1.5	0.5	(15)

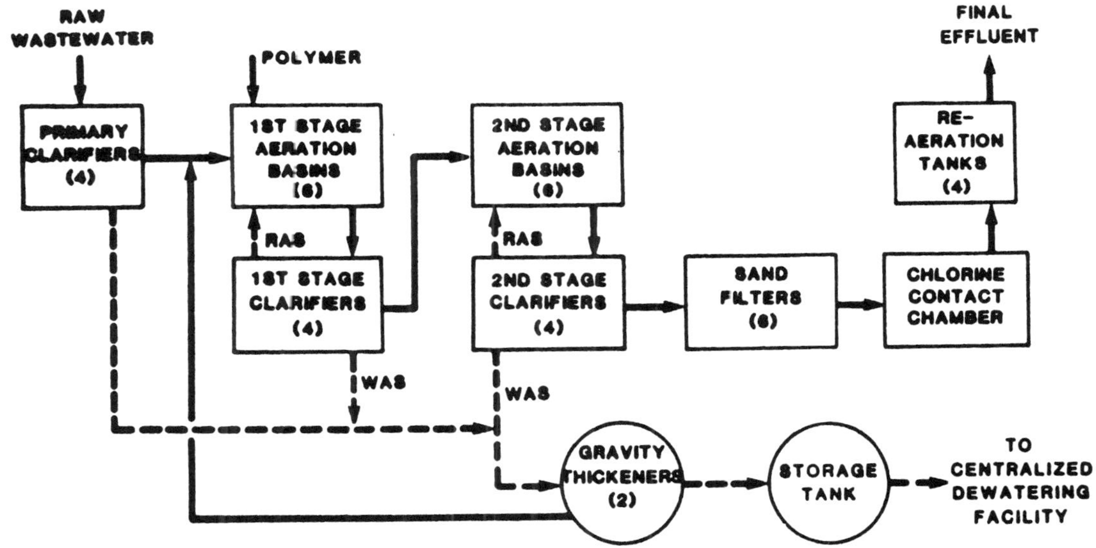

PASSAIC VALLEY WASTEWATER TREATMENT PLANT
Newark, New Jersey

Coping with industrial waste discharges to a 300 mgd POTW in a highly industrialized area is a challenging task. The Passaic Valley Sewerage Commissioners (PCSC) maintain an industrial waste control staff to monitor nearly 400 industries that contribute 20 percent of the wastewater volume and 50 percent of the waste strength. The PVSC performed their first Industrial Waste Survey for database development in 1972, and adopted a set of Rules and Regulations (including local limits) in 1976. By 1982, a comprehensive system consistent with the Federal Clean Water Act of 1977 had been adopted, which established uniform user fees for mass and volumetric loadings in the Passaic Valley plant.

The influent wastewater to the POTW is considered a high-strength waste, with typical BOD and TSS values of 290 and 450 mg/l, respectively. Despite the strength of the influent, the plant is close to meeting the 30/30 NPDES discharge limits, even though the primary clarifiers are not scheduled to go on-line until later this year (1986). The high percentage of industrial flow volume is responsible for the high influent BOD, and hence an interference exists, although the number of industries makes it impossible at this time to determine which are responsible for the interference. The PVSC believes that the addition of primary treatment coupled with the economic incentives for pretreatment created by the user charge system will reduce the effluent to consistently below the limits.

The individual constituents of concern to the PVSC fall into three general categories:

- metals
- flammables
- fibers

The sources of heavy metals are chemical manufacturers, platers and tanneries. One of the smaller (30,000 gpd) chemical companies had been identified as a significant contributor (120 lbs/day) of mercury to the POTW. Although the mercury level of 50 ug/l at the influent was not inhibitory to the activated sludge, the concentration of mercury in the sludge limited the municipality's disposal options. It is anticipated that ocean disposal of sludge will not be permitted much longer, which will require the PVSC to incinerate. The Federal Air Pollution Standards limit the mercury discharge to 3,200 g/day, which translates into a local limit of 0.4 lbs/day in the wastewater from the industry in question. The chemical company responded by isolating the relevant process streams and utilizing a batch recovery system for the mercury, reducing the discharge from 120 down to 5 lbs/day. When ocean disposal is formally eliminated as a disposal option, the company can employ carbon treatment for removal of the remaining mercury.

The oxidation of trivalent chromium to the hexavalent form in a POTW sludge incinerator is a problem caused by the chromium-laden discharge from various industrial users. An additional problem caused by the tanning industrial category is the clogging of local sewers that results from hides being inadvertently discharged from the companies. Similar clogging problems existed at the pretreatment plant due to the balled-up fibers from the pulp and paper manufacturers which close off sludge return lines, orifices and nozzles. This condition improved substantially when the moving-bridge primary clarifiers were placed in service in December, 1985.

The Passaic Valley plant had a unique problem with high concentrations of flammable materials in the influent wastewater. The lower explosive limit (LEL) is defined as the "lowest concentration of a combustible substance in air through which a flame, once ignited, will continue to propogate". When a wastewater approaches 50 percent of the LEL, it is important that it not be discharged into the sewer collection system. The pure oxygen process has a control built into the system which vents all oxygen away from the activated sludge treatment process when high LEL is detected. Since the venting of the oxygen reduces the treatment efficiency it can result in a permit violation as well as creating a health hazard.

The PVSC instituted a three-part program in October of 1984 to mitigate the problems of flammables:

- required industries using or manufacturing solvents which come in contact with discharged wastewater to install LEL detection instruments, and to provide pretreatment to isolate the flammables if high LELs were detected;

- surveyed other industries which used solvents but had no such discharge to determine if a potential existed, requiring necessary control mechanisms; and

- monitored the collection system more closely for illegal dumping of such chemicals.

Representatives of Passaic Valley made it clear that a cooperative attitude on the part of industry was an important factor in successful mitigation of interference problems. In fact, it was the local pharmaceutical manfacturer that conducted the research resulting in the type of LEL instrument recommended by the Advisory Committee when the LEL regulation was adopted.

PASSAIC VALLEY WASTEWATER TREATMENT PLANT
Newark, New Jersey

Design Flow:	**330 mgd**	Location:	**Adjacent to Newark Bay**
Secondary Treatment:	**Activated Sludge (Pure Oxygen)**	Population Served:	**1.5 Million**

INFLUENT WASTEWATER

	Typical (Upset)
Ave. Flow, mgd	250
% Industrial	19
BOD_5, mg/l	290 (500)
SS, mg/l	450 (750)

SIGNIFICANT INDUSTRIES

Industry	Flowrate (mgd)	Problem Pollutants
Pulp and Paper (2)	10.3	Fibers
Pharmaceuticals	6	Xylene, Toluene, Hexane
Tanneries (3)	1.5	Cr
Chemicals (3)	0.5	Cd, Cr, Hg, Pb

PLANT LOADING

Primary Clarifiers	Typical (Upset)
Overflow Rate, gal/sf/day	1,100
Detention Time, hours	2.0
Effluent BOD_5, mg/l	225
Effluent SS, mg/l	125

Aeration Basins	Typical (Upset)
F/M, lbs BOD_5/lbs MLSS/day	0.6
MCRT, days	5
MLSS, mg/l	2,800
Detention Time, hours	1.6
Return Flow, %	35
D.O. Level, mg/l	4-12

Secondary Clarifiers	Typical (Upset)
Overflow Rate, gal/sf/day	480
Detention Time, hours	5
SVI, ml/gm	65

PLANT PERFORMANCE

	Permit Limit	Typical (Upset)
BOD_5, mg/l	30	25 (40)
SS, mg/l	30	25 (60)

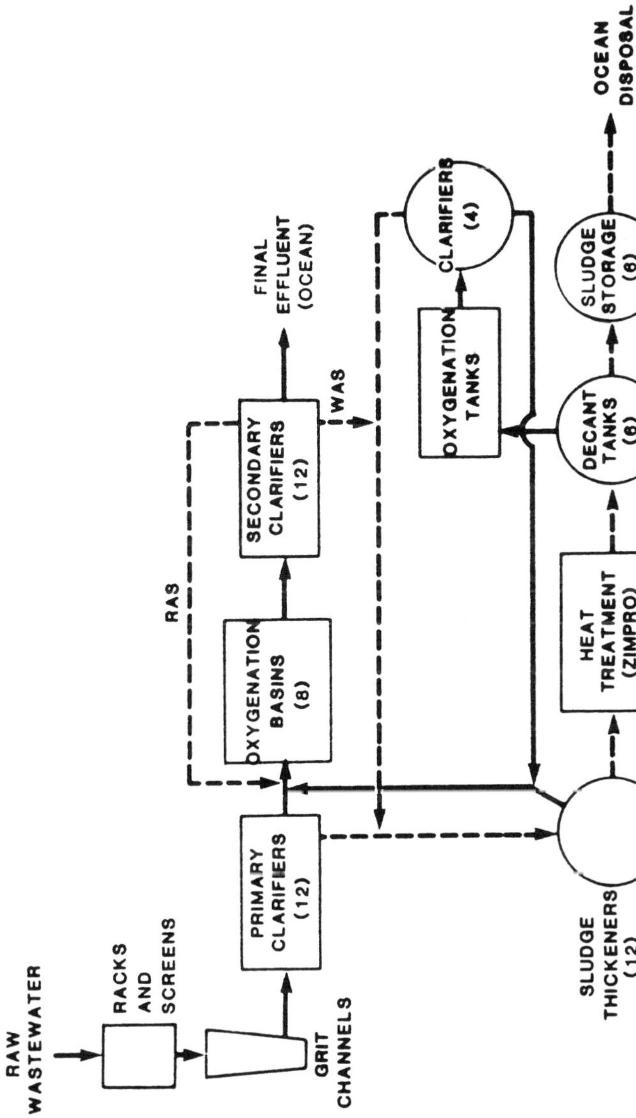
RAW WASTEWATER
RACKS AND SCREENS
GRIT CHANNELS
PRIMARY CLARIFIERS (12)
OXYGENATION BASINS (8)
RAS
SECONDARY CLARIFIERS (12)
WAS
FINAL EFFLUENT (OCEAN)
SLUDGE THICKENERS (12)
OXYGENATION TANKS
CLARIFIERS (4)
HEAT TREATMENT (ZIMPRO)
DECANT TANKS (6)
SLUDGE STORAGE (6)
OCEAN DISPOSAL

SIOUX CITY WASTE TREATMENT PLANT (SCWTP)
Sioux City, Iowa

The Sioux City Waste Treatment Plant (SCWTP) treats a combined industrial and municipal wastewater average flow of 13.5 mgd and discharges to the Missouri River. More than 140 industries were identified by an industrial survey as potential sources of wastewater. Of these, four are categorical metal finishing or electroplating industries and, as of recently, eleven industries contributed significantly to the suspended solids, BOD and oil and grease discharged to the SCWTP. Although the total volumetric load of the industrial wastewater is typically less than 10 percent of the total flow, the industrial organic loads to the plant account for greater than 50 percent of the observed loads.

The SCWTP has experienced two separate incidents in which industrial discharges have interfered with normal plant operations. Isolated slug loads of zinc were experienced by the SCWTP in March and again in April of 1984. Levels as high as 16 mg/l Zn were observed in the treatment plant influent and both slug-load incidents resulted in an upset of the activated sludge process and violations of the NPDES discharge limits. Effluent BOD_5 concentrations exceeded 60 mg/l and effluent suspended solids concentrations in excess of 200 mg/l were observed. The investigation of the first slug load of zinc was somewhat hampered by the lack of in-house capabilities for metals analysis and the first indication of a contamination problem was the process upset itself. Upon confirmation of the nature of the interference, a temporary system for the continuous addition of lime to the primary clarifiers, which would result in the precipitation of subsequent slug loads of zinc, was installed and operated until such time that frequent and periodic monitoring and analysis of the influent for metals could be performed at the SCWTP.

The source of the metal discharge was identified from the City's industrial use survey and from samples of wastewater and solids collected at specific locations in the wastewater collection system. The floor drain at the manufacturing facility through which the zinc discharges occurred was disconnected from the sanitary sewer. In addition to the process upsets, several years accumulation of sludge held in storage lagoons and slated for disposal by land application became contaminated with zinc. Upon receipt of special permitting from the State, the SCWTP was allowed to dispose of the sludge as planned.

In 1985, a pharmaceutical manufacturer came on-line discharging batches of high strength waste without pretreatment. The strength of the waste ranged from 10,000 to 100,000 mg BOD_5/l and the waste contained high levels of salt and sulfite. The average BOD_5 of the waste was 35,000 mg/l and the batch dumps represented 45 percent of the total organic load to the SCWTP. The activated sludge process was severely overloaded and intermittent depressions of the D.O. level occurred. It was possible to operate the activated sludge process to accommodate the severe organic loads, but the process would again be upset during the weekends when the pharmaceutical manufacturer was not discharging waste and the organic loads were reduced. Throughout 1985, the SCWTP experienced severe violations of their NPDES BOD_5 and suspended solids discharge limits. Frequent violations of the pharmaceutical manufacturer's discharge permit occurred with respect to the organic strength and daily mass loading of the waste. The industrial user was placed on a compliance schedule and continued violations of the discharge permit necessitated actions that would result in flow equalization and reductions in the levels of methyl mercaptan, sulfite and sulfide. Presently, all batch waste dumps are transported by bulk to the SCWTP where they are metered, by SCWTP personnel, into the plant influent under controlled conditions.

The upset conditions presented in the following table represent conditions related to the discharge of the pharmaceutical wastewater. The reported upset conditions represent averages for several months of 1985 whereas the typical conditions were based on data for 1984 which spanned nine months and included those months in which the slug loads of zinc were experienced.

SIOUX CITY WASTE TREATMENT PLANT
SIOUX CITY, IOWA

Design Flow:	**30 mgd**	Location:	**Northwest Iowa**
Secondary Treatment:	**Activated Sludge (Conventional)**	Population Served:	**135,000**

INFLUENT WASTEWATER

	Typical (Upset)
Ave. Flow, mgd	13.5
% Industrial	7
BOD_5, mg/l	380 (612)
SS, mg/l	550 (630)

SIGNIFICANT INDUSTRIES

Industry	Flowrate (1000 gpd)	Problem Pollutants
Meat Processing	1,000	BOD_5, oil and grease, SS
Pharmaceutical	70	BOD_5, methyl mecaptan, sulfite
Metal Finishing	20	Zn, Cr, Ni

PLANT LOADING

Primary Clarifiers	Typical (Upset)
Overflow Rate, gal/sf/day	577
Detention Time, hours	2.9
Effluent BOD_5, mg/l	220 (370)
Effluent SS, mg/l	240 (235)

Aeration Basins	Typical (Upset)
F/M, lbs BOD_5/lbs MLSS/day	0.2 (0.3)
MCRT, days	10
MLSS, mg/l	2500
Detention Time, hours	15
Return Flow, %	40
D.O. Level, mg/l	2.5

Secondary Clarifiers	Typical (Upset)
Overflow Rate, gal/sf/day	722
Detention Time, hours	3
SVI, ml/gm	150

PLANT PERFORMANCE

	Permit Limit	Typical (Upset)
BOD_5, mg/l	30	34 (37)
SS, mg/l	30	33 (45)

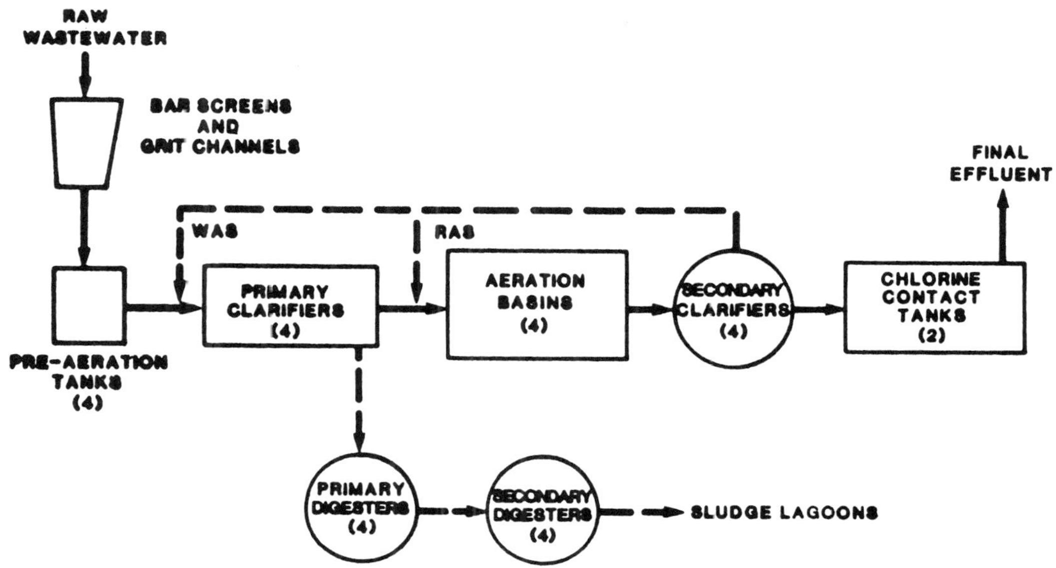

TOLLESON WASTEWATER TREATMENT PLANT
Tolleson, Arizona

The Tolleson Wastewater Treatment Plant (TWTP) is a two stage trickling filter plant that treats a predominantly domestic wastewater from Phoenix, Arizona suburbs. The successful operation of the TWTP is dependent on the one significant industrial contributor to the treatment plant, a meatpacker who processes 1,000 to 1,400 head of beef per day.

The influent to the TWTP could be typified as medium to high-strength municipal wastewater with average BOD_5 and SS levels being 275 mg/l and 225 mg/l, respectively. Approximately 25 to 30 percent of the average organic and solids loading is contributed by the meatpacker average at levels of 1,100-1,600 mg/l BOD_5 and 700-1,200 mg/l SS, for wastewater flows of 0.8-1.0 mgd. In general, the domestic/industrial waste stream BOD_5, and SS can both be treated to below 10 mg/l, well within the 30/30 discharge limits. However, in the past the meatpacker has upset the treatment process by slug discharging blood or other high strength organic slaughter by-products with BOD_5 and SS levels of up to 2,200 mg/l and 1,375 mg/l, respectively. Prior to 1982, these upset conditions would last for several days and result in weekly and monthly effluent suspended solids of 30-40 mg/l, in violation of permit limits.

Treatment upsets have diminished in frequency and intensity since 1982 for two reasons:

- A legal contract with the meatpacker limits flow to 0.8 mgd, BOD_5 to 10,675 lbs per day (1,600 mg/l) and SS to 6,670 lbs per day (1,000 mg/l), and provides for fines or disconnection if these limits are exceeded, and

- Improved treatment plant process monitoring has enabled operators to better detect, and thus act on, a potentially upsetting condition.

The contract with the meat packer attempts to prevent waste blood from being stored for more than about eight hours at a time before discharging to the sewer. Prior practice resulted in blood being held back for up to a week at a time before being discharged all at once.

Primarily through trial and error, the operators of the TWTP have established several operating parameters that help them in detecting upset conditions in the plant. The depth of sludge in the primary clarifiers is monitored closely; a high or rapidly increasing sludge depth is indicative of upset conditions and is caused by the high solids content of the meatpacking waste. The mixed liquor in the solids contact basin following the second trickling filter is monitored closely as well, with levels above 500 mg/l signaling possible problems. Mixed liquor concentrations of 1,500 mg/l generally result in effluent suspended solids of greater than 30 mg/l. To remedy an upset condition, primary sludge pumping rates are manually increased above their normal levels to reduce the solids inventory and prevent escape in the effluent.

TOLLESON WASTEWATER TREATMENT PLANT
TOLLESON, ARIZONA

Design Flow:	**8.3 mgd**	Location:	**South Central Arizona**
Secondary Treatment:	**2 Stage Trickling Filter with Solids Contact**	Population Served:	**65,000**

INFLUENT WASTEWATER

	Typical (Upset)
Ave. Flow, mgd	7.4
% Industrial	14
BOD_5, mg/l	275 (340)
SS, mg/l	225 (280)

SIGNIFICANT INDUSTRIES

Industry	Flowrate (1000 gpd)	Problem Pollutants
Meat Packer	1000	BOD, SS

PLANT LOADING

Primary Clarifiers	Typical (Upset)
Overflow Rate, gal/sf/day	860
Detention Time, hours	1.9
Effluent BOD_5, mg/l	160
Effluent SS, mg/l	95

Intermediate Clarifiers	Typical (Upset)
Overflow Rate, gal/sf/day	735
Detention Time, hours	2.4
Effluent BOD_5, mg/l	30
Effluent SS, mg/l	30

Secondary Clarifiers	Typical (Upset)
Overflow Rate, gal/sf/day	480
Detention Time, hours	7.4

First Stage Trickling Filter	Typical
Hydraulic Loading, gal/sf/day	1,000
Organic Loading, lbs BOD_5/1000 cf/day	45
Recirculation, %	100

Second Stage Trickling Filter	Typical
Hydraulic Loading, gal/sf/day	500
Recirculation, %	100

PLANT PERFORMANCE

	Permit Limit	Typical (Upset)
BOD_5, mg/l	30	9 (25)
SS, mg/l	30	9 (35)

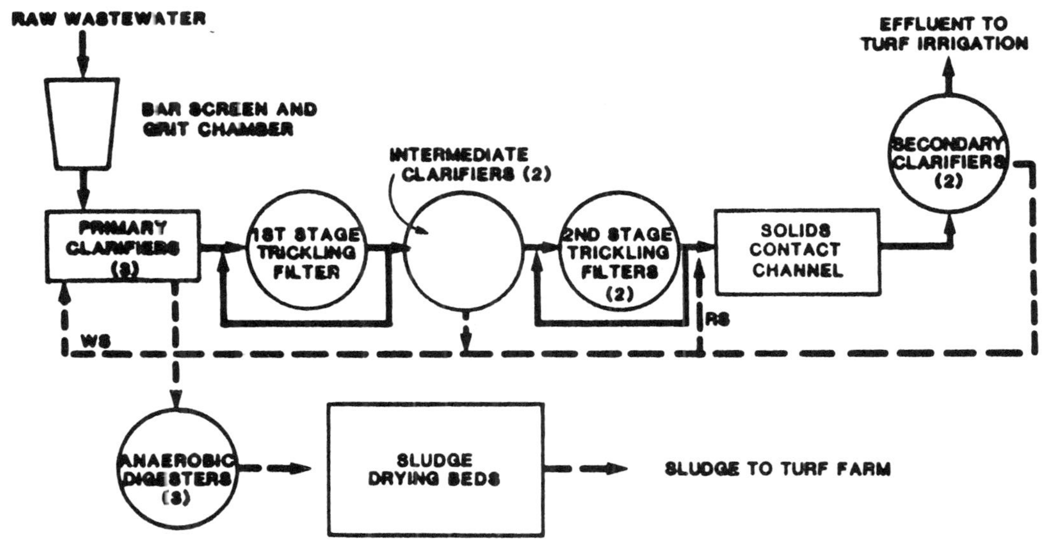

Appendix B: Interfering Substances

CONVENTIONAL

Biochemical Oxygen Demand
Fats, Oil and Grease
pH
Suspended Solids

METALS AND INORGANICS

Alkalinity
Ammonia
Arsenic
Barium
Beryllium
Boron
Cadmium
Calcium
Chloride
Chromium
Cobalt
Copper
Cyanide
Iodine
Iron
Lead
Magnesium
Manganese
Mercury
Molybdenum
Nickel
Nitrogen
Phosphorus
Selenium
Silver
Sodium
Sulfate
Sulfide
Sulfite
Tin
Thallium
Vanadium
Zinc

AGRICULTURAL CHEMICALS

Aldrin/Dieldrin
Chlordane
Chlorophenoxy Herbicides
DDT
Endrin
Heptachlor
Lindane
Malathion
Organometallic Pesticides
Toxaphene

AROMATICS

Benzene
Chlorobenzene
Dichlorobenzene
Dinitrotoluene
Nitrobenzene
PCBs
Toluene
Zylene

HALOGENATED ALIPHATICS

Carbon Tetrachloride
Chloroform
Chloromethane
Dichloroethane
Dichloroethylene
Dichloropropane
Hexachlorobutadiene
Hexachlorocyclohexane
Hexachloroethane
Methylene Chloride
Tetrachlorodibenzodioxins
Tetrachlorodibenzofurans
Tetrachloroethane
Tetrachloroethylene
Trichloroethane
Trichloroethylene
Vinyl Chloride

NITROGEN COMPOUNDS

Acetanilide
Acetonitrile
Acrylonitrile
Aniline
Benzidine
Benzonitrile
Chloroaniline
Dichlorobenzidine
Dimethylnitrosamine
Diphenylhydrazine
Dyes
EDTA
Ethylpyridine
Fluorenamine
Hydrazine
Nitrosodiphenylamine
Pyridine
Trisodium Nitrilotriacetate
Urea

OXYGENATED COMPOUNDS (Acids, Alcohols, Aldehydes, Esters, Ethers, Ketones)

Acetone
Acrolein
Adipic Acid Esters
Allyl Alcohol
Benzoic Acid
Boric Acid
Butanol
Butyl Benzoate
Chlorobenzoate
Chloroethyl Ether
Cinnamic Acid
Crotonol
Cyclohexanecarboxylic Acid
Diethylene Glycol
Ethoxy Ethanol
Ethyl Acetate
Ethylene Glycol
Formaldehyde
Formic Acid
Heptanol
Hexanol
Isophorone
Linoleic Acid
Malonic Acid
Methanol
Methylethyl Ketone
Methylisobutyl Ketone
Octanol
Polyethylene Glycols
Polyvinyl Alcohols
Protocatechuic Acid
Syringic Acid

PHENOLS

Catechol
Chlorophenol
Cresol
Dichlorophenol
Dinitrophenol
Nitrophenol
Pentachlorophenol
Phenol
Trichlorophenol
Trinitrophenol
Vanillin

PHTHALATES

Dimethylphthalate
Disoctylphthalate
Ethylhexylphthalate

POLYNUCLEAR AROMATIC HYDROCARBONS

Anthracene
Benzo (a) Anthracene
Chloronaphthalenes
di-Isopropylnaphthalene
Naphthalene
Phenanthrene
Pyrene

Part II

Literature Review, Telephone Survey, and Case Studies

The information in Part II is from *Interferences at Publicly Owned Treatment Works,* prepared by Edward D. Wetzel and Scott B. Murphy of James M. Montgomery, Consulting Engineers, Inc., for the U.S. Environmental Protection Agency, September 1986.

Acknowledgments

This document was prepared by James M. Montgomery, Consulting Engineers, Inc. Mr. John Grantham was the Project Manager, with Dr. Edward Wetzel and Mr. Scott Murphy acting as Project Engineer and Assistant Project Engineer, respectively. Messrs. Wetzel and Murphy are the principal authors of this report, with contributions from Dr. Roger Stephenson to portions of Section 2 (Literature Review). In addition to the staff listed above, site visits were made by Messrs. David Harrison and Paul Skager and by Ms. Sheila McShane. The contributions of a number of other members of the Pasadena engineering staff to the literature review, and of the support staff to the production of this report, are gratefully acknowledged.

This report was prepared under the technical direction of Dr. Sidney Hannah of the Water Engineering Research Laboratory (WERL), with Mr. Gregory McBrien and Ms. LeAnne Hammer of the Permits Division, Office of Water Enforcement and Permits. Additional guidance and assistance was provided by the other members of the Pretreatment Support Group of WERL, consisting of Messrs. James Kreissl, Dolloff Bishop, Richard Dobbs, Kenneth Dostal and Henry Tabak.

1. Introduction

The purpose of this study is to investigate the impact of industrial discharges on the operation of publicly owned treatment works (POTWs). Specifically, interferences resulting from the presence of toxic organics, heavy metals, or variable-strength conventional pollutants emanating from industrial sources are the focus of this report. Industrial wastewaters are often blamed for process upsets that occur at municipal wastewater treatment facilities. However, in many cases, such upsets may simply be the result of improper operation and maintenance practices. Pajak, et al. (1977) have indicated a lack of definitive information on treatment plant upsets, particularly with respect to the actual causes of violating discharge permit standards. The intent of this document is not to assess blame, but rather to identify cause and effect relationships and to seek techniques for the mitigation of upsets created by the introduction of abnormal contaminants into POTWs.

BACKGROUND

The enactment of the Federal Water Pollution Control Act Amendments (FWPCA, PL 92-500) on October 18, 1972, established the National Pollutant Discharge Elimination System (NPDES) permit program requiring any entity discharging treated wastewater into a receiving water course to obtain a NPDES Permit. Industrial permits issued under PL 92-500 required industries to employ the best practicable technology (BPT) available to limit discharges of conventional pollutants such as BOD, TSS, and fats, oil, and grease (FOG) (Silva, 1981). The FWPCA requirements were amended by the Clean Water Act (CWA, PL 95-217), which was signed into law on December 28, 1977. This legislation brought about significant changes for industrial dischargers, calling on the EPA to focus its attention on controlling toxic pollutants. The EPA Administrator was directed to establish a list of toxic pollutants with effluent limitations for each. Industries with permits would then be required to utilize the best available technology (BAT) economically feasible for the control of toxics and the best conventional technology (BCT) currently available to control the discharge of conventional pollutants (Snyder, 1978).

Industries that discharge to POTWs need not obtain a NPDES Permit; rather they are subject to the General Pretreatment Regulations promulgated on June 26, 1978, and subsequently amended on January 28, 1981 (40 CFR Part 403). The regulations establish both categorical pretreatment standards and prohibited discharge standards for the control of wastewater discharges into POTWs. The categorical pretreatment standards apply to the most significant industrial and commercial sources of toxic pollutants from specific industrial categories. The prohibited discharge standards, which apply to all industrial and commercial sources, prohibit the discharge of pollutants that:

- create a fire or explosion hazard;
- are corrosive;
- obstruct wastewater flow;
- upset treatment processes, or
- increase the influent wastewater temperature above 40°C.

The General Pretreatment Regulations require POTWs with a hydraulic design capacity greater than 5 mgd to establish a formal pretreatment program. POTWs with smaller design flow may also be required to establish such programs if their industries are subject to national categorical pretreatment standards or if there is a history of nondomestic discharge problems or NPDES permit violations. A POTW without a pretreatment program may still develop and enforce local limits for the prevention of specific interference problems as needed to ensure compliance with an NPDES permit or sludge disposal practices.

Problems associated with the implementation and perceived complexity of the pretreatment program slowed the progress of this effort. In response, the Pretreatment Implementation Review Task Force (PIRT) was established on February 3, 1984 by EPA Administrator Ruckelshaus. The task force was composed of 17 representatives from POTWs, States, industry, environmental groups, and EPA Regions. The charge given to the PIRT was to identify problems experienced by industry, States, and POTWs with respect to the pretreatment program, and to make recommendations to the EPA for program improvement.

In their Final Report to the Administrator (U.S. EPA, 1985), the PIRT considered five problem areas affecting implementation. The first problem area was the perception that the pretreatment program is complex and difficult to understand. For this problem, the task force recommended that the EPA sponsor a program of regional workshops, guidance manuals and seminars. One of the specific problems identified by PIRT was the difficulty experienced by POTWs in the recognition, tracking, and mitigation of interferences caused by industrial discharges. Their recommendation was for the EPA to provide guidance to municipalities regarding such interference problems. This report is part of the EPA's response to that recommendation.

SCOPE OF WORK

Impacts of industrial waste discharges on POTWs were broken down by JRB Associates (1981b) into six categories:

- Pass-through and water quality
- Interference and upsets
- Sludge contamination
- Worker health and safety
- Air pollution
- Groundwater pollution

It is the second category, interference and upsets, that is addressed in this study. The U.S. Congress recently redefined (Federal Register, June 19, 1985) interference to generally mean "...the causation of a POTW's noncompliance with its permit or inability to lawfully use or dispose of its sludge". The operative word in this new definition is causation, which replaces the phraseology "contributes to a violation" from the original regulations. Since the interference definition includes sludge disposal considerations, category three above (sludge contamination) is also within the scope of this study. The work associated with the production of this report extended over a 9-month period from mid-July 1985 until the end of April 1986. The overall work plan was divided into three main task assignments:

- Literature review
- Telephone survey
- Case histories

An extensive search of the recent literature pertaining to the treatabilty and impact of industrial wastes in the POTWs was the first step taken to understand the problem. Three hundred articles and reports were selected for detailed review from an initial list of nearly 8,000 titles. The complete review is presented in Section 3 of this report.

State and Federal regulatory agencies, sanitation districts, and major cities and municipalities were contacted by telephone in an effort to uncover specific interference problems in their respective jurisdictions. Surveying each of the 50 state environmental agencies was also a primary source for contacts at POTWs with industrial wastewater problems. Section 4 documents the results of this work effort.

The literature review and telephone surveys helped generate a list of some 120 municipal treatment facilities as potential case studies. Thirty of these plants were selected for one-day site visits based on the criteria discussed in Section 5. This portion of the study was the most productive in terms of developing effective procedures for the correction of industrial waste contamination problems.

Section 6 attempts to synthesize the information presented in the report into a coherent plan of action for POTWs. This section will become the basis for the preparation of an abridged guidance manual for operations and management personnel at municipal treatment facilities.

2. Conclusions

SUMMARY

The literature and case studies discussed in this report reflect the diversity of interference issues and concerns. The literature reviewed in Section 3 consists of all English-language-published work from 1980 through 1985 available to the authors by established searching techniques, pertaining to interference or treatment of industrial wastewaters. A large portion of that work is devoted to the effects of industrial pollutants on activated sludge processes. In general, the quantity of literature available on interferences decreases from suspended growth biological processes (including anaerobic systems) to fixed film processes to physical-chemical treatment. The design and operations concepts that permeated the reviewed literature included:

- **Acclimation.** Biological systems will acclimate to industrial wastes that are discharged at a consistent level. Highly variable industrial discharges cause the greatest inhibition problems.

- **Staged Treatment.** The use of staged or series treatment typically results in better degradation of industrial wastes than if the same treatment units are used in parallel.

- **Operations Modifications.** The POTW operator has operational tools available (particularly at an activated sludge plant) that can mitigate the effects of industrial contaminants.

- **Influent Wastewater Characteristics.** The concentration and form of metal and organic contaminants has an important effect on the treatability of the waste and is dependent on a number of inter-related factors.

The telephone survey discussed in Section 4 served to reinforce the idea of the diversity of industrial interference being experienced by POTWs. The survey was conducted nationwide and consisted of telephone interviews with regional, State, and local officials involved with wastewater treatment and pretreatment programs in all 50 states. The findings of the telephone survey support several general conclusions about POTW interference:

- The major cause of interference is highly variable discharge of conventional pollutants.

- Treatment plants most adversely affected are less than 5 mgd in size with one or two major industrial contributors.
- Interference mitigation is most successful when POTWs have involved industry in source control (pretreatment) program development in a cooperative atmosphere.
- Some POTW interference problems result from poor wastewater characterization before facility design.

The case studies reported in Section 5 represent a wide cross-section of interference mitigation efforts. The POTWs studied reflect diversity in terms of size, treatment process, interfering pollutants and industries, and methods utilized to control interference. The reports served to reinforce the findings of the previous two chapters in addition to presenting interesting methods for implementing source control measures.

Section 6, Results, synthesizes the information presented in the previous three chapters and discusses the necessary aspects of a successful pretreatment program. Monitoring of POTW wastewater streams (influent, side streams, etc.), operations monitoring and industrial user monitoring are discussed. Interference is also discussed with special emphasis on identification and type of interference. A section is included that lists interfering compounds and typical reported inhibiting concentrations. Section 6 also discusses the many aspects of interference mitigation including pretreatment and source controls and treatment plant controls.

RESEARCH NEEDS

The literature review, surveys and site visits performed as part of this project have identified a number of areas where additional bench-scale, pilot-scale and full-scale research would be helpful in identifying and mitigating the impacts of industrial wastewaters on POTWs. For convenience, these research needs have been subdivided into four general areas:

- treatment processes
- plant operations
- toxicity testing
- management aspects

Treatment Processes

The volume of data available on the impact of nonconventional pollutants (particularly the heavy metals) on activated sludge and anaerobic digestion is sizeable. Unfortunately, the results demonstrate the vulnerability of both processes to the discharges from a variety of industrial wastewaters. Design engineers are becoming less inclined toward conventional suspended growth biological processes. This has been largely responsible for the use of powdered activated carbon augmentation of activated sludge for adsorption of priority organics. Additional process modifications such as the use of pure oxygen or anaerobic selector zones should be more thoroughly evaluated relative to their resistance to industrial discharges.

Fixed film treatment had nearly become a forgotten technology for wastewater processing prior to the development and acceptance of plastic media. Since then, RBC and biotower systems have become popular alternatives because of their low maintenance and ease of operation. However, problems of clogging, excess biomass development, odors and reduced process efficiency potentially attributable to industrial wastes have occurred. More research on the impacts of specific substances on RBCs, plastic media trickling filters and fluidized bed systems (aerobic and anaerobic) is needed. Staged systems such as roughing filters followed by activated sludge or the trickling filter-solids contact process should be similarly evaluated.

The final area of research regarding unit operations deals with solid-liquid separation. The impacts of metals and organics on the performance of clarifiers, thickeners (flotation and gravity) and dewatering devices (vacuum filters, belt presses, plate and frame presses and centrifuges) are essentially unknown at this time. Similarly, little consideration has been given to the response of advanced wastewater treatment processes to industrial compounds, except as pertains to the treatment of contaminated sources for drinking water purposes.

Plant Operations

The major research need in the area of plant operations is the identification of specific cause and effect relationships between industrial pollutants and individual unit processes. For example, high levels of surfactants in an influent wastewater stimulates the production of white foam on top of activated sludge reactors; yet the growth of a dark brown foam may indicate a toxicity problem. It would be quite useful to relate the color and degree of foaming to specific classes of toxic compounds. Such knowledge would aid in both the mitigation and source identification efforts following an incident. Unfortunately, responses of complex biological populations to any stimulus tend to be site specific.

A second need in operations is the development of more useful control measures (analogous to MLSS, MCRT and F/M) for fixed film treatment technologies. Although measurements of DO and slime thickness are acceptable performance monitors, little positive control exists to respond to influent wastewater changes. Recycle ratios and feeds are the only adjustments in trickling filter operations, while RBC plants provide no control other than reversal of rotation direction or the use of supplemental aeration for biomass control.

Inorganic chemical salts and polymers are now used extensively in POTWs for improved settleability, sludge conditioning and odor control. The removability of industrial pollutants, especially metals, by chemicals has been widely studied. However, the effect of a sudden discharge of a priority organic on existing chemical feeds is not well known and needs attention.

Toxicity Testing

The use of biological inhibition testing procedures for the identification and tracing of industrial sources to POTWs should continue to grow in the future. Before this technology is widely accepted and utilized by operations personnel, low-cost portable units and in-line instrumentation will need to be commercially available. The automatic detection of inhibitory substances before reaching the POTW would be a significant breakthrough in interference mitigation.

Additional research efforts should continue to be directed at correlating the results of acute toxicity procedures with the more widely accepted fish bioassay methods. Such correlations allow for routine toxicity testing without the logistical problems and high costs associated with traditional biomonitoring.

Management Aspects

The final report of the Pretreatment Implementation Review Task Force (U.S. EPA, 1985) indicated five general conclusions regarding the existing pretreatment program:

1. Pretreatment programs are perceived as too complex.
2. Enforcement and consistent implementation of these programs is critical to their success.
3. Adequate financial resources are needed for proper implementation.
4. Program success depends on a cooperative partnership between the Federal and State government and the POTWs.
5. Specific changes to the regulatory requirements were needed (and have since been implemented).

Treatment plant and sewer authority management must properly implement their pretreatment programs in order to satisfy the requirements of the regulations and to eliminate interferences at their facilities. There is a need to identify the perceived or actual stumbling blocks facing the POTW managers and to develop programs to aid them in implementation and enforcement. A common problem identified during the telephone survey and site visits was the political sensitivity of pressuring industrial waste dischargers who are significant employers and taxpayers in a community. Programs designed to educate both the industry and municipal operator about the regulations and rationale for pretreatment could prove beneficial.

3. Literature Review

To accomplish the efficient and thorough gathering of information on pollutants which interfere with POTW operation, two literature search approaches were utilized. The initial approach involved accessing several computerized technical literature databases. The second approach involved studying pertinent literature:

1. reviewed annually by the Water Pollution Control Federation (WPCF);
2. presented at applicable technical conferences;
3. categorized at year end in the Journal of the Environmental Engineering Division (JEED) of ASCE; and
4. compiled in the Revised Pretreatment Guidelines (JRB Associates, 1981a) and in a study by the Franklin Research Center (Geating, 1981).

Prior to starting the literature review portion of the project, it was necessary to establish a time period for the literature contained in the computer databases. Such a constraint served to keep the number of citations down to a manageable level, while at the same time did not duplicate the efforts of previous researchers. The two documents listed in Item No. 4 above contain sizeable bibliographies pertaining to the treatability and impact of pollutants in municipal treatment facilities. Both reports utilized computerized searching techniques, and in the case of the Franklin Research Center report, considered literature dating back to 1914. A check on the breadth and depth of the references indicates that the bibliographies contained in the abovementioned reports provide complete coverage of the literature prior to 1980. The current effort therefore limits the computerized database search to the time period from 1980 to 1985.

A list of eighteen appropriate databases available through the Dialog network was developed and prioritized. The first five databases were accessed and the suitable reference abstracts were obtained and screened. The number of citations generated by each database was as follows:

National Technical Information Service	932
Compendex	666
Engineering Meetings	1,054
Aqualine	100
Dissertation Abstracts	60
Total	2,812

With a large number of duplicate references being noted by several of the accessed databases, and with no pertinent unique databases left unaccessed, further computerized database searching was not pursued.

The second part of the literature review was designed to locate additional references of interest to the study and to validate the computer-based search. Cross checks were made to see if citations of interest obtained from the WPCF, technical conferences and JEED searches were also included in the computerized databases. Approximately 5,000 references were screened in the second approach to the literature search.

As a final check of the literature search process, the resulting references were scanned for the presence of several journal and conference articles known to be important to this study. Based upon the positive checks, it was decided that the literature search was successful and complete. The combination of all literature search efforts resulted in reviewing nearly 8,000 reference titles. As a result of "key word" limitations in the searching procedures, a large number of the 8,000 references reviewed were either not applicable to this project or were not deemed sufficiently important to warrant futher investigation. Through careful scrutiny, some 400 articles/reports were selected for potential comprehensive review. Approximately 25 percent of these documents proved to be inappropriate to this study upon receipt of the full article, resulting in a final bibliography numbering just over 300 references.

The selected articles and reports were distributed for review among ten members of the Wastewater Department in JMM's Pasadena, California headquarters. The literature review form shown in Appendix B was developed and utilized by the reviewers for each reference read. This approach generates more consistent information about each article, and provides a more efficient transfer of data to the authors than if critiques were independently written by the reviewers. Each document reviewed was coded, so that the literature review authors could easily refer back to the reference if they sought more information than the form provided.

The literature review is divided into several categories and subcategories according to treatment process. The major process categories with the number of references reviewed in each are given below:

General Wastewater	32
Primary Treatment	5
Secondary Biological Treatment	
General	7
Suspended Growth Processes	134
Fixed Film Processes	48
Advanced Treatment	
Biological Nutrient Removal	11
Physical/Chemical Process	35
Sludge Processing	
Treatment	15
Disposal	13

GENERAL WASTEWATER

It is generally assumed that the sources of the contaminants causing process upsets and interferences in municipal treatment are industrial wastewaters. In an analysis of numerous surveys conducted over the last ten years at POTWs across the country, JRB Associates (1981b) concluded that process upsets and O&M problems were typically associated with industrial wastewaters. Specifically:

- Approximately 80 percent of POTWs with O&M problems receive industrial wastes.
- Over 70 percent of the POTWs receiving industrial wastes implicated the industrial flows as a source of the O&M problems.
- 65 percent of the operators at plants with permit violations attributed those violations to industrial discharges.
- 84 percent of the POTWs sustaining upsets claimed the industrial wastes contributed to the upsets.

Other conclusions were that smaller facilities (less than 5 mgd) were more susceptible to interferences from industrial contaminants than larger plants, and that a high incidence of upsets were attributable to wastewater discharge from the food, electroplating, mechanical products and textile industries. Similar conclusions could be drawn from a series of unpublished inspection reports investigating the cause(s) of POTW permit non-compliance (JRB Associates, 1982-1984).

Some authors argue that industrial discharges are used as scapegoats by municipal treatment plant operators for their O&M problems. Gray, et al. (1979) ranked industrial loadings fifth on their list of ten factors contributing to O&M problems in a survey of 120 Eastern, U.S. plants. In a similar survey of 50 western, U.S. facilities, Hegg, et al. (1980) concluded that plant operations problems resulting from poor process control were unjustifiably blamed on toxic inputs. The discrepancy between these conclusions and those of JRB Associates (1981b) may stem from the fact that JRB chose POTWs with significant industrial contributions, while Hegg, et al. and Gray, et al. focused their efforts on organically and hydraulically underloaded facilities with generally insignificant industrial waste loads. In both of the latter studies, performance limiting factors at activated sludge plants mainly resulted from inadequate process control, while trickling filter problems commonly resulted from design deficiencies.

Some studies have focused on the characterization of the wastewater at specific locations. Ongerth and DeWalle (1980) analyzed the influent at Metro Seattle and found that although industries were contributing only 10 percent of the wastewater flow volume, it represented 20 percent of the organic load. They also discovered that the concentration of toxic pollutants in the influent increased in proportion to the industrial flows. Similar results were discovered at the Moccasin Bend Plant in Chattanooga, Tennessee. Seventy-two of the priority pollutants (including cyanide, zinc, chromium, phenol and toluene) are present in the influent, representing 21 of the 37 EPA designated industrial categories. Industrial wastewater constituted 50 percent of the dry weather flow at the plant, but 65 percent of the influent BOD. Iannone, et al. (1984) studied the wastewater influent to 12 New York City plants where the industrial contribution ranged from less than 1 percent to 8 percent by volume. In contrast to the Seattle example, the authors found that only 13 of the 114 organic priority pollutants were present in greater than 10 percent of the samples, and that the major source of these organics was from domestic or stormwater sources. Similar wastewater characterization was performed by Levins, et al. (1981) at Cincinnati, St. Louis, Atlanta and Hartford. They found that although industrial sources dominated the loading on the POTW, few toxic pollutants were found in the wastewaters, and those present generally existed at low concentrations.

Two studies were funded by the U.S. EPA in an effort to quantify the contribution of toxicants from non-industrial sources. Hathaway (1980) identified common household uses for the priority pollutants, including a breakdown of sources for each within the home (i.e., toilet flushing, laundry, kitchen sink, etc.). The appendix contains lists of abbreviations, synonyms and trade names for each of the 129 compounds on the priority pollutant list. E.C. Jordan Company (1984) analyzed a series of storm events and discovered that high concentrations of heavy metals (> 1 mg/l) and toxic organics (0.5 mg/l) were sometimes present in combined sewer flows. Comparisons of these data with dry weather flow analyses indicates that the stormwater contribution is significant, but that the concentrations are still less than those present in typical sanitary wastewater.

The U.S. EPA has previously sponsored research on the treatability and effects of pollutants at POTWs using a combination of literature and actual plant data. The three-volume Federal Guidelines for State and Local Pretreatment Programs (U.S. EPA, 1977) included a summary of treatability and impact for a variety of metals, conventional pollutants and organics. Wastewater processes considered include primary sedimentation, activated sludge treatment (including nitrification), trickling filters and anaerobic digestion. A series of useful figures is presented indicating the concentration levels at which specific contaminants produce no effect, are inhibitory or cause upsets in activated sludge, nitrification and digestion. These guidelines were later updated and consolidated into a two-volume, unpublished EPA report (JRB Associates, 1981a).

Perhaps the most comprehensive work done to date on toxic pollutants was the publication of a five-volume Treatability Manual (U.S. EPA, 1981). This project utilized the literature and treatment plant records to produce separate volumes on treatability, industrial processes and descriptions, and treatment technologies. Excellent coverage of the literature pertaining to biodegradibility and toxicity was provided by Geating (1981, Volume I). Volume II of that same document develops a "permutated index" of chemicals in water from over 600 references published from 1974-79. While the index was not particularly useful to this project, the extensive bibliography provided 150 additional relevant references to be considered for this review.

Process Comparisons

Six different treatment processes were subjected to 21 organics plus background concentrations of five metals in order to compare removal efficiencies of the five alternatives with the conventional activated sludge process acting as a control (Hannah, et al., 1985). Suspended growth systems (lagoons) proved to be the best alternatives, followed by high rate trickling filters and then physical-chemical processes. Smith, et al. (1981) subjected two complete-mix activated sludge units (one with powdered activated carbon addition) and a rotating biological contactor (RBC) to shock loadings of pesticides. The RBC unit proved to be more stable with lower oxygen requirements, and not subject to the sludge bulking which occurred in activated sludge treatment. However, the testing was performed using non-acclimated bacterial cultures.

Some research compares suspended growth and fixed film processes for their reliability and stability in treating specific wastewaters. Stracke and Baumann (1977) set up a side-by-side pilot scale comparison of trickling filter and activated sludge treatment at a chemical manufacturing industry which was causing a POTW to violate their NPDES Permit for BOD, TSS and NH_3. Although the trickling filters were easier to operate and not as susceptible to upsets from shock loads, the long detention times required for treatment made the activated sludge system a more practical design alternative.

Coal derived wastewaters containing phenols, cyanides, thiocyanates and ammonia have been analyzed by Holladay, et al. (1978) and Medwith and Lefelholz (1982). In the former reference, a continuous stirred-tank bioreactor (CSTBR), a packed-bed bioreactor (PBBR) and a fluidized-bed bioreactor (FBBR) used acclimated cultures to degrade the waste. The results indicate:

- good biodegradation of phenols and thiocyanate in all reactors
- phenol degradation rates decreased from FBBR > PBBR > CSTBR
- the CSTBR is most vulnerable to shock loadings
- the PBBR tended to develop excess biomass.

The later study tested a bench-scale, hybrid, suspended growth-fixed film reactor using coal dust as an inert support medium in a suspended culture of organisms acclimated to the wastewater. Simultaneous carbonaceous and nitrogenous oxidation occurred, with improved sedimentation resulting from the coal dust addition. Use of powdered activated carbon (PAC) in lieu of the inert coal did not improve the treatment.

Austin, et al. (1981) found the oxygen activated sludge process to be more stable and flexible, with better settling characteristics than air activated sludge when treating Carson, California wastewater in pilot reactors. Both systems met discharge standards for all parameters except chromium, nickel and zinc. The air system accomplished nitrification (the oxygen system did not), but suffered from sludge resuspension due to denitrification in the final clarifier.

Gaudy, et al. (1982b) compared the behavior of 24 organic compounds dosed at 5 mg/l into settled municipal sewage in batch reactors. Only pentachlorophenol and 2-chlorophenol caused increases in effluent COD. Eight of the compounds were then tested in continuous flow reactors with no increase in COD, but phenol and methylene chloride caused effluent TSS to rise. When four of the organics were treated by extended aeration, improved effluent quality existed in each case, with no evidence of pass through.

Conventional activated sludge and extended aeration were compared by Kincannon, et al. (1981) in bench scale reactors treating varying dosages of phenolics. At 5 mg/l there was no increase in effluent COD, but slightly higher TSS values in conventional treatment. However, as the phenol concentrations increased, the effluent COD and TSS also increased in the effluent. The extended aeration process was affected at all levels of dosing, but generally produced a higher quality effluent than the conventional system.

Niku, et al. (1981) studied over 40 activated sludge wastewater treatment plants for general reliability, stability and variability. The following generalities for activated sludge process variations were found: (1) step feed and step aeration processes provide the best year-round BOD removal efficiency, (2) conventional activated sludge provides the best total suspended solids removal, (3) completely mixed activated sludge performance is lower than the previous three variations, and (4) contact stabilization provides the lowest performance of all.

Some research efforts have compared RBCs with other biological processes. Lytle (1984) determined that operating costs were lower for RBCs than either trickling filters or activated sludge systems when treating photofinishing wastewaters. An acclimated RBC fed liquid detergent effluent (MBAS = 2.8 mg/l) provided comparable treatment to that of an aerated lagoon, as long as the wastewater temperature remained above 15°C (Lense, et al., 1978).

Toxicity and Inhibition

The most relevant literature for this study are those references pertaining to the impact of pollutants on POTW operations. Pajak, et al. (1977) lists the six general ways in which biological treatment process upsets are manifested as:

- direct toxicity to organisms
- inhibition of biological processes
- exertion of BOD after treatment (e.g., in receiving stream)
- high effluent COD caused by refractory material
- disruption of sludge treatment
- high oxygen demands resulting in a poor bioreaction environment

The report further develops a "Hazardous Materials Effects Matrix", including an alphabetical listing of chemicals with inhibition/toxicity and treatability data.

Russell, et al. (1983) reviewed the literature on inhibition of the activated sludge, nitrification and anaerobic digestion processes by metals and organics. They noted some disagreement with previous publications concerning threshold inhibitory effect concentrations. The authors concluded that generally organic toxic pollutants do not occur at concentrations that would affect POTW operations, but that metals are often found above threshold levels in influent wastewaters. The form of the metals is considered as important as total concentration when assessing the impact on treatment processes. In the literature survey section of another article (Yost, et al., 1981), trickling filters where shown to be less sensitive to heavy metal shock loadings than activated sludge systems. In their work at the Kokomo, Indiana plant, the authors noted high removals of metals, particularly for chromium (98 percent), iron (98 percent) and lead (95 percent). The metals were determined to be predominantly from industrial point sources throughout the sewer system, but the highly variable levels did not affect their removal.

An advanced treatment pilot plant was used to treat Indianapolis wastewater containing between 0.15 and 1.0 mg/l of a variety of organics in a study performed by Cain, et al. (1983). Phenols at 1 mg/l caused a 10 percent to 20 percent reduction in nitrification, and 0.15 mg/l of pentachlorophenol destroyed the nitrification process, but no other inhibitory or toxic effects were realized. A number of typical cause and effect relationships, shown on Table 1, were highlighted by JRB Associates (1981b) based on their 77 POTW study.

A Trainee Notebook (Environ. Res. Train. Ctr., 1979) has been developed from a Field Manual (Culp, Wesner, Culp, 1978) to accompany a troubleshooting course for O&M problems at wastewater treatment plants. A portion of the course presents a variety of appropriate responses to process upsets caused by hydraulic overloads, organic overloads and toxic inputs to both activated sludge and trickling filter (addenda). processes. Some valuable troubleshooting tables are presented in both documents, outlining a list of possible negative impacts to treatment processes, particularly activated sludge and anaerobic digestion.

PRIMARY TREATMENT

The recent literature contains very little about primary sedimentation of toxicants, and the articles that have been written generally relate to the removability of contaminants in the clarifier as opposed to the effect of those compounds on the process.

Metals -- In a review of the literature on metals removal by primary treatment, Lester (1983) found few consistent patterns other than:

- Removal was largely a function of the form of the metal (soluble or insoluble).

- The extent of removal typically varied among the common metals present as Pb > Cu > Zn > Cd > Cr > Ni.

Patterson, et al. (1983) developed similar conclusions in their bench scale work to those of Lester, with metals removals ranging from 14 to 41 percent. A significant contradiction, however, is their ranking of metals removability which they show to decrease from Ni > Fe > Pb > Cr > Al > Cd > Cu > Zn. Having

TABLE 1
COMMON TREATMENT PLANT UPSETS
from JRB Associates (1981b)

Cause	Effect
Low pH	Corrosion of pipes and equipment
Solids or viscous pollutants	Flow obstruction
High concentrations of volatiles	Explosion/fire hazards
Organic or hydraulic overloads	Process upsets
Heated discharges	Altered biological activity
Presence of toxics	Inibition of biological activity

most readily removed in primary treatment is inconsistent with nearly all other authors, who typically indicate nickel to be one of the least removable.

Pilot scale testing on the removal of 13 metals from the Mill Creek (Cincinnati) wastewater was conducted by Petrasek and Kugelman (1983). Removals ranged from 15 to 40 percent for most elements except arsenic, calcium and magnesium, which demonstrated negligible partitioning in the clarifier. Relative concentrations of metals found in the waste activated sludge were two to five times greater than in the primary sludge, indicating stronger removals of metals in secondary treatment. Nielsen and Hrudey (1983) studied the fate of metals at the Gold Bar Plant in Alberta (Canada), and found that with the exception of cadmium, most of the removal took place in primary sedimentation. This apparent contradiction with the work of Petrasek and Kugelman results from the particulate form of the metals in the Gold Bar influent wastewater. Low overall removal of zinc (< 55 percent) contradicts results obtained at other facilities.

Organics -- A few authors have considered the fate of priority organics in primary clarifiers. Ongerth and DeWalle (1980) observed that the high molecular weight compounds were more easily removed than the smaller compounds, presumably due to their adsorption onto the settling solid particles. Petrasek, et al. (1983) noted that removals tended to be variable between classes of compounds, but consistent within a particular class. Two references (McIntyre, et al., 1981b; Gutierrez, et al., 1984) have looked at the behavior of PCB's and other organochlorine insecticides in primary sedimentation. Both papers indicated a direct relationship between suspended solids removal and the removal of these organics. McIntyre, et al. found that comparable removals (50-75 percent) of suspended solids and insecticides existed at the Sandford Works in Oxford, England. Gutierrez, et al. discovered that optimizing suspended solids removals in a pilot scale application also resulted in optimum removals of PCB and other organochlorines. The authors also noted that organics removals were inversely proportional to hydraulic loading, but were not affected by varying concentrations of suspended solids, surfactants (MBAS) or fats, oils and greases (FOG). In contrast to the abovementioned results with insecticides, Hill, et al. (1985) found insignificant removals of chlorophenoxy herbicides in primary clarification during pilot scale testing.

Case Studies -- In a 30 day study (E.C. Jordan, 1982) of the Moccasin Bend Wastewater Plant, organics and metals removals averaged less than 20 percent in primary treatment, compared with 30 percent removal of suspended solids.

A one-week study (U.S. EPA, 1977b) of the 80 mgd T.E. Maxson plant in Memphis, Tennessee uncovered numerous problems with primary treatment. High influent BOD and TSS values resulted from the discharges from 162 industries, particularly brewery and food-processing wastes. Sludge buildup in the clarifiers produced septicity, bubbling and excessive solids carryover.

Lowry and Chwirka (1983) reported on the impact of papermill wastewater on a 3 mgd combined treatment facility in Brewer, Maine. The industrial flows were responsible for large flow and TSS fluctuations, plus occasional high temperature flows resulting from mill boilouts and washups. Such variations resulted in stratification and short-circuiting in the primary clarifier. Polymer addition to the clarifier influent and by-passing the domestic fraction of the wastewater around the primaries directly to the aeration basins have proven successful in mitigating the upsets.

SECONDARY BIOLOGICAL TREATMENT

This section of the literature review is divided into two major treatment categories: suspended growth and fixed film processes. Activated sludge is the dominant secondary biological treatment method studied in the literature, with the addition of powdered activated carbon to activated sludge, lagoons and anaerobic reactors also considered in this review. Despite their relatively recent entrance into the marketplace, more research has been performed on the treatability and impact of various pollutants on rotating biological contactors (RBCs) than any other fixed film treatment method. Other processes covered are trickling filters (including biotowers), aerated submerged filters and anaerobic filters.

Some studies have considered the biodegradability of pollutants irrespective of the treatment system. Bedard (1976) focused his attention on the inherent treatability of numerous organic compounds based on the results of bench-scale testing. Compounds were assigned a numerical refractory index (R.I.) indicating their potential for total biodegradation, or a biological interference value (B.I.V.) based on the degree of interference with the biological system. Oxygenated compounds such as the glycols demonstrated a high degree of biodegradability (high R.I.), while the PCB's and halogenated aliphatics (notably vinyl chloride) show low R.I. values. Similarly, aliphatics (chloroform) and phenols (dichlorophenol) produced the largest B.I.V.'s, indicating a tendency toward causing interference. Tabak, et al. (1981) used static culture flask biodegradation screening methods to determine degradability and rate of acclimation for 96 compounds. The phenolics, phthalate esters and naphthalenes demonstrated significant degradation and rapid acclimation, while PCB's and organochlorine insecticides were poorly degraded with variable acclimation rates.

Baxter, et al. (1975) exposed prepared cultures of microorganisms to varying concentrations of numerous PCB compounds alone and in combination with one another. They observed that:

- the presence of PCB reduced the rate of biodegradation.
- the degradation rate is inversely proportional to the number of chlorine atoms per molecule.
- the degradation rate may increase when the PCB compounds are present in a mixture of compounds.

Two studies considered the biodegradation of polyethylene glycols. Haines and Alexander (1975) determined the rate of degradation to be inversely proportional to molecular weight. The prepared cultures used in the research were able to fully degrade ethylene glycol in 2 days, di, tri, and tetra-ethylene glycol in 5 days, and the polyethylene glycols (PEG) up to a molecular weight of 20,000 only after extended time periods. The mechanism for PEG degradation is apparently the depolymerization of the compounds by extracellular enzymes, at which time the simpler molecules can be absorbed through the cell membrane. Similar results were obtained by Watson and Jones (1977) using bacterial strains isolated from sewage. The three strains used successfully degraded PEG up to a molecular weight of 4000. The authors also noted that biodegradation was frequently more rapid in the presence of other nutrients.

DiGeronimo, et al. (1979) studied the utilization of chlorobenzoates by sewage bacteria. They observed that all configurations of chlorobenzoate and 3, 4-dichlorobenzoate were readily degraded, but that 2, 4-dichlorobenzoate and 2, 3, 6-trichlorobenzoate were not metabolized.

Suspended Growth Processes

Activated Sludge --

In one of the earlier studies on the degradability of organics, Pitter (1976) outlined the three factors affecting treatability as:

- physical-chemical (temperature, pH, solubility, DO level)
- biological (microbial culture)
- chemical (pollutants)

The research tested 94 aromatic, 15 hydroaromatic and 14 aliphatic compounds in a bench scale adapted activated sludge.

The mechanisms for removal of organics in the activated sludge process were outlined by Kincannon, et al. (1983) as:

- air stripping (volatilization)
- biodegradation
- adsorption onto sludge flocs
- chemical oxidation

The authors performed bench scale testing on a synthetic wastewater spiked with varying concentrations of 15 toxic pollutants. The acclimated cultures present in the complete-mix system were able to achieve greater than 99 percent removal for all compounds except tetrachloroethane, nitrobenzene and 2, 4-dichlorophenol. In general, biodegradation was the prevalent mechanism for the oxygenated, nitrogen and phenolic compounds, while the halogenated hydrocarbons were removed by stripping. Aromatic compounds were removed by a combination of stripping and biological oxidation, leaving sorption as an insignificant mechanism. The authors also investigated the effects of ozone on treatability, and found it can have a significant but inconsistent impact. For example, 1, 2-dichloropropane became biodegradable in the presence of ozone, while acrylonitrile demonstrated reduced biodegradability. In similar bench scale studies (Lawson and Siegrist, 1981) run at food-to-microorganism (F/M) ratios less than 1.0, biodegradation rates were greater than those observed at higher F/M values. Air stripping rate constants were found to be sensitive to temperature changes, and sludge sorption was a significant mechanism for dioctyl phthalate only.

A few studies have measured removals of contaminants by the activated sludge process. Bishop (1982) determined the total removal (including primaries) to be greater than 90 percent for organics, and between 60 and 80 percent for metals. In E.C. Jordan's (1982) study at Moccassin Bend, the activated sludge removed 74 percent of the metals, 80 percent of volatile organics, 69 percent of the acid extractables and 62 percent of the base neutral compounds. Neiheisel, et al. (1982) tested for acute aquatic toxicity on the influent, primary effluent and final effluent of a plug flow activated sludge pilot plant. Fifty ug/l of 16 priority organics were added to a combined municipal/industrial wastewater feed, but they did not alter the toxicity. The activated sludge reduced the toxicity level from the primary effluent to the final effluent, while insignificant changes in toxicity occurred through the primary clarifier.

Conventional Pollutants -- A study by Manickan and Gaudy (1983) compared the response of an activated sludge system to organic, hydraulic and combined organic-hydraulic shocks of the same total mass loading. After dosing six-fold mass loading increases, it was determined that the three types of shocks produced three different operational disruptions. Hydraulic and organic-hydraulic shock produced equal increases in total effluent COD, but the disruptions were primarily due to decreased suspended solids removal and soluble COD leakage, respectively. Organic shock resulted in the greatest effluent COD increase, but in terms of total mass leakage, hydraulic shock was most disruptive. Predictive kinetic equations did not adequately describe the disruption of solids or substrate concentration. Selna and Schroeder (1978) reported similar changes to kinetic parameters for organic shocks of five hour durations. Only minor increases in effluent suspended solids were reported, but noticeable increases in effluent COD values were observed for increased organic loading (three times normal) at mean cell residence times of five and ten days.

George and Gaudy (1973b) showed that a bench-scale activated sludge unit could withstand substantial pH shock and concluded that full scale units could accommodate a change of 1.0 pH unit without a noticeable decrease in organic removal efficiency. A study of the effect of the dissolved oxygen (DO) level on the activated sludge process showed no relationship between DO concentrations and COD removal efficiency, respiration rate or solids yield (Thabaras and Gaudy, 1969). The response of the activated sludge process to organic shock loads was independent of DO as well. From a baseline temperature of 25°C, the activated sludge process responded more favorably to increases in temperature than decreases (George and Gaudy, 1973a). In once-through reactors, eight hour hydraulic residence times were more favorable than four hour residence times.

In utilizing a synthetic substrate (BOD=200mg/l) which included readily biodegradable synthetic surfactants, inconsistent COD removal was observed until 2-5 mg/l of phosphate was added (Painter and King, 1978). Before addition of the phosphate, the activated sludge exhibited signs of bulking as well. Tokuz and Eckenfelder (1979) showed that their bench-scale activated sludge system was not inhibited by the addition of sodium chloride and sodium sulfate salts up to the levels 35 g/l and 30 g/l, respectively. It appeared that the reactor solids level and the number of filamentous organisms were proportional to the salt concentration, while the number of protozoa was inversely proportional to the salt concentration. However, no sludge bulking was observed in the range of salt concentrations mentioned above. During the acclimation of an activated sludge unit to a substrate with the salinity of sea water, the sludge volume index (SVI) decreased, but at the same time an increase in the number of pinpoint flocs caused serious degradation of the supernatant (Imai, et al., 1979).

Studies of the biodegradation of cationic surfactants in activated sludge were conducted by Fenger, et al. (1973) and Sullivan (1983). Fenger showed that tetradecyldimethylbenzyl-ammonium chloride (TDBA) was reduced by up to 75 percent at influent levels as high as 20 mg/l in acclimated cultures, but unacclimated activated sludge was inhibited by TDBA levels as low as a few mg/l. Sullivan showed that ditallowdimethyl-ammonium chloride (DTDMAC) was biodegradable in conventional and extended aeration activated sludge, with ultimate biodegradation being more rapid in extended aeration. DTDMAC removal also appeared to be the result of sorption/precipitation mechanisms. Removal of a nonionic surfactant (NEODOL 45-7, Shell Chemical Corporation) was 90 percent complete in a full scale contact stabilization activated sludge treatment plant with influent concentrations as high as 10 mg/l (Sykes, et al., 1979). Cook (1979) investigated another nonionic surfactant (DOBANOL 45-7) that showed rapid and complete biodegradation by activated sludge micro organisms at concentrations up to 500 mg/l.

In a study of biodegradability by prepared activated sludge cultures, Gerike, et al. (1978) concluded that quaternary ammonium salts could be degraded in acclimated activated sludge treatment plants in concentrations up to 15 mg/l. In another biodegradation test, Novak and Kraus (1973) reported that the degradation of saturated long chain fatty acids was proportional to their solubility. The degradation of unsaturated long chain fatty acids was independent of chain length or degree of saturation.

The T.E. Maxson plant in Memphis suffers from high influent BOD, TSS and FOG, the presence of pesticides and other toxic organics, plus rapid pH fluctuations (U.S. EPA, 1977b). Consequently, a one-week EPA study discovered numerous problems with the contact stabilization process used for secondary treatment, including:

- low DO in the contact basins
- high F/M
- growth of filamentous bacteria
- poor sludge settling
- foul odors

Their recommendations were to add an additional air compressor to maintain the DO levels between 1.0 and 2.0 mg/l during peak flow periods, and to increase the SRT.

Phenols and Cyanides -- A number of studies reported generally successful treatment of phenolic bearing wastewaters by acclimated activated sludge (Dechev and Matveena, 1977; Neufeld and Valiknae, 1979; Khararjian, et al. 1979; Luthy and Jones, 1980; Kim and Armstrong, 1981; Luthy, 1981; DeWalle, et al., 1982; Gaudy, et al., 1982a; and Rozich and Gaudy, 1985). DeWalle, et al., presented records from 25 treatment plants (locations not specified), nearly all of which employed the activated sludge process. Activated sludge was shown to exhibit generally good removal of phenolics, with methylated phenols showing greater removal than chlorinated phenols. Particularly poor removal was shown for 2-nitrophenol, 2-chlorophenol and 2,4-dichlorophenol.

The response of an activated sludge process utilizing a phenolic substrate to quantitative shock loadings was studied by Rozich and Gaudy (1985). An acclimated pilot-scale plant showed no substantial changes in operation when the phenol concentration was stepped from 500 mg/l to 1,000 mg/l, but eventual washout occurred in stepping from 1,000 mg/l to 2,000 mg/l. Radical changes in biomass characteristics occurred during the second shock, the biomass becoming viscous, clumpy and highly dispersed. The first indication of process failure was a white foam which collected at the top of the aeration basin.

Dechev and Matveena (1977) showed that the rate of oxygen uptake is proportional to the phenol concentration in activated sludge. Khararjian, et al. (1979) demonstrated that the oxidation rate was independent of SRT in the range of 3 to 15 days, but that high SRTs were required to treat low COD and phenol concentrations. The authors also showed effective removal of phenol to below 0.1 mg/l, with good sludge settling characteristics.

A study of thiocyanate inhibition of phenol biodegradation showed that the effect was particularly significant when low effluent phenol levels were desired (Neufeld and Valiknae, 1979). A design and operation equation was developed relating sludge age (SRT), effluent phenol level, thiocyanate level and temperature. In a treatability study, the substrate utilization rate (r_{su}) for a phenol substrate decreased as pH deviated from neutral and as salinity increased (Kim and Armstrong, 1977). In contrast, nitrogen and phosphorus levels did not affect r_{su}. Phenol decomposition resulted in a considerable decrease in pH, so that the buffering capacity of the wastewater was the most important factor in determining treatability.

Capestany, et al. (1977), treating a mixture of phenol, toluene and benzoic acid observed good phenol and BOD removal (>97 percent) only after sufficient sulfur was added to satisfy a BOD:N:P:S ratio of 1000:5:5:5. Three studies investigated the activated sludge treatment of chlorophenols. Broecker and Zahn (1977) showed that activated sludge, treating an industrial waste, successfully acclimated to the addition of 5 mg/l of 3,5-dichlorophenol (DCP) with only a slight, temporary lowering of wastewater biodegradation. Higher DCP levels were not treated as successfully and the authors suggested 5 mg/l as the toxicity limit. Edgehill and Finn (1983) and Moos, et al., (1983) both addressed the activated sludge treatment of pentachlorophenol (PCP). In the former article, activated sludge (without recycle) required seven days to acclimate to 40-120 mg/l PCP and showed poor response to PCP shock in an acclimated culture. The latter article reported that PCP concentrations as low as 250 ug/l inhibited PCP biodegradation, but that the activated sludge could be acclimated to 20 mg/l PCP or higher. PCP biodegradation was proportional to biomass concentration, while sorption and volatilization were not significant removal mechanisms.

Lordi, et al. (1980) studied six wastewater treatment plants for cyanide problems in the Metropolitan Sanitation District of Chicago. They found variable removal efficiencies with influent cyanide concentrations varying from a low of 0.07 mg/l in domestic plants to 0.40 mg/l in industrial plants. In general, 0.67 to 0.80 pounds of cyanide were removed per million gallons of waste treated. Gaudy, et al. (1982a) estimated a maximum of 60 percent of the cyanide removal was by air stripping at influent levels of 20 mg/l. Raef, et al. (1982) ranked the cyanide removal mechanisms in decreasing importance at neutral pH as: air stripping > biological metabolism > bio-adsorption > chemical reaction with substrate. Neufield, et al. (1981) showed virtually complete thiocyanate removal by acclimated activated sludge at thiocyanate levels of 1,000 mg/l in the pH range of 5-7. At higher pH values or sludge ages less than six days, significant process instability resulted. The critical sludge age for thiocyanate biodegradation in this study was 4.14 days.

A substantial amount of work has been done on the treatment of coal coking and coal gasification wastewaters, which are characterized by high concentrations of phenols, cyanide and thiocyanates. Luthy (1981) reviewed the literature concerning treatment of these wastes by activated sludge. It was indicated that under proper conditions of acclimation, good COD, BOD, phenolics, ammonia-nitrogen and cyanide removal could be expected. Low microbial yield was a typical characteristic of coal byproduct wastewater treatment, resulting from inhibitory constituents in the wastewater. Single stage nitrification was possible under controlled conditions.

Jones, et al. (1984) studied the biological treatment of a high-strength coke plant ammonia liquor where operation of the activated sludge system led to filamentous growth. A pH increase, exposure to hydrogen peroxide, 4 percent wasting of the mixed liquor and elevated operation temperatures (42°C) were all attempted to mitigate filamentous effects. Only the last response was successful, although long term effects were not studied. Luthy and Jones (1980) utilized a similar wastewater in a completely-mixed activated sludge unit, resulting in good removal of phenolics, thiocyanate and COD at a mean cell residence time of 40 days. In the absence of substantial nitrification, the effluent ammonia-nitrogen was higher than the influent due to degradation of the thiocyanate and release of ammonia. Another study of a similar waste showed good phenol and

COD removal for an SRT of only 6.4 days when preceded by lime precipitation, ammonia stripping and solvent extraction (Hung, et al., 1981).

Metals -- Numerous studies dealing with the relationship of metals and the activated sludge process have been reported. Concentrations which reportedly inhibit the activated sludge process and reduce metal removal efficiency by activated sludge vary widely for most metals, but are remarkably similar in some cases. The wide range of reported values can be attributed to a number of factors including varying metal forms and solubilities, metal complexation and competition, characteristics of the substrate and activated sludge biomass, and process variations and conditions.

A number of references deal primarily with removal efficiencies of metals in the activated sludge process (Nielsen and Hrudey, 1983; Rossin, et al., 1982; Sterritt and Lester, 1981; Petrasek and Kugelman, 1983; Lester, 1983; and Patterson, et al., 1983). In general, the studies agree on the broad removal efficiency classifications presented in Table 2.

In most cases, the studies used to construct Table 2 utilized acclimated activated sludge and showed that the activated sludge process is at least as important as primary sedimentation in metals removal. Due to sludge recycle, metals concentrations are consistently higher in activated sludge than in the primary clarifier sludge.

Sterritt and Lester (1981) and Rossin, et al. (1982) showed that heavy metals removal varied with SRT for those metals effectively removed by activated sludge (Cd, Cr, Cu, Pb, Zn). Maximum affinity of the metals for the biomass occurred at SRTs of 9-12 days, but maximum removal occurred from 12-16 days, indicating the increased mixed liquor suspended solids (MLSS) more than offset the metal's lower biomass affinity. Both studies observed Pb removal to be proportional to suspended solids (SS) removal, as was Cr, Ni and Zn removal according to Rossin. Sterritt and Lester reported Ag, Cu, Ni, and Zn removal to be proportional to COD removal. The study by Rossin, et al., and another by Nielsen and Hrudey (1983) indicated the ratio of soluble metal to total metal increased as most metals passed through the activated sludge plant, with the effluent being predominantly in soluble form.

Studies by Nelson, et al. (1981) and Patterson, et al. (1983) both showed that adsorption isotherms adequately represented the distribution of metals between the solid biomass phase and the solution phase. In the Patterson, et al. study, sorption of seven metals in activated sludge mixed liquor ranged from a low of 8 percent to a high of 98 percent with the following sorption ranking: Fe < Ni < Zn < Cd < Cr < Cu < Pb. A metals distribution model and full system metals removal model were developed. Nelson, et al., determined that the equilibrium distribution of metals between solids and solution was affected primarily by physical-chemical factors and not by biological transport processes. pH was the single most important factor influencing the chemical speciation of metals and their solids-solution distribution.

After studying the inhibition of the activated sludge process by several metals, Kang, et al. (1981) found that chromium was a fairly sensitive overall indicator

Table 2
Metal Removal Efficiency by Activated Sludge

Metal	Removal Efficiency
Ag	Fair
Al	Poor
As	Poor
Ca	Poor
Cd	Good
Co	Poor
Cr	Good
Cu	Good
Fe	Good
Mg	Poor
Mn	Fair
Mo	Poor
Ni	Poor
Pb	Fair-Good
Zn	Fair-Good

Poor = < 20% Fair = 20-50% Good = >50%

Table 3
Metal Concentrations Inhibiting the Activated Sludge Process

Metal	Concentration (mg/l)
Cd	4.0 - 5.0
Cr	1.0 - 20
Cu	0.7 - 1.0
Ni	1.0 - 5.0
Pb	0.1 - 1.0
Zn	2.5 - 20
Hg	1.0 - 2.5
Co	<5.0
V	20

of the occurrence of heavy metals in a waste stream. Additional studies on treatability of metals and metal inhibition of activated sludge were also conducted by Sujarittanonta and Sherrard (1981), Bagby and Sherrard (1981), Weber and Sherrard (1980), Petrasek (1981a, 1981b, 1981c), Bieszkiewicz and Hosezowski (1978), Robins and Green (1974), and Kunz, et al. (1976). Volesky, et al. (1977) reviewed the literature concerning the effects of metallic species on the performance of biological treatment systems. Table 3 presents a compilation of metals inhibition data from the previous ten references. Inhibition is variously defined as causing an increase in effluent COD or suspended solids, decreasing sludge settleability, or a combination thereof. As defined, inhibition does not include inhibition of nitrifying organisms. The values reported in Table 3 are in general higher than values listed as inhibitive by EPA (JRB Associates, 1981a).

In reviewing the literature, Volesky, et al. found the following general relationships to be true:

- For a given metal ion concentration, a decrease in sewage strength results in lower organic removal efficiency.
- Increased metal ion concentrations result in lower organic removal efficiency and longer process recovery times.
- Inhibitory concentrations of metals are proportional to the MLSS concentration.
- Some metals appear to behave independently of others (when in mixtures) in the activated sludge process.
- The order of increasing toxicity to activated sludge is: Cr(VI) < Co < Zn < Cd < Cr(III) < Cu < Ni.

Lester (1983) drew many of the same conclusions in another literature survey. He concluded that metal toxicity depends on the following six factors: metal concentration, the form of the metal, species of organisms, suspended solids concentration, sludge age and concentration of other cations.

Morozzi and Cenci (1978) compared the toxicity levels of some metals to that of their tetracyanide salt complexes in activated sludge. The results showed: cadmium and zinc complexes to be more toxic than the elemental form, nickel metal toxicity was higher than the corresponding complex and no difference in the effect of mercury and the corresponding tetracyanide complex. In contrast, Cenci and Morozzi (1977) found no evidence of inhibitory differences between cadmium and its corresponding tetracyanide complex at the 0.5 to 2.0 mg/l level. This inconsistency was suggested to be a result of higher MLSS in the 1978 study.

Webber, et al. (1977) showed that acclimated activated sludge could accommodate continuous boron concentrations up to 150 mg/l with no decrease in COD removal efficiency. The fate of the boron was not discussed, but boron pretreatment guidelines at the time suggested only 1 mg/l of boron as being inhibitory. The organotin compound, tributyl tin oxide (TBTO), was treatable in acclimated and unacclimated activated sludge at levels up to 100 μg/l and 25 μg/l respectively (Argaman, et al., 1984). Activated sludge exposed to TBTO showed less biomass diversity compared to control sludge.

Two studies concerning the effect of nickel on the activated sludge process found that increased substrate concentration and increased SRT were mitigating responses to inhibiting nickel concentrations (Sujarittanota and Sherrard, 1981; Bagby and Sherrard, 1981). The latter article showed the same relationship for cadmium. Kao, et al. (1982) found that in the presence of the chelating agent EDTA, cadmium uptake into the biomass was an order of magnitude less than without EDTA, and the soluble portion of cadmium increased. A pilot-scale study of cadmium using municipal sewage showed some cadmium breakthrough into the effluent at influent concentrations as low as 2 mg/l, but activated sludge inhibition did not occur until influent concentrations were greater than 4 mg/l (Petrasek, 1981c). At higher cadmium concentrations (>8.0 mg/l), large decreases in protozoa populations and sludge settleability occurred.

Casey and Wu (1978) found that cadmium and copper (and presumably other metals) were more readily adsorbed to activated sludge biomass under phosphorus-limited conditions than under normal operating conditions. Optimum COD to phosphorus ratios depended on the metal involved.

Henney, et al. (1980) documented an accidental spill of 225 kg of chromium (VI) to the Mankato, Minnesota wastewater treatment plant that resulted in a maximum aeration basin concentration of 10 mg/l. By the day after the spill, the respiration rate had dropped by a factor of three, free swimming protozoa disappeared from the aeration tank and stalked ciliates were noticeably reduced. Recovery was fairly rapid however, with effluent suspended solids levels returning to normal after another day. BOD and COD removal efficiency decreased for three days (from 89 percent to a 76 percent low) and returned to normal in three weeks. The greater part of the Cr(VI) passed through the plant, being removed only to the extent that it was reduced to Cr(III). The highly insoluble Cr(III) remained largely adsorbed to the biomass but did not appear to be harmful to the activated sludge in moderate amounts. The first indication of plant problems after the nightime spill was a characteristic (chromium) yellow coloration of the plant wastewater.

Muttamara and Islam (1983) studied the activated sludge treatment of a tannery waste high in chromium. They found COD removal will remain unaffected if the food to microorganism ratio (F/M) is $\geq$0.25, with a metal to biomass ratio $\leq$ 0.005. They found lime addition improved sludge settleability and reduced toxic effects by causing precipitation of chromium.

Petrasek (1981b) found that mercury partitions strongly to the activated sludge mixed liquor. Design and operation equations were developed correlating Hg levels in the influent and effluent with return activated sludge concentrations, MLSS and primary sludge concentration. In a similar study, lead was also found to partition strongly to the mixed liquor (Petrasek, 1981a).

A study by Mirzadeh, et al. (1977) investigated the effect of sodium bentonite (clay) and ferric chloride on the activated sludge treatment of wastewater. It was concluded that additions of sodium bentonite (0-500 mg/l) and ferric chloride (0-200 mg/l) in the presence of sodium bentonite (100 mg/l) were ineffective in improving an unstable activated sludge system.

Other Organics -- A comprehensive investigation of the fate of 22 toxic organic compounds in an activated sludge treatment plant was conducted by Petrasek, et al. (1983). Four pesticides, three phenols, six phthalates and nine polynuclear aromatic hydrocarbons were dosed in municipal sewage at approximately 50 ug/l each. Ninety-seven percent removal occurred for all compounds except lindane, bis-(2-ethylhexyl)-phthalate, di-n-butyl-phthalate and pyrene.

Klečka (1982) showed that methylene chloride rapidly degraded to CO_2 and Cl in acclimated activated sludge. The rate of biodegradation was determined to be twelve times greater than the rate of volatilization. Unacclimated activated sludge did not successfully degrade methylene chloride. Vaicum and Eminovici (1974) looked at the effect of Lindane on activated sludge characteristics and found no effect at or below 5 mg/l (Lindane), but found reduced COD removal, biomass protein levels and enzymatic activity at Lindane levels of 10 mg/l.

Ishikawa, et al. (1979) investigated the removal of organic acids by acclimated activated sludge. Acetic, propionic and butyric acid were removed at the rate of 35.9, 23.1 and 29.3 mg acid/gm MLSS/hour, respectively. Lau (1978) studied process stability and removal efficiency of an activated sludge system treating various aromatic acids. Two or three stage aeration provided much better stability in dealing with quantitative shocks than did conventional single-stage aeration.

Methyl alcohol and ethylene glycol did not affect an acclimated activated sludge process at concentrations up to 1500 mg/l and 200 mg/l, respectively (Bieszkiewicz, et al., 1979). Concentrations above those levels caused reduced biodegradation and poorly settling sludge. Aircraft de-icing fluid (the main component being glycols) was successfully treated at concentrations less than 193 mg/l (as carbon) with a total BOD load of 480 mg/l (Jank, et al., 1974). Organic loadings were kept to less than 0.15 kg BOD/kg MLSS/day to control bulking, thus operating in the extended aeration range. Cox (1978) reviewed the literature concerning polyethylene glycol (PEG) biodegradation and found successful results for molecular weights less than 400. Biodegradation was generally inversely proportional to molecular weight.

Four adipic acid esters were treated by acclimated activated sludge in a study by Saeger (1976). In the range of concentrations studied (20-56 mg/l), the esters were substantially degraded with no inhibition of the activated sludge process. Ultimate biodegradation did occur. In a study of the activated sludge treatment of a combination of methanol, acetylene, vinyl acetate, acetic acid and formaldehyde, BOD removal was 99 percent at influent levels up to 1000 mg/l (Schwartz, 1984). A two-stage (series) activated sludge unit provided more process control and a low rate (extended aeration) activated sludge unit was not as easily upset as conventional activated sludge.

Avendt and Avendt (1983) reported that an unspecified organometallic pesticide dosed at levels up to 75 µg/l in a municipal wastewater did not affect the performance or stability of a full scale plant. Due to sludge recycle, pesticide concentrations up to 80 times the raw feed concentration were observed in the activated sludge mixed liquor. Iannone, et al. (1984) found that pesticides "passed through" New York City wastewater treatment plants utilizing activated sludge rather than being treated. Another study utilizing a pesticide mixture observed significantly lower BOD removal at the 0.5 mg/l pesticide level and some decrease in removal efficiency at 0.1 mg/l (Smith, et al., 1981). At both levels, however, the pesticides were significantly removed by the activated sludge.

Kaneko, et al. (1976) reported no change in BOD removal efficiency at polychlorinated biphenyl (PCB) concentrations up to 10 µg/l, however there were resulting changes in microflora, and increased aldolase and oxygen uptake activity. PCBs were concentrated in the activated sludge by a factor of 10^3-10^4, without showing any appreciable decay. Herbst (1977) found no organic degradation of PCBs as well, but did find some volatilization occurring. Furukawa (1978) and Liu (1980) studied the biodegradation of PCBs by microorganisms isolated from activated sludge cultures. Varying rates of biodegradation were reported, with the following two relationships being evident:

- Degradation is inversely proportional to the level of chlorination.
- PCBs containing two Cl atoms in the ortho position (2, 2- or 2, 6-) showed great resistance to degradation.

Baird (1977) and Barth and Bunch (1979) reported that benzidine inhibits aerobic digestion in unacclimated cultures, but the latter article reported no interference to acclimated activated sludge in concentrations up to 30 mg/l. Hydrazine had no effect on the activated sludge process and underwent complete treatment at the 1 mg/l level (Farmwald and MacNaughton, 1981). At levels greater than 5 mg/l, COD removal seriously degraded and hydrazine in the effluent was > 0.4 mg/l. Non-acclimated activated sludge handled slug doses of hydrazine up to 44 mg/l without serious impairment of COD removal.

Trisodium nitrilotriacetate (NTA) was 80-90 percent biodegradable in acclimated activated sludge at levels up to 30 mg/l (Renn, 1974). Treatment was poor using unacclimated cultures. Nay, et al. (1974) reported successful biodegradation of trinitrotoluene (TNT) with no substantial effects on acclimated extended aeration activated sludge up to concentrations of 20 mg/l. At higher TNT levels, biodegradation is inversely proportional to the TNT concentration. Contact stabilization with extended aeration is suggested for best biological treatment. Chow and Ng (1983) studied the biodegradation of n-methyl-2-pyrrolidone at 100 mg/l. The compound was 95 percent biodegradable but produced an intermediate carbonyl metabolite with a significant COD. Vaicum and Eminovici (1974) found that 50 mg/l of trinitrophenol had no effect on acclimated activated sludge processes while 200 mg/l caused substantial reduction in COD removal efficiency.

Activated sludge treatment efficiency was not affected by various organic dyes at 10 mg/l concentrations (Dohanyos, et al., 1978). The removability of dyes by activated sludge is proportional to the number of hydroxyl groups, the number of nitro groups, the number of azo groups, the length of the molecule, increasing electrokinetic potential and is inversely proportional to the number of sulfo groups. Gledhill (1975) found rapid activated sludge acclimation to 200 µg/l of 3, 4, 4'-trichlorocarbanilide (TCC). TCC was also substantially degraded.

Miscellaneous industries -- A wide variety of aqueous wastes are generated by industrial concerns and these wastes can result in a range of operational difficulties in an activated sludge system. Hegg, et al. (1980) conducted a survey of operational and maintenance problems at 50 POTWs. Problems such as toxic pH conditions due to an acid discharge, turbid supernatant due to crankcase oil, and loss of sludge inventory due to a suspected metal plating waste were reported. A survey of German activated sludge plants (Wagner, 1984) revealed that wastes from paper, milk processing, vegetable processing, hide and glue, fruit processing, and distillation industries may exhibit a tendency to cause bulking sludge associated with filamentous organisms because of unbalanced nutrient loadings (high organic carbon and nitrogen, low phosphorus). Slaughterhouse, metal processing and textile wastes were felt to be less likely to cause bulking problems. Appropriate responses for sludge bulking could be pretreatment denitrification, phosphorus addition or precipitation, and/or lime addition for maintenance of favorable nutrient levels and mixed liquor pH.

Interferences with activated sludge processes may also be the result of particulate types of contaminants. Wilson, et al. (1980) reported poor process performance resulting from the introduction of filamentous organisms into the system from a pharmaceutical manufacturer -- a condition that was corrected by increasing the SRT and MLSS of the system.

Wastes from petrochemical industries can contain degradable, nonbiodegradable, and toxic organic compounds. The effects of urea and dichloromethane, two possible constituents of refinery wastewater, on activated sludge treatment were investigated by Gerber, et al. (1979). Urea was removed, but high ammonia nitrogen concentrations resulted. Dichloromethane was removed by stripping and it had a detrimental effect on nitrification prior to acclimation of the system. Matsui, et al. (1975) examined the degradability of organic substances found in wastewater from petrochemical industries. Thirty-five substances were evaluated, the following fifteen of which were found to be highly resistant to biodegradation:

4-chloro-2-methyl aniline
p-toluene-m-sulfonic acid
2-chloro-5-amino-p-toluene sulfonic acid
dinitrotoluene
β-naphthol
β-oxynaphthoic acid
4,4-dimethyl-1,3-dioxane
polyethylene glycol
PVA
silicon surfactant
methyl cellulose
2-chloroethanol
higher alcohol surfactant
acetoacetic-o-anilide (3A-4)
acetoacetic-m-xylidide (3A-5)

For wastewaters from individual industries, activated sludge COD removals (after 24 hours of aeration) ranged from 14 to 81 percent.

Stover and Kincannon (1983) experimentally investigated the treatability of 12 organic compounds using a bench-scale, activated sludge system. The results of their thorough investigation included the removal efficiencies and biokinetic constants for the individual compounds and various combinations, and the percent removals attributed to volatilization. Kincannon, et al. (1982) evaluated the treatability of two semiconductor-industry wastewaters containing a wide variety of organic compounds, concluding that the two types of wastewaters examined were compatible with the activated sludge process.

Monnig, et al. (1981) conducted investigations on the suitability of various pesticide manufacturing wastewaters for discharge to municipal activated sludge treatment facilities. The wastewaters examined were characterized only by the type of pesticide from which they were generated and the results were described qualitatively. Carbaryl, maneb, mancozeb, and dazomet wastewaters at dilutions ranging from 0.1 to 10 percent were degraded with no adverse effects on COD removal. Glyphosate and atrazine wastewaters were nondegradable with no adverse effects while dinoseb and oryzalin wastewaters disrupted treatment (even at low dilutions) and mancozeb, dazomet, and carbaryl wastewaters inhibited nitrification.

Food processing wastewaters are generally amenable to activated sludge treatment as illustrated by the work of Heddle (1979). Interferences with POTW operation can occur due to organic overloading. Gerardi (1981) reported on the interference of a fatty acid derivative waste from a manufacturer of food emulsifiers on the performance of a POTW. This particular waste resulted in an excess of filamentous organisms and poor sludge settleability. The problem was resolved by equalization of flow and pretreatment of the manufacturing wastewater. Other remedial measures attempted had no effect. Tolaney (1976) characterized citrus processing wastewaters, described pilot-scale testing, and presented design criteria for the high purity oxygen activated sludge treatment of such a wastewater. An operational problem encountered was the result of excessive filamentous organism growth within the system. This was a result of a high influent temperature which was remedied by including a cooling tower in the process scheme.

Wilson (1981) reported several problems associated with the discharge of a potato chip processing wastewater to a small POTW. Very poor primary sedimentation performance was possibly due to oil and grease, caustic process water, and/or nonsettleable starches. Pretreatment of the industrial wastewater stream was the appropriate response. Poor sludge setteability in the secondary clarifier was the result of low amounts of phosphorus with respect to other nutrients and substrate. Identification of the causes of the poor performance of this POTW was confounded by the occurrence of hydraulic surging.

Radik (1984) documented the experiences of more than five years of operation of a dairy industry wastewater treatment plant. Spills from various processes resulting from equipment malfunctions and operator errors, and their effects on the wastewater treatment process, were documented in addition to other experiences related to both the diary product and wastewater treatment facilities.

Textile manufacturing facilities produce wastewaters with significant contaminants. The discharge of textile fibers to the wastewater stream can result in fouling and plugging of aerators and pumps and dense blanket formation on clarifiers (Troxler and Hopkins, 1982). In general, however, such textile wastewaters have a higher BOD and lower TSS than domestic wastewater.

Contaminants such as kerosene, polyvinyl alcohol and textile dyes may be present which have various degrees of removal by activated sludge (Porter and Snider, 1976). Polyvinyl alcohol can be removed by activated sludge treatment without inhibiting the process (Hahn, et al., 1977). Gaffney (1976) reported that chloro-biphenyls, while removed by activated sludge treatment, caused an unbalanced microbial population and poor overall process performance. Textile dyeing wastewaters were tested by Horning (1978) using bench-scale biological and physical-chemical processes. The author discovered that dye bath components were not inhibitory to carbonaceous BOD removal, but could inhibit nitrification. The optimum treatment turned out to be staged biological plus physical-chemical processing, as not even acclimated bacterial seeds could metabolize the color-producing dye molecules.

Shock loads of textile mill wastewater can also interfere with POTW process performance. Blevins (1982) reported that following such fluctuations in loading, variations in an activated sludge microbial population occurred, resulting in poor sludge settleability.

Dagon (1973) characterized photoprocessing wastewaters and presented a discussion of the possible effects of the various constituents on the activated sludge process. Experiments were also conducted in which it was found that the presence of a photoprocessing wastewater at dilutions of 20 percent or less with municipal wastewater had no adverse effects on an activated sludge process. Bard, et al. (1976) considered the effects of various forms of silver in photoprocessing effluents on activated sludge treatment. It was concluded that insoluble silver or silver thiosulfate would exhibit no effects while silver nitrate and silver chloride would inhibit the activated sludge process.

Reinbold and Malleville (1975) investigated the activated sludge treatability of process water from the steaming of timber. Such waters are characterized by a wide range of soluble and insoluble organic compounds and bench-scale experimentation revealed that they are treatable provided that sufficient nitrogen and phosphorus are added. Lowry and Chwirka (1983) reported that the activated sludge treatment of combined municipal and papermill wastewater was hampered by nutrient deficiencies and paper mill wastewater dumping, typically characterized by high flow, extreme pH and fluctuating temperature. Dissolved oxygen fluctuations resulted from the papermill wastewater variability which in turn resulted in excessive filamentous growths and the attendant sludge bulking. A buildup of papermill clay in the mixed liquor suspended solids was also observed.

Kashinaya and Yoshimoto (1980) characterized tannery wastewater with respect to the specific tanning processes that were involved. Pure oxygen activated sludge treatment of combined municipal and tanning wastewater was also successfully demonstrated and design criteria were presented.

Cummins (1981) experimentally evaluated the effect of sanitary landfill leachate on the activated sludge process. His primary result was that at the higher leachate loadings, poor settleability of the sludge solids occurred. It was recommended that further studies be conducted with respect to possible system acclimation and process control strategies.

Powdered activated carbon/activated sludge process--
The ability of an activated sludge process to handle organic contaminants can be enhanced by the addition of powdered activated carbon (PAC). Bauer, et al. (1981) compared the treatment of coke plant wastewater by the RBC and PAC enhanced activated sludge (PAC/AS) processes. Both processes performed comparably, but the PAC/AS process had a better capability for nitrification on a continuous basis, independent of the COD loading. Weber, et al. (1983) found that lindane could be removed by the PAC/AS process. Other studies have similarly examined the applicability of the PAC/AS process for treating a variety of wastewaters and contaminants (Soderberg and Bockrath, 1985; Janaczek and Lamb, 1983; and Shaul, et al., 1983).

Variables of importance related to PAC enhancement of activated sludge include the carbon dosage and the degree of mixing--both of which can govern the contaminant removal efficiency of the process (Weber, et al., 1983).

From bench-scale experimentation of the pretreatment of chemical production wastewaters, Leipzig and Hockenburg (1980) found that the PAC/AS process had the capacity to handle organic shock loads in addition to allowing nitrification to occur despite the presence of inhibitory materials. Sundstrom, et al. (1979) found that periodic addition of PAC to an activated sludge system was effective in removing shock loads of phenol but glucose shock loads were moderated only slightly.

Ferguson, et al. (1979) evaluated PAC used in conjunction with contact stabilization activated sludge. Shock loadings of trichlorophenol were evidenced by a loss of COD removal efficiency only when the PAC dosage was low. Excessive carbon dosages were, however, also detrimental to the process.

Several studies of full-scale PAC/AS systems have been published. Dunn and Hutton (1983) reported that organic and heavy metal removals of 50 to 60 percent could be obtained. Hutton and Temple (1979) found that a PAC/AS process could outperform a conventional activated sludge system — particularly in the removal of extractable organics. Pitkat and Berndt (1981) discussed the treatment of textile waste at a municipal PAC/AS treatment plant and presented operational data for both the overall treatment process and for the wet air oxidation-carbon recovery process.

Lagoons --
Little research has been conducted on pollutant interference of municipal wastewater lagoons, however some characteristics of an aerated lagoon or oxidation pond are similar to extended aeration activated sludge. Facultative lagoons cannot be as easily likened.

Abeliovich and Azov (1976) found that ammonia and methylamine concentrations over 2.0 mM at pH levels greater than 8 inhibit photosynthesis and therefore oxygen levels in an oxidation pond. Increasing the detention time of the oxidation pond lowered the ammonia concentration and restored proper operating conditions. In another oxidation pond study, Miller. et al. (1977) studied the impact of increased carbon supply on algae growth. Growth of algae (and therefore oxygen production) was enhanced due to the additional nutrients of an added sludge stream.

A three-cell, 0.40 mgd lagoon system in Eudora, Kansas was studied by McKinney (1977). Most of the biodegradation was accomplished in the first cell, but inadequate settling of the dead microbes in cells two and three resulted in an effluent TSS which exceeded the 30 mg/l permit limit.

The performance of an aerated lagoon treating a refinery waste was improved after the addition of a selective bacteria (Nyer and Bourgeois, 1981). Belly, et al. (1975) showed that the ferric chelate of EDTA was biologically degraded by a mixed population of aerated lagoon microbes.

Meel, et al. (1976) investigated the treatment of combined potato processing and municipal wastewater by anaerobic and aerobic lagoons. Various combinations were studied and best results were obtained by staging the lagoons in an anaerobic/aerobic series. Low pH and temperatures had little effect on the process. In another study, Klein (1974) found 90 percent steady state removal of trisodium nitrilotriacetate (NTA) in an aerobic stabilization pond.

Anaerobic Reactors --
Engineers have historically been reluctant to design anaerobic treatment systems for industrial wastewaters containing toxicants. The belief that methanogens are inherently more sensitive to contaminants than aerobic or facultative organisms is disputed by Parkin, et al. (1983). The authors tested four pollutants (nickel, sulfide, formaldehyde and ammonium) in batch, semi-batch and continuous flow reactors, and discovered that acclimated cultures could tolerate toxicant concentrations 2.4 to 12 times higher than unacclimated systems with no upsets. The use of high SRTs promotes such acclimation, mitigating the effects of decreased gas production resulting from process inhibition. The study also noted that extended periods of zero gas production may not indicate total process failure, but rather the system will often start up again after the toxicant is purged from the system.

Additional evidence promoting the use of anaerobic treatment with acclimated cultures is provided by other authors. Pearson, et al. (1980) note the inhibitory effects of formaldehyde, zinc sulfate, phenol and ammonium in bench-scale testing of recreational vehicle tank waste, but observed greater stability of acclimated over non-acclimated systems at the 50 percent kill dosages. Sulfide and sulfite inhibition of anaerobic treatment in both continuous flow and batch modes was studied by Eis, et al. (1983). The impact of these contaminants on methane production and COD removal efficiency was minimized when the reactors were acclimated to sulfur.

There is some discrepancy in the literature regarding the anaerobic degradation of phenol. Healy and Young (1978) report complete degradation of phenol to methane and carbon dioxide following an 18-day acclimation period plus an additional 14-days of biodegradation. The authors even cite the first reported case of anaerobic degradation of catechol, using a 32-day acclimation period. Neufeld, et al. (1980b) indicate the need for a 40-day SRT for effective and stable phenol decomposition to organic acids, but no methane production. At phenol concentrations above 686 mg/l, they note distinct substrate inhibition to non-methanogenic decomposition. Later work by Healy and Young (1979) produced results similar to their phenol study for eleven simple aromatic compounds. Anaerobic decomposition to methane was achieved with long acclimation and treatment periods.

Fixed Film Processes

Trickling Filters --
An overview of trickling filter performance was provided by Haugh, et al. (1981) in a survey of 11 plants in 6 states with flows ranging from 0.5 to 33.7 mgd. They found that increased effluent BOD generally resulted from:

- increased flow rates.
- increased influent BOD.
- pH out of 6.5 to 8.5 range.
- the presence of toxics.

Most of the recent literature has studied the treatment or effects of specific industrial wastes on trickling filter processes. Two references have shown that using trickling filters as a first-stage preceding activated sludge can control bulking sludge problems resulting from filamentous organisms. Anderson (1980) reported on a pharmaceutical wastewater in Lincoln, Nebraska containing high concentrations of sugars in a nitrogen deficient environment. The addition of an equalization basin and a biotower mitigated the effects of shock loadings on the activated sludge system, resulting in overall BOD and COD removals of 94 to 99 percent and 86 to 91 percent, respectively. The use of a pilot-scale plastic media trickling filter preceding an oxidation ditch for the treatment of brewery wastes was analyzed by Biesinger, et al. (1980). High organic loadings (400 to 1200 lb BOD/d/1000 ft^3) to the fixed film reactors serves as a population selector for non-filamentous organisms, eliminating their presence in the oxidation ditch.

High hydraulic and organic loadings coupled with high concentrations of fats, oils and grease make slaughterhouse wastewaters troublesome. Li, et al. (1984) found that employing air flotation for oil and grease removal ahead of biofiltration was an effective treatment system. Combined domestic and tannery wastewater was treated by a combination of pilot-scale systems in Alexandria, Egypt (Hamza, et al., 1982). High concentrations of metals, organics, sulfide salts and ammonia result in a highly toxic wastewater. A combination of physical/chemical, biofiltration and post-filtration processes were successfully employed to remove conventional pollutants and reduce the toxicity of the wastewater.

Chloro-biphenyls are present in dye carriers used in the carpet industry which severely inhibit secondary biological treatment processes. Trickling filters subject to such compounds have produced thin slimes, a red color, black deposits, large populations of filamentous bacteria with no attached ciliates or rotifiers in northwest Georgia (Gaffney. 1976). These facilities showed poor treatment of conventional pollutants, but were removing the biphenyls. Fox and Merrick (1982) studied a PCB spill contamination at an Alcoa facility treating combined domestic and industrial waste in Lafayette, Indiana. In this case, the authors noted that 84 percent of the PCBs were removed in the clarifiers, with only an additional 10 percent taken out by the trickling filters. Upsets were prevented by optimizing the polymer feed to the clarifiers which minimized the effect on the biological process.

Nitrification inhibition by fertilizer constituents was analyzed by Beg, et al. (1981). A packed-bed, bench-scale reactor was fed synthetic wastewater containing a fixed ammonia concentration with varying dosages of sodium fluoride, sodium arsenate and potassium dichromate. Inhibition, defined as a 50 percent reduction in biological activity, occurred at 50, 292 and 1218 mg/l for Cr, As and F, respectively.

Rotating Biological Contactors --
The rotating biological contactor (RBC) process has only been commercially available in this country since the early 1970's, yet more research has been performed regarding industrial pollutant treatability and impact on this process than all other fixed-film processes. Such high levels of activity are partly due to the notoriety of RBCs and partly resulting from the availability of suitable pilot-scale equipment.

Conventional pollutants-- Pilot-scale studies (Fry, et al., 1982; Ouyang, 1984) have analyzed the impact of hydraulic and organic shock loads on the treatment of domestic or synthetic wastewater. At three times the normal flow rate, the dissolved oxygen level in the basins decreased and nitrification was inhibited. At still higher flows, BOD and COD removals may also become inhibited. No adverse effects were noted with organic shock loadings until they reached three to four times the design loads. Hammer (1983) reported on the organic shock loadings from a cheese processing wastewater discharged to the Newman Grove, Nebraska municipal treatment plant. One to two hour flows containing 35,000 mg/l BOD would cause the effluent BOD to increase from 10 to 60 mg/l for a two day period. Attempts to equalize organic loads throughout the RBC trains by altering the staging arrangements proved unsuccessful in this case.

Organic overloads are a common problem at numerous municipal RBC facilities. The Kirksville, Missouri plant witnessed excessive biomass growth (and ultimately a shaft failure) and decreased BOD removal due to organic overloads (Newbry, et al., 1982). Supplemental air at 70 cfm/shaft increased sloughing, thereby reducing shaft weight, but with no noticeable improvement in BOD removal. Similar results were obtained by Surampalli, et al. (1984) at Ames, Iowa in their use of supplemental air to eliminate Beggiatoa and reduce biomass thickness. RBC system staging and configuration has also been shown to influence organic loading capacity (Zogorski, 1984). An eight shaft, eight-stage (one shaft/stage) train arrangement could not withstand the same organic loadings as an eight shaft, four-stage (4-2-1-1 configuration) system.

RBC treatment of munitions manufacturing wastewater at the Radford, Virginia Army Base is subject to shock organic loadings and severe pH fluctuations (Smith and Greene, 1982). The organic shocks cause an initial decrease in BOD removal, with a subsequent growth of biomass to accommodate the loading. Similar BOD removal reductions occur with pH shock, but the system quickly recovers following pH correction. Attempts to mitigate these effects with adjustments in the equalization basin have met with modest success.

Metals -- A few studies have considered the effects of metals on RBCs. Copper has no impact on the process up to 10 mg/l, (Chang, et al., 1985; Smith and Moore, 1984) but reduces the organics removal efficiency by 10 to 15 percent in the 25 to 50 mg/l concentration range. Cadmium has been shown to limit organics reduction by 10 percent over the entire range from 5 to 20 mg/l (Chang, et al., 1985). Hexavalent chromium levels of 1.4 mg/l resulted in a slight reduction of both BOD and 1, 1, 1-trichloroethane removal over a wide range of applied loadings for these parameters (Blumenschein and Helwick, 1983).

Pretreatment -- Numerous industrial wastewaters have been tested in RBC pilot systems for treatability in pretreatment applications. Coal gasification wastewater was treated in a single-stage RBC by Turner, et al. (1984), who observed BOD and phenol removals of 94 percent and 100 percent, respectively, 30 to 35 percent removal of COD and TOC, with negligible treatment of ammonia, cyanide or thiocyanate. The results indicate that these refractory organics are not inhibitory to biological processes. Bracewell, et al. (1980) noted 40 to 80 percent reductions for COD, phenols and formaldehyde in an acclimated RBC pilot system.

Two separate studies by Landon-Arnold and Chan (1982, 1984) have demonstrated the treatability of aqueous film forming foam (AFFF), a combination of fluorochemical surfactants, hydrocarbon surfactants, ethylene glycol and water. The results indicate significant BOD reduction (97 percent) up to 2000 mg/l AFFF concentrations, provided pH adjustments are made and nutrients are added if not mixed with domestic sewage. At higher concentrations, foaming problems may occur.

Landfill leachates have been successfully treated in a couple of pilot-scale RBC applications. Coulter (1984 observed significant degrees of COD, BOD, NH_3, phenols, cyanide and iron removal, with limited toxicity reduction. Smith and Moore (1984) obtained similar results using synthetic leachate and pesticides. Eighty-five to 95 percent pesticide removals were recorded for influent concentrations of 100 mg/l and less. COD and BOD removals remained high, with pesticide concentrations as large as 2000 mg/l.

Pilot-scale tests were performed comparing the addition of air with pure oxygen to an RBC system treating high strength starch waste (Li, et al., 1982). At loading rates of 38 and 95 g COD/m^2/d, the air system experienced foaming and septic conditions, while no problems were encountered in the oxygen-fed system.

Watt and Cahill (1980) exceeded 80 percent BOD and COD removals when treating synthetic wastewaters containing phthalate esters and other oxochemicals with bench-scale RBCs. When COD loadings were kept below 40 g/m^2/d, removals consistently topped 90 percent. The treatment of relatively high strength (1600 mg/l BOD) synthetic wastewater was not affected by the presence of explosives (RDX, HMX, TNT), as BOD removals were at least 95 percent (Chesler, 1980).

Slaughterhouse industrial wastewater discharging to a 16-shaft POTW was studied by Stover and Rakness (1984). The soluble BOD loading of 5 lb/d/1000 ft^2 resulted in a dark black, thick biomass covered with a white, stringy growth. Staged treatment at the municipal site was deemed the best alternative for achieving NPDES permit compliance.

Aerated Submerged Filters --
Comparatively little research has been reported in the literature on aerated submerged filters (ASF), particularly regarding treatment and effects of industrial contaminants. Most of the information appears in manufacturer's literature or in-house reports. Like any emerging technology, objective analysis will follow as such systems become popular and are installed in treatment facilities. References that do exist in the literature have focused on treatability of specific contaminants or comparisons with existing technologies, with nothing written on process upset or mitigation.

Kao and Kang (1984) treated soda pulp wastewater with a bench-scale submerged biofilter, and found it to be more efficient than the reported performance of activated sludge plants over the range of BOD loadings tested. Rusten and Thorvaldsen (1983) compared the performance of a pilot-scale ASF with an activated sludge unit for the treatment of a combination of food processing wastes. The authors noted that the desired effluent COD concentration could be achieved by the ASF at an organic loading rate four times greater than the activated sludge system.

A fluidized bed reactor was employed that successfully removed cadmium, copper and zinc from a synthetic industrial wastewater (Remacle and Houba, 1983). The metals were present in soluble form, and were subject to a two-day detention period at 150 percent bed expansion with a dissolved oxygen concentration of 4 mg/l.

Biodegradation of phenol (50 to 200 mg/l) in synthetic wastewater was accomplished by Richards, et al. (1983) in a bench-scale, fluidized-bed apparatus. Inert coal was used as the attached medium with oxygen bubbled into the reaction flasks. The kinetics of phenol degradation were described on the basis of a competitive inhibition model. Phenol treatability was also studied (Olthof and Oleszkiewicz, 1983) in a pilot-scale packed bed reactor fed coke plant wastes. Over a range of hydraulic loadings, there seemed to be a random pattern of phenol degradation with organic loading.

Anaerobic Filters --
Anaerobic processes have become popular among a number of researchers because of their process stability and the methane produced as a by-product of the reactions. Kennedy and Berg (1982) demonstrated that their bench-scale anaerobic filters maintained satisfactory removal efficiencies over a wide range of loadings and temperatures when treating three different wastewater streams.

Smith (1980) noted that an anaerobic RBC was not very sensitive to overload conditions when properly acclimated to the wastewater.

Acclimation was further stressed by Wu, et al. (1982) in a comprehensive literature review of submerged anaerobic reactors. General conclusions of their work were that:

- Influent metals resulted in decreased gas production and increased effluent COD.
- Acclimation to high concentrations (specified) of sulfide, sodium, formaldehyde, acrylic acid and acrolein was possible.
- Cyanide, ammonia and nickel toxicity was reversible, but was non-reversible for chloroform, formaldehyde and sulfide.

Rivera (1983) used acclimated cultures in an upflow reactor to treat zinc levels ranging from 100 to 1000 mg/l, resulting in decreased COD removal and gas production.

Slonim, et al. (1984) studied anaerobic/aerobic treatment of 4, 6-dinitro-o-cresol (DNOC). Their results indicate that DNOC is toxic to anaerobic processing, unless a readily degradable carbon source (sucrose) is also present in the waste stream, in which case DNOC is converted into simpler, biodegradable compounds. Although anaerobic treatment did not remove BOD, the conversion of DNOC in the anaerobic step made the waste amenable to COD reduction in the aerobic process.

Two studies analyzed the impact of phenolics on bench-scale anaerobic activated carbon filters used for industrial treatment (Khan, et al., 1981; Suidan, et al., 1981). Long detention times (140 days) were required for acclimation, but once acclimated, the cultures were resistant to shock loading effects. Stover, et al. (1984) found that high loadings of fuel alcohol wastes resulted in increased volatile acid levels and subsequent reduction in methane production and treatment efficiency.

A fluidized, anaerobic, activated carbon filter was used by Skrinde and Bhagat (1982) for denitrification of a mixed food-processing wastewater. The authors noted that when the system was pre-acclimated with methanol, no inhibition would occur when the feed switched to the industrial wastes. Lower hydraulic detention times were required for this process than for other denitrification systems, although as denitrification increased, so did the effluent BOD.

ADVANCED TREATMENT

Advanced treatment includes the physical, chemical, and biological wastewater treatment processes that are used for the removal of nutrients or as a polishing step following secondary treatment. In this section of the review of the literature, the effects of various pollutants and process conditions on biological nutrient removal and physical/chemical processes used in wastewater treatment are considered.

Biological Nutrient Removal

The removal of nitrogen and phosphorus can be achieved biologically through the processes of nitrification/denitrification and phosphorus uptake. Biological nutrient removal is subject to interferences due to contaminants, or other conditions, which inhibit the metabolic processes of the bacteria. Several recent articles examine the effects of various wastewater contaminants on the nitrification of ammonia to nitrate. An additional article compared the ability of several processes to achieve nitrification in the presence of a suspected contaminant. A dearth of information exists on the subjects of interferences to denitrification and biological phosphorus uptake in the recent literature.

Nitrification --
Conditions which can influence nitrification are temperature, pH and organic loading. Low pH was found to retard nitrification in a second-stage RBC in a study conducted by Stratta, et al. (1982). The system under study was for the nitrification of a high-rate, trickling filter effluent with nitrogen concentrations of 19-21 mg/l. Low alkalinity reduced nitrification in the long-term for pH values less than 8.5. Short-term conditions of pH values from 7.0 to 8.5 had little impact. An interference such as this could be remedied by pH monitoring and adjustment. Neufeld, et al. (1986) studied the influence of elevated temperatures on nitrification biokinetics. The authors found an apparent maximum rate at 30°C for stable nitrification. Rates of nitrification decreased on either side of this optimum. Ito and Matsuo (1980) found that high organic loadings reduced nitrification in an RBC system. This would be expected since nitrification is a function of the relative amounts of carbonaceous and nitrogenous substrates. Yu, et al. (1984) examined RBC performance at various wastewater treatment plants and established loading criteria for which nitrification could occur.

Metals -- The influence of cadium, chromium, and/or nickel on biological treatment has been the subject of a number of investigations. Sujarittanonta and Sherrard (1981) found that while carbonaceous substrate removal was not significantly affected in their bench-scale, completely-mixed activated sludge system, a nickel concentration of 5 mg/l caused nearly complete inhibition of nitrification. The nitrifying organisms were thus more sensitive to nickel interference than were the heterotrophic bacteria. Similar activated sludge system behavior in the presence of nickel and cadmium was evidenced by the work of Bagby and Sherrard (1981). Weber and Sherrard (1980) found that at cadmium concentrations of 5 and 10 mg/l, nitrification was reduced by 30 percent and 60 percent, respectively. Nitrification in an RBC secondary treatment system was studied by Kang and Borchardt (1982). Steady levels of chromium caused suppression of nitrification. Shock loads at 50 mg/l of chromium resulted in immediate reductions of both COD removal and nitrification. Evidence was also presented supporting the concept that inhibition is the result of the adsorption of chromium onto active organisms.

Huang and Skeikhdeslami (1983) investigated the inhibition of bench-scale nitrification in the presence of metals. Chromium, nickel, and zinc were found to be inhibitory to both the ammonia and nitrite oxidation steps, although ammonia oxidation was the most sensitive. Chromium reduced the rate of ammonia oxidation by 40 percent at a concentration of 0.065 mg/l Cr. Nickel concentrations of 2 mg/l or greater completely inhibited ammonia oxidation, while at lower concentrations, zinc was found to be the most lethal metal with respect to ammonia oxidation.

Braam and Klapwijk (1981) investigated copper inhibition of nitrification in a bench-scale, fill-and-draw activated sludge process. They examined the form of copper (soluble versus insoluble) as governed by the prevailing chemical conditions and they found that the rate of nitrification varied inversely with the ratio of insoluble copper to mixed liquor suspended solids. The effects of pH on the rate of nitrification were therefore due, at least in part, to the effect of pH on the amount of insoluble copper present for a given total copper concentration. It is also interesting to note that the addition of nitrilotriacetic acid, (NTA) cancelled the inhibitory effect of copper when the NTA was added within 24 hours of the addition of the copper.

Beg, et al. (1982) investigated the use of a fixed-film process for the nitrification of a fertilizer manufacturing wastewater containing chromium, arsenic, and fluoride. All three constituents were toxic with chromium having the most significant effect and fluoride the least. Beg and Atiquallah (1983) found that for a given concentration of fluoride or arsenic, the degree of inhibition due to the weak inhibitor increased with increasing chromium concentration up to some maximum degree of inhibition. As a result of their work, chromium, arsenic and fluoride were all classified as reversible, non-competitive inhibitors.

Miscellaneous industries --- Stafford (1974), Greenfield and Neufeld (1982), Neufeld, et al. (1980a), and Yurovskaya (1982) investigated the effects of coking plant wastewater constituents on nitrification. All three investigations involved bench-scale testing in which nitrification was achieved using populations of nitrifying organisms to which no organic carbon sources were fed (other than the compound under study). Their results indicated that phenols and other heterocyclic compounds interfered with the nitrification process by inhibition of the Nitrosomonas bacteria. Thus, the oxidation of ammonia to nitrite was inhibited, whereas the biological oxidation of nitrite to nitrate was unaffected. Stafford (1974) found that inhibition occurred for phenol concentrations exceeding 3 mg/l, and further suggested that upon acclimation of the nitrifying organisms, the sensitivity to shock loadings could be minimized. Neufeld, et al. (1980) examined the mechanism of phenol inhibition and presented a "modified non-competitive" kinetic model.

Stafford (1974) also examined the effects of five pyridine compounds and found that either nitrification step (ammonia to nitrite or nitrite to nitrate) could be inhibited depending upon the specific compound. Thiocyanate, another constituent of coking plant wastewater, also inhibits nitrification, but only at much higher concentrations than those observed for phenol inhibition (Greenfield and Neufeld, 1982). Likewise, unionized ammonia in coking plant wastewater also has been found to be an inhibitor (Neufeld, et al., 1980).

Other work by Neufeld, et al. (1986) found the toxic inhibition to biological nitrification decreased in the order of free cyanide, coal tar acid phenolics (derived from a coke plant), phenol, 2,3,6-trimethylphenol, 2-ethylpyridine,

2,4,6-trimethyphenol, complexed cyanides and thiocyanate. All substances except free cyanide appear to follow a "shoulder effect" where low levels of inhibitor had no influence on rates of biological nitrification while higher levels had profound effects. The concentrations up to which there is no effect on the maximum rate of nitrification are 1.2 mg/l for the tar acid phenolics, 1.4 mg/l for reagent phenol, 4.9 mg/l for 2,3,6-trimethylphenol, 10.0 mg/l for 2 ethylpyridine, 30.0 mg/l for 2,4,6-trimethylphenol, 69.0 mg/l for complexed cyanide and 236 mg/l for thiocyanate. Based upon the experimental observations, it appears that free cyanide levels greater than 0.20 mg/l must be avoided for stable operation of biological nitrification. Inhibitory effects of moderately toxic substances such as complexed cyanides, thiocyanates and methylated phenolics can be compensated for by increasing the sludge age.

Ganczarczyk (1979) evaluated a two-stage activated sludge system with second-stage nitrification for the treatment of coking plant wastewater. Dephenolization was achieved in the first-stage, and despite significant fluctuations of ammonia, thiocyanate, and phenol concentrations, second-stage nitrification was achieved. Both stages were operated with long hydraulic residence times and relatively high mixed liquor suspended solids concentrations.

An additional organic nitrification inhibitor that has been investigated is aniline -- a compound that is known to severely interfere with nitrification. Joel and Grady (1977) found that nitrification occurred in a completely - mixed activated sludge system of 7-day solids residence time that was fully acclimated to aniline. When plug flow reactor conditions were simulated, however, nitrification did not occur for the same SRT, although the aniline substrate underwent degradation to less-than-toxic levels.

Panzer (1982) compared a conventional extended aeration nitrification system with a combined nitrification/denitrification (N/D) extended aeration process for the secondary treatment of cattlehide tannery wastewaters. Both systems produced good COD and TKN removals (> 90%), but the N/D process was able to handle a 40 percent greater COD loading rate while producing 28 percent less sludge.

Hill and Gelman (1978) evaluated the effect of saline wastewater, as may be generated by certain food and cheese processing operations, on nitrification. They used a bench-scale activated sludge process with second-stage nitrification. Chloride concentrations of 10,000 and 20,000 mg/l as Cl^- reduced the rate of nitrification by 70 percent or more at temperatures of 15°, 25°, and 35°C. The three previous citations indicated that the maintenance of high mixed liquor suspended solids concentrations can lessen the impact of those contaminants.

Process comparisons -- Sampayo and Hollopeter (1979) reported on the bench- and pilot-scale testing that was conducted for process selection for the treatment of a combined domestic and industrial waste. Several alternatives (conventional activated sludge, sludge reaeration, and rotating biological contactors) were eliminated from consideration because of insufficient and/or inconsistent nitrification. The absence of nitrification with these alternatives was felt to be due to the presence of some unidentified, volatile organic contaminant. A two-stage, high-purity-oxygen activated sludge system with second-stage nitrification similarly failed to provide ammonia oxidation. An intermediate stripping-stage was added prior to the second-stage so that air-stripping of the suspected volatile contaminant could be achieved. Some nitrification was achieved when dissolved oxygen levels of 6 mg/l or greater were maintained within the stripping stage. Additional improvement in process performance was achieved when the second-stage, high-purity-oxygen process was replaced by an open reactor operated at a minimum dissolved oxygen level of 5 mg/l. Although consistent and acceptable nitrification was not realized, carbonaceous BOD removals occurred. The final process evaluated was a single-stage, powdered-carbon, activated sludge process followed by filtration. With this process, consistent ammonia reductions of 70 percent and greater were achieved. The results of the field testing did indicate the presence of some volatile nitrification inhibitor. An on-site gas analysis revealed 6 chlorinated hydrocarbons, of which 1, 2-dichloroethane was the major contaminant, at a concentration of two milligrams per cubic meter.

Phosphorus Removal —
As previously stated, the recent literature contains little information on interference in biological phosphorus removal. Yall and Sinclair (1971) found that 10^{-3}M concentrations of 2,4-dinitrophenol resulted in anywhere from 20 percent to 80 percent inhibition of phosphorus uptake by activated sludge. At 10^{-2}M, the inhibition was 100 percent. A literature review was conducted by Elliot, et al. (1978) in which proposed mechanisms of biological phosphorus removal in the activated sludge process were examined. The conclusions were in the form of recommendations for maximizing phosphorus removal and as such, they may be useful in handling an interference to wastewater treatment that results from high phosphorus loadings. The recommendations were the following:

- Prevent anaerobic conditions in the secondary clarifier.
- Separately treat nutrient rich return flows, supernatants, and filtrates.
- Maintain a minimum mixed liquor pH of 6.

Physical/Chemical Processes

The physical/chemical treatment processes of adsorption, flotation, precipitation, ion exchange, filtration, and oxidation can be used alone or in various combinations as tertiary treatment following secondary biological treatment, or as processes for the selective removal of organics or metals. The emphasis of the recent literature has been on the treatability and removal of wastewater contaminants by various physical/chemical processes. In this section, the removals of organic and inorganic pollutants by physical/chemical processes will be presented followed by an examination of the performance of several advanced wastewater treatment processes for the simultaneous removal of organic and inorganic pollutants.

Organic Pollutants —
The use of activated carbon for the removal of organic pollutants has been widely investigated. El-Dib and Badaway (1979), Wood, et al. (1983), and McManus, et al. (1985) demonstrated the effectiveness of granular activated carbon for the removal of a wide variety of organic pollutants. The operational

variables of surface loading rate and pH can have significant effects on adsorption efficiency (Wood, et al., 1983). Additionally, the presence of multiple pollutants can reduce the efficiency of the process, the desorption of pollutants from the carbon can be significant, and activated carbon adsorption appears selective for organics over metals (McManus, et al., 1985).

Ion exchange can be used as an alternative to activated carbon for the removal of organics. Fox (1979) surveyed the available literature and concluded that ion exchange resins could perform better than activated carbon. Concentration reductions of chlorinated pesticides, phenols, and other compounds of three orders-of-magnitude and greater were reported.

Beds of ion exchange resins are subject to fouling due to a number of causes which result in decreased capacity, reduced effluent quality, and operational difficulties (Pelosi and McCarthy, 1982a, 1982b). Cationic resins are subject to fouling by numerous cations. For example, iron, manganese and copper fouling can result in channelling, decreased capacity and disruption of cross-linkages within the resin -- problems that can be corrected through washing and chemical treatment. Both cationic and anionic resins are subject to fouling because of the accumulation of oil and grease, silt and clay, and microorganisms.

Organics removals can also be achieved using membrane filtration processes. Phenolic compound removals by bench-scale membrane processes were demonstrated by Klemetson and Scharbow (1979) and by Bhattacharyya, et al. (1983). Important variables that must be considered include the membrane characteristics and pH. Hrubec, et al. (1983) found that reverse osmosis followed by activated carbon adsorption was effective in the removal of micropollutants from treated municipal wastewater. Removals of trihalomethanes (THM's) and chlorinated solvents were not, however, obtained.

The chemical oxidation of organic pollutants can be a feasible way of removing those pollutants from the process stream. Rice (1981) described the known aqueous reactions of ozone with both organic and inorganic hazardous chemicals. Ozone oxidation is dependent upon conditions of pH and the specific compounds undergoing oxidation. For example, unsaturated aliphatic compounds generally are readily reactive with ozone, while saturated aliphatic hydrocarbons are unreactive toward ozone. The formation of intermediate oxidation products must also be considered -- some of which may be hazardous. Ozonation of a wastewater can also influence the biodegradation of organic constituents. Medley and Stover (1983) examined the effects of ozone on the biodegradability of three organic compounds with varying results. Acrylonitrile was not oxidized by ozonation and adverse effects on biodegradation were apparent. 1,2-dichloropropane was readily oxidized with small ozone doses and although not oxidized, the biodegradability of 2,4-dinitrophenol was increased.

Oxidation of organics by ozone may be promoted in the presence of ultraviolet (UV) radiation (Nall, 1980). Termed photo-ozonation, this process was found to result in higher reaction rates than for the oxidation of phenols with ozone (Otake, et al., 1979). Organic compound oxidation can be conducted in other ways as well. DeLuca, et al. (1983) experimentally investigated the oxidation and removal of toxic organics using a two-step process of oxidation with potassium ferrate followed by alum coagulation. One-hundred-percent removals of trichloroethylene and napthalene were acheived while other results showed little process effects. Other workers examined the oxidation of phenols with hydrogen peroxide in the presence of an iron catalyst (Butler and Nandam, 1981).

McCarty and Reinhard (1980) investigated the removal of 25 trace organics by a full-scale, advanced wastewater treatment (AWT) plant (Water Factory 21, Orange County, California). The treatment scheme consisted of lime clarification, ammonia stripping, recarbonation, filtration, activated carbon adsorption, and reverse osmosis. Treatment plant records spanning three years were examined in detail and very high organics removal efficiencies had been achieved. The data also revealed a significant improvement in organics removal upon the near simultaneous change from trickling filters to activated sludge preceding the AWT process and a reduction in the industrial component of the wastewater. Although the separate effects of each change could not be ascertained, they were both felt to be of possible significance.

Inorganic Pollutants --
As with organic pollutants, several physical and chemical processes can be used for the removal of metals and other inorganic pollutants from wastewater. Metals can be readily precipitated from water as described by Jenke, et al. (1983). Upon the addition of lime (or other strong base), metals can precipitate out of solution in the form of metal hydroxides which can then be removed by conventional solid-liquid separation processes. Kim and Amodeo (1983), however, reported on an instance in which the presence of chelating agents or synthetic fat in the waste interfered with hydroxide-precipitation treatment of an industrial wastewater. The interference was reduced by the precipitation of heavy metal sulfides following the addition of calcium sulfide.

Christensen and Delwiche (1982) reported a pilot-scale investigation on the removal of metals by hydroxide precipitation, flocculation, and ultrafiltration. In this instance, the presence of cyanide interfered with hydroxide precipitation because of an increase in copper solubility resulting from the formation of copper-cyanide complexes. It was felt that the interferences from cyanide would be reduced at higher process temperatures due to reduced stability of cyanide complexes, or upon chemical destruction of the cyanide by chlorination.

Other processes have been proposed for the removal of metals which involve dissolved air flotation. Chavalitnitikul and Brunker (1981) conducted bench-scale testing of foam flotation. With this process, soluble metals combine with added surfactants and are removed with the float. Percent removals of 60 to 75 for several metals including cadmium, chromium, and lead were achieved. A similar process termed colloid foam flotation was investigated by McIntyre, et al. (1983). In this process, soluble metals adsorb onto the surface of inorganic coagulant flocs and can then be removed with the float. Good removals of copper, chromium, and zinc were achieved and suggested design and operational data were presented. Benjamin and Leckie (1981) investigated the mechanism of heavy metal removal by adsorption onto amorphous iron oxyhydroxide which, while serving as the basis for the flotation process investigated by McIntyre, et al. (1983), would also serve as a basis for metals removal by conventional coagulation/sedimentation/filtration processes.

The removal of metals from wastewater can also be accomplished with ion exchange or membrane filtration processes. Huang and Wirth (1981) demonstrated the removal of cadmium to concentrations less than 0.1 mg/l using an aluminosilicate ion exchange material. Although a nonmetal, cyanide is also subject to removal by the ion exchange process (Chanda, et al., 1983). Bhattacharyya, et al. (1979) demonstrated the feasibility of heavy metals, sulfate, and suspended solids removal from a scrubber blow-down water using ultrafiltration.

Beckwith, et al. (1983) investigated the removal of particulate mercury in three activated sludge POTWs. They concluded that the addition of aluminum sulfate or sodium aluminate had no measurable effect on the fate of mercury within the process. Chen, et al. (1984) investigated the removal of organic and inorganic forms of mercury by a sulfide precipitation/activated carbon/neutralization process. Reductions from greater than 100 mg/l to less than 1 mg/l were realized, and it was felt by the authors that the process was superior to precipitation by either carbonate or sodium borohydride addition.

As noted by Rice (1981), metals of lower oxidation states are readily oxidized to their highest state and inorganic anions (e.g. cyanide, sulfide, and thiocyanate), can be destroyed by ozonation and converted to carbon dioxide, sulfate, and nitrogen. Bench-and pilot-scale investigations conducted by Sukes, et al. (1984) demonstrated cyanide destruction by ozone oxidation. As a result of their work, they formulated a lengthy list of conclusions regarding the process and a set of design criteria were established.

The destruction of cyanide by a copper-catalyzed, sulfur dioxide/air oxidation process was demonstrated on a laboratory-scale by Nutt and Zaidi (1984). The process was proposed for the treatment of cyanide-containing wastewaters from gold mills and metal finishing plants.

Miscellaneous Processes —
Osantowski and Hendricks (1982) investigated the treatment of coke plant wastewater using several advanced processes. The most efficient process examined consisted of a two-stage activated sludge process followed by activated carbon adsorption and granular-media filtration. Excellent volatile organics removals with incomplete metals removal were obtained. Zwikl, et al. (1982) investigated the treatment of a similar wastewater using activated carbon followed by alkaline chlorination. Effective removals of organic and inorganic contaminants, except phenol, were achieved.

Vuceta, et al., (1979) examined the removals of inorganic and oily waste contaminants by various physical and chemical processes including the chemical destruction of cyanide and reduction of chromium, followed by precipitation, flocculation and sedimentation, and filtration. The bench-scale tests demonstrated that the various pollutants could be effectively removed.

Rooney and Wu (1982) reported on the performance of a full-scale tertiary treatment facility for the joint treatment of municipal and meat-packing wastewaters. After pretreatment in anaerobic lagoons, the industrial wastewater was combined with domestic waste to further biological treatment in an aerated pond. The tertiary treatment step consisted of coagulation, flocculation, clarification, filtration, and chlorination. With the exception of ammonia, final effluent standards were met. Although recommendations for further ammonia reductions were presented (increase detention times of ponds and/or breakpoint chlorination), supporting data were not made available.

SLUDGE PROCESSING

Interferences in sludge processing operations at POTW facilities can be of a primary or secondary nature. Primary interferences are those in which a contaminant directly inhibits or otherwise decreases the efficiency of a sludge processing operation. Secondary interferences are those interferences in sludge processing which are the result of a change in the characteristics of the sludge which accompanies an interference of a preceding wastewater treatment process.

For purposes of this literature review, sludge processing has been defined to include biological and physical-chemical sludge treatment operations and the ultimate disposal of the processed waste sludge solids.

Sludge Treatment

Digestion--
The focus of recent literature on the inhibition of biological digestion processes has been on anaerobic digestion. This perhaps is the result of the known, sensitive nature of the process. It would, however, be logical to conclude that interferences to aerobic digestion can occur. Additionally, the interferences may be similar to those experienced by the activated sludge process with the exceptions that the typically long solids residence times and high biological solids concentrations of aerobic digestion may serve to reduce the effects of some contaminants.

Metals -- Interferences of the anaerobic digestion process can be caused by the presence of metals and other inorganic contaminants. However, noted by Barth and Bunch (1979), metal toxicity in full-scale POTWs may be primarily related to shock loading situations, since the continuous sulfide precipitation of heavy metals within a digestor can preclude the adverse effects of the metals.

The toxicities of zinc, cadmium, copper, iron and chromium to bench-scale anaerobic digestion have been investigated (Mosey and Hughes, 1975; Matsumoto and Noike, 1978; Parkin and Miller, 1982). Suggested "safe" levels of metal ions for anaerobic digestion of sludge have also been presented (Barth and Bunch, 1979). But, it is apparent that the effect of a given concentration of a metal toxicant will be a function of the temperature and solids residence times of the digestor (Parkin and Miller, 1982) and of the conditions within the digestor with respect to pH and the sulfide and carbonate concentrations (Mosey and Hughes, 1975). The latter chemical conditions control the solubilities of the various metals and their possible precipitation and removal from the supernatant.

Henney, et al. (1980) documented the effect of an accidental spill of chromium on the performance of a municipal wastewater treatment plant. Forty-one kilograms of chromium (VI) gradually entered the anaerobic digestors with the maximum concentration reaching 17.4 mg/l. There was, however, no noticeable effect on the digestion process.

Kobylinski and Bell (1983) reported that the light metals calcium and sodium could inhibit volatile solids destruction at relatively high metal concentrations. Ammonia inhibition of anaerobic digestion of poultry manure has also been documented (Ripley, et al., 1984). In this case, very long acclimation periods resulted in a biomass that was tolerant to high ammonia concentrations.

Organics -- Chou, et al. (1978) evaluated the effects of 52 petrochemical compounds on a bench-scale, anaerobic fermentation process. The testing was done using acclimated microbial populations and inhibition or toxicity was indicated by reduced methane production. Relationships between the chemical structure of the contaminants and their toxicities were examined. Aldehydes, for example, were found to be toxic and the toxicity varied with the reciprocal of the carbon chain length. Upon acclimation of the system, the aldehyde toxicity was reduced. The presence of amino, hydroxyl, and carboxyl functional groups in various compounds were found to result in a decrease in the observed toxicity.

Parkin and Miller (1982) evaluated the response of bench-scale methane fermentation to the continuous addition of toxic compounds. The effects of process temperature and solids residence time were also evaluated. For inhibition by chloroform, formaldehyde, and hydrazine, the longer solids residence times (SRT) of 25 to 50 days tended to lessen the severity of the inhibition through the acclimation of the bacteria to the toxicant. A process temperature of 35°C also appeared to optimize the acclimation process.

Parkin and Speece (1982) reported the results of a similar bench-scale investigation of the effects of shock loading of various toxicants, including formaldehyde, on methane fermentation. Temporary reductions in methane production were observed. It was found that by increasing the SRT of the system, the recovery process was promoted. Also, models for predicting the rate of recovery were presented.

The inhibition of anaerobic digestion by 24 organic pollutants was investigated by Johnson and Young (1983). Of those compounds exhibiting toxicity at concentrations of 100 mg/l (50 percent or more reduction in gas production), ring-structure compounds containing nitro-groups were distinguished by the reversal of inhibition following the biological reduction of the nitro-groups to amines. Hexachloroethane was also degraded, at least in part, as evidenced by apparent dechlorination of this compound. Other compounds were removed from solution not by degradation, but by adsorption onto the biological solids. As demonstrated by other investigators, inhibitory effects were again found to be lessened by biological acclimation and longer SRTs. It was also hypothesized, but not substantiated, that inhibitory effects should vary directly with the hydraulic residence time of the system.

Choate, et al., (1982) reported on the less-than-expected capacity of an anaerobic digestor used to pretreat a high-strength, soluble starch waste prior to discharge to a municipal treatment system. The diminished capacity of the digestor was, however, the result of a loss of solids with the supernatant. By retaining the solids using ultrafiltration, an acceptable effluent was achieved.

The bulk of the literature has dealt with toxic effects, and not treatibility of a contaminant during anaerobic digestion. An exception to this is the work of Moore and Barth (1976) who found that a salt of nitrilotriacetic acid (NTA) was biodegradable and did not inhibit digestion. This could be significant in view of the use of NTA as a complexing agent for minimizing the effects of metals in biological treatment (Braam and Klapwijk, 1981).

Dewatering and Thickening--
The review of the recent literature yielded no published work on the interference of sludge dewatering and thickening operations by contaminants. Interferences of a secondary nature can be inferred. For example, Gerhardi (1981) reported on an interference of an activated sludge process that was caused by a food manufacturing waste. The overall result was the production of biological solids of poor settleability and such a change in the sludge characteristics could be a secondary interference that would affect the handling, thickening, and dewatering characteristics of the waste activated sludge.

Disposal

The primary interference to sludge disposal which results from pollutants entering the POTW process stream is contamination of the sludge which may preclude lower cost sludge disposal options. JRB Associates (1981b) reported that of the POTWs with known sludge contamination, the sludge disposal method was affected in two-thirds of those facilities.

The presence of metallic and organic contaminants in waste sludges has been documented by Bailey and Zimomra (1981), Clevenger, et al. (1983), Fricke, et al. (1985), and Strachan, et al. (1983). The occurance of heavy metal contaminants has been primarily related to industrial sources (Coker and Mathews, 1983; and Kock, et al.,1982).

Numerous studies have been conducted regarding the possible effects of land disposal of contaminated sludges. (Naylor and Loehr, 1982a, 1982b; Beckett and Davis, 1982; Coker and Matthews, 1983; Chang, et al., 1981; Weber, et al., 1983; Baxter, et al., 1983; Jeffus, 1981; and Theis and Podgett, 1983). These studies have considered such factors as crop uptake of pollutants, crop damage, and leaching phenomena.

McIntyre, et al. (1981a) found that the distribution of PCBs and organochlorine insecticides between the solid and liquid phase in sludge was not altered by common chemical conditioning and dewatering operations. Thus, the uptake of pollutants by biological adsorption could result in a sludge disposal problem.

A possible remedial measure for the removal of metals from sludge is acid treatment to lower the pH and subsequent leaching of the metal contaminants (Woyniak and Huang, 1982; Jenkins, et al., 1981; and Button and Gaudy, 1982). Cornwell and Westerhoff (1980) demonstrated that metals could be removed from sludge using a liquid ion exchange solvent. The tests were, however, of a preliminary nature and additional work would be required before such a technique could be fully evaluated. Metals removal by chelation has also been examined by Jenkins, et al. (1981), but they concluded that neither method of metals removal and recovery would be economically attractive. However, such methods may be necessary when no other disposal option exists.

Werthman, et al. (1982) presented a case study in which several alternatives for the disposal of a contaminated sludge were evaluated. Included in the evaluation were co-disposal with municipal refuse, separate landfilling, incineration, and a re-evaluation of sludge dewatering operations in light of the increased disposal costs resulting from the "hazardous" classification of the sludge.

SUMMARY

The literature concerning process upsets and interferences at municipal treatment facilities is not equally distributed among the various process operations. Clearly, the activated sludge processes have received the bulk of the attention by researchers in bench, pilot and full-scale studies. Activated sludge treatment rests on a firm theoretical foundation, is widely used in POTWs and is relatively easy to model using pilot-scale reactors -- so it is not surprising that its' database regarding interferences is more substantial than other processes. In general, the quantity of literature available on interferences decreased from suspended growth biological systems (including anaerobic degradation) to fixed film processes to physical-chemical treatment. In fact, the use of physical-chemical advanced treatment in POTWs is so infrequent that industrial wastewater treatment literature must be consulted to properly review these processes.

The above literature review investigated the treatability and impact of conventional and toxic pollutants on a process-by-process basis. However, several design and operations concepts permeate the literature across arbitrary process distinctions.

Acclimation

The most important concept in the resistance of biological treatment processes to upsets caused by priority organics (and to a lesser extent metals) is the acclimation of the microbial system to the contaminant. Suspended growth and fixed film systems alike have demonstrated a two to twenty-fold increase in tolerance of shock loads when switching from unacclimated to acclimated treatment. Some evidence suggests that acclimation to a particular compound may provide a measure of resistance to similar compounds. Biological systems in POTWs will acclimate to industrial organic wastes that are discharged at a consistent level. It is therefore the occasional or highly variable industrial discharges that create toxic inhibition problems, which must be handled by at-the-source pretreatment or flow equalization.

Staged Treatment

Staged treatment can be defined as the use of separate biological systems linked together in series for the removal of soluble organics and metals, and possibly nitrification. A common example is the use of biotowers (roughing filters) preceding either activated sludge or RBC treatment. Such sytems have been shown to be more resistant to contaminant upset and not as susceptible to classical problems such as filamentous bacteria in an aeration basin or Begiattoa in an RBC biofilm. The use of step-feed or step-aeration activated sludge represents a compromise between single-stage and two-stage operation. Staged treatment also helps protect the sensitive autotrophic organisms responsible for nitrification, as contaminant shock loads are mediated by the first stage treatment system.

Operations

The literature has consistently demonstrated a number of operational tools which may prove valuable in mitigating the effects of contaminants. Most of these tools are available to operators of activated sludge plants, as this is where the research has been focused and such facilities generally provide the most flexibility. Maintaining high mean cell residence times (for both secondary treatment and digestion) and high mixed liquor suspended solids concentrations helps to resist process upsets. Increasing the SRT and MLSS at the first indication of a toxic dump will also help weather the storm. The addition of powdered activated carbon to activated sludge aeration basins has been shown to improve priority organic treatability and reduce the potential for process inhibition. Adjustment of the wastewater pH to provide a more suitable environment for microorganisms or to help control metal solubility is a very important operational parameter.

Influent Wastewater Characteristics

The concentration and form of metal or organic contaminants can play a significant role in downstream processing. The effects of metals on POTW operation is largely dependent on whether they exist in a soluble or insoluble form when they enter the plant. The metals-to-microorganism ratio has been shown by some authors to be just as important to the fate of metals as F/M is to organic degradation.

Obviously the higher the influent concentrations of pollutants, the more detrimental to treatment plant operations. However, the literature indicates no consistent relationship between influent concentration of organics and effluent concentration, treatability or toxicity. Some authors have indicated consistency in particular classes of compounds, but not among organics with widely varying properties. Pesticides and metals have demonstrated the least consistent results of the contaminants considered by this review. The literature offers widely varying conclusions concerning the treatability and impact of these compounds by wastewater treatment systems.

In general, metals either enter POTWs at concentrations near their inhibition levels, or concentrate in sludges to critical values. Priority organics, on the other hand, enter at non-toxic concentrations and then are biodegraded by acclimated aerobic or anaerobic systems. Hence, metals contamination is generally more troublesome to POTWs than organics. Exceptions to this rule would be the presence of non-biodegradable organics or large diurnal fluctuations of BOD or TSS not mediated by equalization. It is these exceptions which provide insight into why specific industries, such as textile producers or certain food processors, are undersirable indirect dischargers who should install pretreatment and/or equalization facilities.

4. Telephone Survey

A nationwide telephone survey was conducted to identify occurrences of POTW interference and investigate the circumstances surrounding such events. While the literature documented a few cases of POTW interferences, most cases resulting from industrial waste discharges were identified through contacts with regional and state water quality officials. This section describes the telephone survey procedure and approach, follows with a summary of general survey results, and closes with discussions concerning each state, grouped by geographical region.

PROCEDURE

Initial telephone contacts were made at the state level. Personnel in state water quality agencies were interviewed, using the form in Appendix B as a general outline. In many instances it was necessary to interview more than one state official in order to obtain all the required information. Typically, discussions were conducted with personnel in both "permits and compliance" and "pretreatment". EPA has yet to delegate authority to some states for NPDES permit writing, enforcement or pretreatment programs, and in such cases, state and regional EPA officials were also contacted and interviewed following the same outline used for state officials. The purpose of these interviews was two-fold:

- to gain an understanding of pretreatment programs and interferences on a general, state-wide basis; and
- to identify specific POTWs experiencing interferences.

Over 90 state and regional wastewater officials were contacted in all.

Information explaining the circumstances surrounding an identified POTW interference was obtained from state personnel. In all instances, when sites were identified as interference cases, the name of a plant operations or management contact was obtained.

Follow-up telephone contact was made with POTWs identified by state personnel as experiencing interferences. Interviews were conducted with POTW personnel, which generally allowed for POTW characterization into one of three categories:

1. POTWs which are currently violating their NPDES Permit or are experiencing sludge contamination as a result of industrial discharges, and are working to correct the problems;

2. POTWs which had previously violated their NPDES Permit or had experienced sludge contamination as a result of industrial discharges, and have mitigated the problems to the point of consistent permit compliance; and

3. POTWs with past or present NPDES Permit violations or sludge contamination, not principally due to industrial wastes.

For POTWs belonging in category 1 or 2 above, additional data or documents to further describe the interference and the resulting mitigation efforts were requested. Documentation typically consisted of engineering reports, operating data or correspondence files.

In addition to locating POTW interferences through state contacts, the 25 largest cities in the United States were contacted independently to identify any industrial related treatment plant operations problems. The premise of such direct contact was that large metropolitan areas potentially treat wastewaters with substantial industrial contributions, however, very few POTW interferences were identified through this approach. Interviews were conducted using the form presented in Appendix B, and when interference occurrences were identified, additional information was obtained. Similarly, in October, 1985, the Association of Metropolitan Sewerage Agencies (AMSA) published an article in the AMSA Monthly Report, at the request of the EPA-Office of Water Enforcement and Permits, asking member cities to contact EPA if they have experienced POTW operational problems associated with industrial discharges. Although there was limited response to the article, several AMSA members were indentified and contacted.

In all, over 160 municipalities or sanitation districts were identified and contacted concerning POTW interference.

GENERAL SURVEY RESULTS

The findings of the telephone survey support several general conclusions about POTW interference:

- The major causes of interference can be ranked in decreasing order as highly variable discharge of conventional pollutants (BOD, SS, NH_3), followed by metals and then toxic organics.

- Food processing, textile, and chemical/pharmaceutical industries are most commonly responsible for treatment plant upsets.

- Treatment plants most adversely affected are small in size ($\leq$5 mgd) with one or two major industrial contributors.

- The most successful interference mitigation efforts have occurred when POTWs and industry have cooperatively worked to solve problems.

- Some POTW interference problems result from poor wastewater characterization prior to facility design.

Despite identification of numerous problems relating to toxic pollutants, the majority of the POTW interferences were associated with highly variable discharges of conventional pollutants. In some cases, flow equalization by industry or by the municipality has alleviated or substantially reduced the problems relating to BOD, SS, pH and NH_3. Interferences associated with metals or toxic organics usually result from "shock loadings" as well, giving the biological treatment system insufficient time to acclimate.

The industries identified with POTW interferences were placed into one of six categories. Since this project is not limited to interferences created by "categorical industries", the industrial descriptions shown below are not necessarily consistent with the EPA Standard Industrial Codes. The percentage of plants surveyed which identified industries in these categories as interference contributors are:

Metal plating/finishing and electronics	39%
Food and meat processing	38%
Chemicals, pharmaceuticals and dyes	27%
Textiles	12%
Wood and paper products	8%
Leather Tanning	7%

Metal plating/finishing and electronics industries generally cause interference with sludge disposal due to biomass metals partitioning, although they can also cause treatment upsets. The remaining industries primarily contribute large loadings of conventional pollutants. The types of industries causing interferences were consistent with those identified through the literature survey.

The telephone survey revealed that wastewater treatment plants experiencing interferences are generally less than 10 mgd in size (average dry weather design flow), with those smaller than 5 mgd being most severely impacted. This fact can be attributed to two factors. First, small to medium size treatment plants are more abundant than large ones. Secondly, dilution is an important factor in larger plants, as domestic/industrial wastewater flow ratios tend to be higher. Industrial waste streams which would impact small to medium sized plants are treatable when diluted with large amounts of domestic wastewater in large metropolitan plants, some in excess of 1 bgd. Most of the small to medium sized plants experiencing interferences fall into two general categories: plants of less than 5 mgd with a few key, identifiable industrial dischargers, and medium sized plants (5-10 mgd) with several industrial dischargers.

State water quality representatives and local treatment plant personnel frequently stressed that cooperation between industries and municipalities was an important element in eliminating POTW interferences and implementing effective pretreatment programs. A simple formula or standard plan for achieving this end does not exist, but the first step towards this goal must necessarily come from the municipality or sanitation district. Industries may not realize that practices they employ are detrimental to the municipal treatment plant's operation. Through heightened awareness, municipalities can cause industries to employ minor housecleaning or process modification steps to the benefit of both the industry (reduced sewer charges or increased efficiency) and the POTW (mitigated interference problems).

Finally, some POTW problems are the result of poor wastewater characterization by design engineers, resulting in treatment plants which experience interference problems shortly after start-up. The problems may be the result of organic or hydraulic overloads, improper plant equalization/pretreatment or a lack of process variability and control. Plant personnel occasionally overstate design deficiencies in an attempt to explain away problems created by operation and maintenance inadequacies, but the problem appears to exist at a level deserving attention.

REGIONAL RESULTS

For the purpose of this study, the country was divided into the five geographical regions shown in Figure 1. Discussions for each region and state follow.

Northeast

Consistent with its reputation as an industrialized area, the greatest number of POTW interferences were reported for the Northeast. In addition, this region is the most diverse in terms of POTW interference as measured by plant size, secondary treatment process and industrial contributors. All of the six major categories of industrial users are represented as causing some POTW interference, with chemical/pharmaceutical industries topping the list.

Connecticut, Delaware, Maine, Massachusetts, New Hampshire and Vermont --
POTW interference in these states is primarily confined to small facilities (< 5mgd) with a single, significant industrial discharger. The industries include metal plating, leather tanning, textiles, and food and fish processing. In some cases pretreatment programs have yet to be implemented and enforced while in other cases industries continue to violate pretreatment limits, occasionally resulting in litigation. Extended aeration activated sludge and aerated lagoon treatment systems are common biological processes at the POTWs identified.

New Jersey --
Chemical industries, including pharmaceutical and paint manufacturers, are the principal cause of POTW interference in New Jersey, affecting both fixed film and suspended growth treatment facilities. A secondary plant with a treatment capacity of 300 mgd experiences interference due to paint manufacturing waste, and was the largest site identified through all telephone contacts.

New York --
New York experiences POTW upsets and interferences primarily due to industries discharging highly variable quantities of conventional pollutants. Dairy, paper products and pharmaceutical manufacturing wastes are among the most difficult to treat, due to the changing nature of the waste products. Industry water conservation and process modifications have been successful in dealing with some interference problems.

Pennsylvania --
All identified POTWs experiencing interferences in Pennsylvania are 6 mgd or smaller in size. In most cases, a single significant industry discharges to the municipal system, some causing chronic POTW effluent discharge limit violations

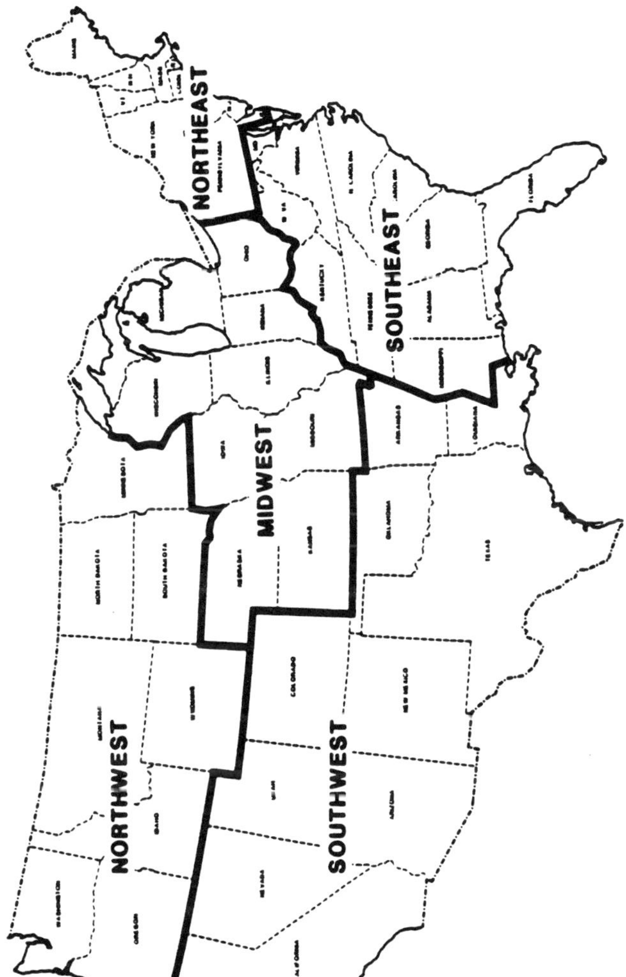

FIGURE 1
GEOGRAPHICAL REGIONS

and others causing periodic treatment plant upsets. As is typical of the region, a wide variety of pollutants and industrial sources are responsible for interferences.

Southeast

Like the Northeast region, the states of the Southeast region reported POTW interferences being caused by a wide variety of industries. The textile, wood and paper products, and leather tanning industries were mentioned as interference causes nearly as often as the metal plating and food and meat processing industries. No facility smaller than 1 mgd was identified as experiencing interference.

Alabama and Mississippi --
Isolated POTW interferences are experienced in Alabama and Mississippi as a result of food processing and timber products industries. Highly variable discharges of conventional pollutants make for difficult treatment situations; chicken processing is the major "problem" industry. In Mississippi, most local pretreatment programs are administered and enforced by the state, rather than by municipalities. State officials report this arrangement to be working satisfactorily.

Florida --
NPDES Permit violations as a result of industrial discharges are infrequent and difficult to identify according to regional U.S. EPA officials. The U.S. EPA retains control over NPDES Permits and pretreatment programs in Florida, while the state enforces water quality standards. Since the great majority of municipalities requiring pretreatment have existing programs, interferences are not common. Problems are typically the result of toxic metal "dumps" or sludge disposal limitations caused by metals accumulation.

Georgia, North Carolina, South Carolina --
The textile industry is a major economic force in these three states and at the same time causes some POTW interferences. Textile mill operations often involve batch processes that result in large, fluctuating wastewater flows at the POTW if pretreatment/equalization is not employed. Flow equalization is sometimes the only step necessary to keep POTWs from experiencing interferences. Metal plating and food processing industries were also reported as causing interferences.

Kentucky and Tennessee --
Leather tanning, metal plating and steel industries create the majority of the POTW interferences in Kentucky and Tennessee. The reported problems occur when the industries are located in small towns with POTW treatment capacities less than 10 mgd. In addition to BOD, SS and metal loadings, tanning operations can cause colored water which presents a pass-through problem. Pretreatment program implementation has significantly curtailed the occurrence of interferences. Better monitoring efforts at POTWs and in collection systems has enabled wastewater officials to mitigate the effects of occasional toxic spills and determine the source of contaminants.

Maryland --
Typical of many states, Maryland possesses myriad industries, some causing POTW interference. In contrast to other states however, two large plants (greater than 60 mgd) in a major metropolitan area experience interferences because of metal plating, chemical and paint manufacturing industries. Smaller facilities report fats, oils and grease (FOG) and BOD associated problems.

Virginia and West Virginia --
Very few POTW interferences were reported for these two states, with high conventional pollutant concentrations creating the biggest industrial wastewater problem.

Midwest

The Midwest region is composed of many states with a mixture of agricultural (food and meat processing) industries and more traditional heavy industries. As such, interferences experienced at POTWs are diverse. POTW operation upsets are not widely occurring due to industrial contaminants, but are still of considerable concern to most state water quality agencies.

Illinois --
Industrial interference problems in Illinois were portrayed by state officials as being difficult to identify. It was stated that non-compliance could be directly attributed to industrial discharges in only a few cases. In these instances, both conventionals and metals were involved as the interfering pollutant, primarily originating from food processing and electroplating industries, respectively. In contrast to other states, most facilities identified as experiencing interferences were medium to large in size ([10 mgd).

Indiana --
Indiana is presently experiencing most of its industrial related POTW upsets in plants with] 2 mgd hydraulic capacity. Small facilities have difficulties meeting effluent ammonia limits (particularly in trickling filter plants) and high conventional pollutant loadings on the part of meat and food processors can periodically inhibit plants to the point of non-compliance. Two cases of interference problems in larger plants were identified as sludge contamination by metals, eliminating land application as a disposal option.

Iowa and Nebraska --
Neither Iowa nor Nebraska are heavily industrialized states, with most existing industry being confined to several of the larger cities. Isolated industrial interferences exist, primarily due to meat and food processing or small electroplating facilities. In some small towns, meat and food processing industries contribute as much as 50 percent of the total wastewater flow and strength. Municipality/industry cooperation and pretreatment programs have corrected and/or controlled most of the potentially serious situations.

Kansas --
Typical of many Midwestern and Southwestern states, greater than 70 percent of municipal wastewater facilities in the state are lagoon/pond systems, often with no direct discharge. Consequently, the permit compliance rate is high and state

water quality officials indicate very few POTW interferences. Municipalities with past interference problems have established pretreatment programs and now maintain properly operating facilities.

Michigan --
State wastewater officials indicated that few POTW interferences are presently occurring in the state. Metal and paper industries have caused operational problems in the past, but long established pretreatment programs have virtually eliminated POTW interferences. In at least one instance, a municipality has not limited industrial pollutant concentrations through a pretreatment program, but instead has opted to adjust it's facilities to accommodate all industrial wastewater, with considerable success.

Missouri --
Missouri is typical of other Midwestern states, as it experiences a small number of conventional and metal-caused POTW interferences. Most interferences occur in small to medium sized towns with a single industrial contributor. Industries identified by state personnel as causing operational problems include poultry processing, cheese production and circuit board manufacturing. Many of the interferences are related to non-equalized hydraulic and organic loading.

Ohio --
Ohio has the largest number of municipal wastewater treatment facilities of any Midwestern state, yet reported fewer than six cases of plant process upsets due to industrial wastewaters, most of these being isolated. Process upsets result mainly from heavy, fluctuating conventional pollutant loadings. Occasional heavy metals contamination of sludge was reported, but metals did not interfere with POTW biological process operation. No POTW interference problems were reported for the larger, heavily industrialized cities of Ohio.

Wisconsin --
Like Ohio, Wisconsin has a large number of POTWs, but state officials report very few interference problems. Because of recent facility upgrades and new facility construction, Wisconsin wastewater treatment plant flows and organic loadings are generally under design capacities, possibly hindering identification of wastewaters that might otherwise cause interferences. Past problems have been connected with the state's dairy industry.

Southwest

POTW interferences in the Southwest region are primarily due to food and meat processing industries and occasionally are due to metal plating and electronics manufacturers. Metals and chemicals are not causing interferences in expanding high-technology areas, presumably because of sophisticated wastewater pretreatment and reuse facilities. Warm temperatures and ample space make non-discharging treatment facilities popular for small communities, thus reducing the likelihood of traditional interferences.

Arizona --
Arizona has fewer than 20 discharging NPDES wastewater treatment facilities in the entire state. Only two POTW interference cases were reported, one involving metal contamination in sludge, and another involving meat processing.

Arkansas --
Discharge permit non-compliance resulting from interferences occur at approximately 2.5 percent of the regulated POTWs. Most major problems are associated with the state's poultry industry, with processing plants often located in small towns. The state has experienced some difficulty in working with POTWs and industries to correct the problems.

California --
Despite its large population and supporting economic base, California does not experience many interferences at wastewater treatment facilities. Most interferences appear to be isolated events, associated with "midnight dumps" of toxic chemicals. Interferences caused by food processors may have been a problem in the past but appear to be corrected. Well implemented and enforced pretreatment programs prevent interferences in the larger metropolitan areas.

Colorado, Nevada, New Mexico, Oklahoma and Utah --
These five states have sparse populations and limited industrial bases. Oklahoma, New Mexico and Utah state personnel essentially reported no interference problems. Colorado and Nevada reported isolated interferences, principally associated with periodic metal toxicity. Media publicity of toxic upsets and improved pollutant detection equipment have helped limit the occurrences and mitigate the negative impacts of toxic pollutant dumps.

Hawaii --
Interferences were not reported by Hawaii state personnel. Interference from seasonal fruit processors may be a potential problem when primary facilities are upgraded to full secondary treatment.

Louisiana --
Interferences are not significant at POTWs. Most notable industries are direct dischargers, while others are encouraged to pretreat by stiff user surcharges.

Texas --
Texas reported a larger number of POTW interferences than any other state in the region. Food and meat processing was the most often cited cause of plant upsets. The discharge of these industries is typified by high BOD, SS and NH_3 concentrations in addition to heavy grease loadings. Cooperative efforts on the part of industry and treatment plant personnel have resulted in some remarkable turnarounds in plant performance and effluent discharge compliance.

Northwest

The Northwest region is principally composed of non-industrialized states, and consequently experiences fewer POTW interference events than do other regions.

Alaska, Idaho and Montana --
These three states reported very few POTW interference problems. None of the three states has a substantial industrial base and no food processors were reported as causing interferences. Some municipalities are developing pretreatment programs, not in response to present industrial dischargers, but in anticipation of future industrial development.

Minnesota --
State water quality personnel report a small number of POTWs experiencing interference problems, primarily due to highly variable discharges of conventional pollutants originating in the food processing and dairy industries. The state appears to keep close watch on non-complying facilities and is involved in litigation as a result of interfering industrial dischargers. Most problems are associated with industries that are significant contributors to small (] 5 mgd) treatment facilities.

North Dakota, South Dakota and Wyoming --
These three states experience very few POTW interference problems. A large majority of the facilities are lagoon treatment systems, some of which are non-discharging. Occasional problems are experienced in all three states with variable discharges of conventional pollutants from food or meat processing plants to small treatment facilities.

Oregon --
Oregon reported infrequent POTW interferences. State and regional personnel noted only one recent incidence of interference, where a small municipal facility (2 mgd) was periodically upset by toxic metals. An agreement between the city and the metal plater has reduced the likelihood of such occurrences in the future.

Washington --
The state experiences some problems with POTW interference, but has implemented pretreatment programs to prevent these occurrences in most instances. Interference potential exists in the central portion of the state where apple processing is prevalent. Many of the chemicals used in apple processing can cause toxic upsets and therefore cooperation between processors and treatment plant personnel is required to prevent such events. Occasional upsets occur when food processing waste volume or strength is suddenly increased at smaller treatment facilities. Many communities discharging to Pacific Coast or Puget Sound waters have primary treatment facilities which must be upgraded to secondary treatment in the near future. The increased susceptibility of biological treatment to upset could result in interference causation by some presently non-interfering waste streams.

5. Case Studies

A list of potential POTW interference case studies was developed as a result of the telephone survey discussed in Section 4. Potential case studies identified through the literature search (Section 3) were also followed up with phone calls, adding a small number of sites to the list. A total of 117 potential case studies were identified and they are presented in Figure 2 and in Tables 4 through 8 for the Northeast, Southeast, Midwest, Southwest and Northwest regions respectively. Forty-two (42) potential case studies were identified in the Northeast, the largest number of any region, while the smallest number of potential case studies, 5, were identified for the Northwest region. Thirty (30) potential cases were identified in the Southeast region, 21 in the Midwest region and 19 in the Southwest region. The design average flow, the secondary treatment process, the type of pollutants involved and the responsible industries are indicated in Tables 4 through 8 for each potential case study.

SELECTION PROCESS

The scope of this project prohibited making 117 case study site visits and so it was necessary to reduce the number of case study sites to a manageable level. The selection process involved several steps, the first of which was to evaluate each potential POTW considering the following:

- Has the interference problem been successfuly mitigated or has the discharge permit compliance rate substantially improved?
- Were treatment plant officials receptive to inquiries concerning this project and interested in providing further assistance?
- Has there been adequate documentation of the interference and its correction?

Affirmative answers to the first two questions were necessary for the POTW to be considered further as a case study site. Concerning the third question, the interests of this study are best served by documenting successful interference mitigation techniques not previously studied in this manner.

As a group, case study sites were chosen to reflect diversity in terms of:

- geographical location
- secondary treatment process
- treatment plant size
- industrial contributors

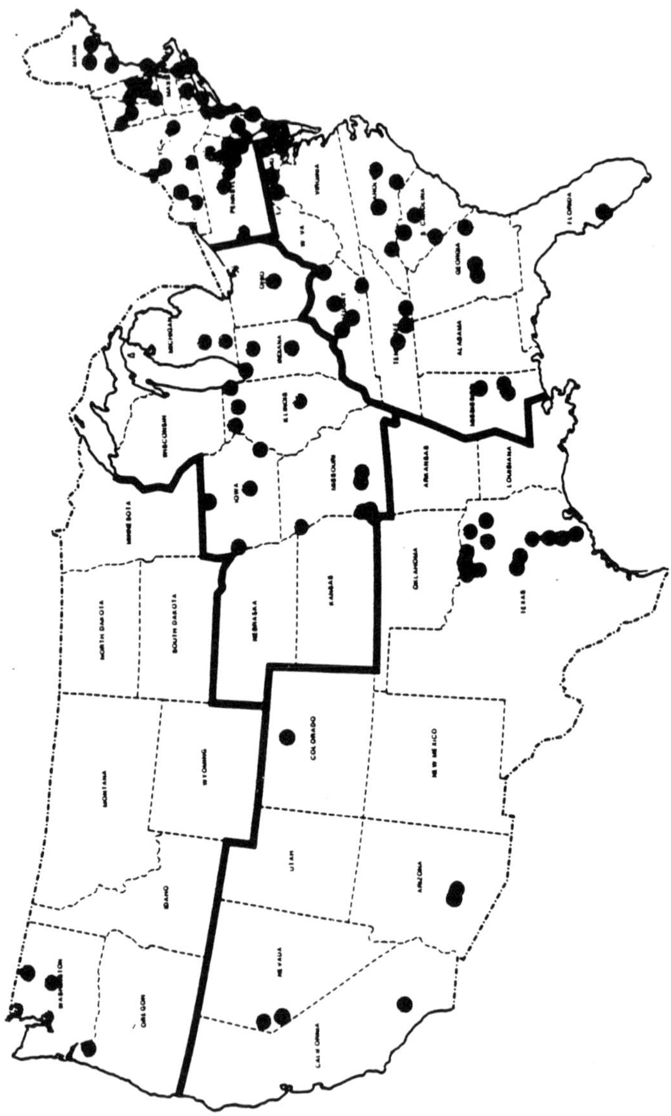

FIGURE 2
POTENTIAL CASE STUDIES

TABLE 4

POTENTIAL CASE STUDIES - NORTHEAST

Location	Design Flow (mgd)	Secondary Treatment	Contaminants Con.	Met.	Org.	Industries
Connecticut						
Branford	3	Oxygen AS	•		•	Textiles, Laundries
Vernon	6.5	PAC		•	•	Textiles, Metal Plating
Delaware						
Bridgeville	0.8	Trickling Filters	•			Food Processing
Maine						
Brewer	3	Complete Mix AS	•		•	Paper, Food Processing
Corinna	1.2	Ext. Aeration AS			•	Textiles
Lisbon	1.5	Ext. Aeration AS	•	•	•	Metal Plating, Paper, Textiles
Massachusetts						
Amesbury	1.9	Ext. Aeration AS		•		Metal Plating
Fall River	30	Oxygen AS	•	•	•	Textiles, Chemicals, Metal Finishing
Gloucester	15	Primary Plant	•			Fish Processing
South Essex	41	Primary Plant		•		Tannery
New Hampshire						
Ashland	1.6	Aerated Lagoons	•		•	Textiles
Derry	1.2	Aerated Lagoons		•		Metal Plating
Merrimack	4.2	TF/AS	•			Brewery
Penacook	4.2	Conventional AS	•	•		Tannery
Pittsfield	0.4	Aerated Lagoons	•	•		Tannery
Troy	0.3	Aerated Lagoons	•		•	Synthetic Materials

TABLE 4 (Continued)

POTENTIAL CASE STUDIES - NORTHEAST

Location	Design Flow (mgd)	Secondary Treatment	Contaminants Con.	Met.	Org.	Industries
New Jersey						
Bay Shore	8	Complete Mix AS			•	Fragrances
Hamilton Township	9.5	TF/RBC	•			Pharmaceuticals
Passaic Valley	300	Oxygen AS		•		Paint
Perth Amboy	10	Primary Plant			•	Chemicals
New York						
Beacon	6	Staged Aeration AS	•			Dairy (cheese)
Binghamton/Johnson City	18	Contact Stab. AS		•	•	Paper
Canandaigua	6.5	RBC	•			Winery
Gloversville/Johnstown	9.5	TF/AS	•			Tannery
Hornell	4	Complete Mix AS	•			Dairy
Niagara Falls	60	Physical/Chemical		•	•	Chemicals
Oswego (East)	3	Conventional AS	•			Paper
Rensselaer County	24	Step Aeration AS	•		•	Dyes, Pharmaceuticals
Syracuse Metro	86	Conventional AS	•		•	Pharmaceuticals
Pennsylvania						
Danville	3.2	Contact Stab. AS		•		Metal Finishing
Hatfield Township	3.6	Complete Mix AS/AWT	•		•	Electronics, Chemicals
Lackawanna River Basin (Moosic)	1	Ext. Aeration AS			•	Paper
Lower Lackawanna Valley	6	Contact Stab. AS			•	Wood Preservatives
Maiden Creek	0.4	Aerated Fixed Film	•			Food Processing

TABLE 4 (Continued)

POTENTIAL CASE STUDIES - NORTHEAST

Location	Design Flow (mgd)	Secondary Treatment	Contaminants Con.	Met.	Org.	Industries
Pennsylvania (continued)						
Monaca	1.2	Conventional AS		•		Glass, Steel
Quakertown	4	Trickling Filters		•		Metal Plating
Telford	0.7	Ext. Aeration AS	•			Poultry
Williamsport	2.4	Conventional AS			•	Chemicals
Rhode Island						
Warren	2	Conventional AS		•		Metal Plating
Warwick	5	Conventional AS		•		Electrical Switches, Metal Finishing
Vermont						
Hinesburg	0.3	Aerated Lagoons	•			Dairy (Cheese)
Williamstown	0.2	Aerated Lagoons	•	•	•	Laundry

TABLE 5

POTENTIAL CASE STUDIES - SOUTHEAST

Location	Design Flow (mgd)	Secondary Treatment	Contaminants Con.	Met.	Org.	Industries
Florida						
Tampa	60	Oxygen AS	•	•	•	Brewery, Food Processing, Metal Plating and Finishing
Georgia						
Macon (2 Plants)	14, 14	TF, AS	•	•	•	Paper, Metal Plating
Sylvania	0.5	AS		•	•	Textiles, Metal Plating
Kentucky						
Ashland	11	Oxidation Ditch	•	•	•	Steel, Tannery
Campbellsville	4.2	Oxidation Ditch		•	•	Textiles, Metal Plating
Elizabethtown	4.5	RBC		•	•	Ink and Dye, Metal Plating
Middlesboro	3.0	Conventional AS	•	•	•	Tannery
Nicholasville	2.8	RBC		•		Metal Plating
Maryland						
Baltimore (Back River)	180	TF/AS	•	•	•	Metal Plating, Paint, Liquid Waste Hauler
Baltimore (Patapsco)	70	Oxygen AS	•		•	Chemicals
Elkton	3	RBC	•			Paint, Metal Finishing
Frederick	7	Trickling Filters	•	•		Dairy, Metal Plating

TABLE 5 (Continued)

POTENTIAL CASE STUDIES - SOUTHEAST

Location	Design Flow (mgd)	Secondary Treatment	Contaminants Con.	Met.	Org.	Industries
Mississippi						
Laurel (2 Plants)	2,2	Conventional AS	•			Food Processing
Meridian	20	Conventional AS	•		•	Timber Products
North Carolina						
Asheboro	4	Trickling Filters	•	•		Chemical, Textiles
Columbus	0.9	Ext. Aeration AS	•			Textiles
Raleigh	23	AS	•	•		Metal Plating, Dairy
Raeford	3	Ext. Aeration AS	•			Poultry, Fragrances, Textiles
South Carolina						
Chester (2 plants)	2, 1	Ext. Aeration AS	•	•		Chemicals, Metal Plating
Gaffney (3 plants)	1.5, 3.2, 3.6	AS	•		•	Food Processing, Textiles
North Augusta	20	Ext. Aeration AS	•			Textiles
Tennessee						
Chattanooga	65	Oxygen AS	•	•	•	Various
Murfreesboro	10	Air/Oxygen AS	•	•		Dairy, Metal Plating
Tullahoma	5	Batch Ext. Aeration AS	•	•		Tannery, Metal Plating
West Virginia						
Martinsburg	3	Trickling Filters	•			Fruit Processing

TABLE 6

POTENTIAL CASE STUDIES - MIDWEST

Location	Design Flow (mgd)	Secondary Treatment	Contaminants Con.	Met.	Org.	Industries
Illinois						
Dakota	0.05	Contact Stab. AS	•			Dairy (Cheese)
Decatur	27	AS/Ponds	•			Food Processing
Gurnee (N.S.S.D.)	20	Step Feed AS		•	•	Metal Plating, Printing, Misc.
Rockford	60	Complete Mix AS		•		Metal Plating
Indiana						
Indianapolis	125	TF/Oxygen AS		•	•	Metal Plating, Pharmaceuticals
Plymouth	1.5	Conventional AS	•			Food Processing
East Chicago	20	Conventional AS		•	•	Steel, Chemicals, Metal Plating
Iowa						
Lake Mills	0.4	TF/Ponds		•		Electronics
Marshalltown	7.5	Conventional AS	•	•		Meat Packing, Metal Finishing
Muscatine	13	Oxygen AS	•			Food Processing
Sioux City	30	Step Feed AS		•	•	Metal Plating, Pharmaceuticals
Michigan						
Kalamazoo	54	PAC	•		•	Paper, Pharmaceuticals
Grand Rapids	66	Conventional AS		•	•	Metal Plating, Misc.

TABLE 6 (Continued)

POTENTIAL CASE STUDIES - MIDWEST

Location	Design Flow (mgd)	Secondary Treatment	Contaminants Con.	Met.	Org.	Industries
Missouri						
Joplin (3 plants)	15,12, 8.5	Conventional AS	•	•		Food Processing, Metal Plating
St. Josephs	27	Conventional AS	•			Food Processing
Springfield (2 plants)	6.5, 30	Conventional AS		•		Electronics
Ohio						
Newark	8	Conventional AS	•		•	Dairy, Fiberglass
Cleveland (Westerly)	50	Physical/Chemical		•	•	Metal Plating, Chemicals

TABLE 7

POTENTIAL CASE STUDIES - SOUTHWEST

Location	Design Flow (mgd)	Secondary Treatment	Contaminants Con.	Met.	Org.	Industries
Arizona						
Phoenix (91st Ave)	120	Complete Mix AS		•		Electronics, Metal Plating
Tolleson	7.5	TF/Solids Contact	•			Meat Packing
California						
Victor Valley	4.8	Modified Step Feed AS			•	Military Installation
Colorado						
So. Fort Collins	1.5	AS		•		Metal Plating
Nevada						
Minden-Gardnerville	1.5	Trickling Filters		•	•	Electronics
Reno-Sparks	24	AS (Phostrip)	•	•		Metal Plating, Electronics, Bottling Plants
Texas						
Corsicana (2 plants)	3, 1	Conventional AS	•			Food Processing
Denisen (3 Plants)	2, 1.5, 1	Ext. Aer. AS, TF	•			Food Processing
Huntsville	1	Ext. Aeration AS		•	•	Metal Plating, Paint
Montgomery County	0.3	Contact Stab. AS			•	Photo Processing
Mt. Pleasant	1	Ext. Aeration As	•			Meat Packing
North Greens	0.3	Contact Stab. AS	•			Meat Packing
Paris	4.2	Ext. Aeration AS	•			Food Processing
Park Ten	1	Conventional AS	•	•		Food Processing, Metal Plating
Sherman	8	TF/AS	•		•	Food Processing
Sulphur Springs	4	Conventional AS	•			Food Processing, Dairy

TABLE 8

POTENTIAL CASE STUDIES - NORTHWEST

Location	Design Flow (mgd)	Secondary Treatment	Contaminants Con.	Met.	Org.	Industries
Oregon						
Newberg	2	Conventional AS		•		Electronics
Washington						
Seattle	190	Primary Plant	•	•	•	Metal Finishing, Bottling
Lynden	1.7	ABF/Oxidation Ditch	•			Food Processing
Oroville	0.4	Ext. Aeration AS			•	Fruit Processing
Brewster	0.4	Ext. Aeration AS			•	Fruit Processing

The final consideration for site selection involved budgetary constraints. Twenty-five to thirty-five (25-35) case study site visits were considered feasible provided the POTWs could be clustered into approximately ten trips. Final site selection involved several iterations to satisfy the previously described criteria.

The 29 selected case study sites are listed in Table 9 and shown in Figure 3. As can be seen, eight sites were chosen from the Northeast region, six from the Southeast, six from the Midwest, seven from the Southwest and two from the Northwest. Table 10 shows the distribution of sites by secondary treatment process, Table 11 shows the distribution of sites by treatment plant size, and Table 12 shows the distribution of sites by major industrial contributor.

SITE VISIT LOGISTICS

The task of making the site visits was divided among five members of the Wastewater Department of JMM's Pasadena Office. Case files were prepared containing all telephone notes, letters, operating data or other information concerning a selected case study. Sites were assigned to individuals such that they could schedule two to four visits in any one trip.

Supervisory personnel at each POTW were contacted by the appropriate JMM site visitor and a tentative visit date was arranged by telephone. After dates were arranged for all sites to be visited in one trip, letters were sent to the same supervisory personnel confirming the visit dates and detailing some of the project objectives. The letters offered a suggested agenda and encouraged the participation of POTW and municipal personnel from operations, laboratory and pretreatment groups in the site visit discussions. The following is a typical agenda for the one day visit:

10:00	-	Introduction and General Discussion
11:00	-	Operations
12:00	-	Lunch
1:00	-	Plant Tour
2:00	-	Laboratory and Records

Interview discussion was conducted in an informal fashion—the length and depth of the interview varied depending on the size of the facility, the number of personnel involved and the pre-visit organization of the contact person. Typically, in addition to talking with the plant superintendent/manager contact, an individual from operations and/or another from pretreatment were involved in the discussions. Laboratory personnel occasionally participated in the discussions as well.

An interview usually began by JMM personnel giving a brief overview of the goals of the project, the work done to date, the purpose of the site visits and a description of the site selection process. The discussions then typically transitioned to the interference issues at the POTW. As a guide to insure that the desired information was obtained on each visit, a standard site visit interview form was developed prior to making any visits. The form was intended to remind interviewers about some specific information that was required from each site, some of which might have been overlooked in the natural course of

TABLE 9

SELECTED CASE STUDY SITES

REGION	STATE	LOCATION
NORTHEAST	New Jersey	Bayshore (Union Beach) Hamilton Township (Trenton) Passaic Valley (Newark)
	New York	Binghamton/Johnson City Canandaigua East Side (Oswego)
	Pennsylvania	Hatfield Township (Colmar) Maiden Creek (Blandon)
SOUTHEAST	Georgia	Rocky Creek (Macon)
	Maryland	Baltimore (Back River, Patapsco)
	North Carolina	Raeford Neuse River (Raleigh)
	South Carolina	Horse Creek (North Augusta)
MIDWEST	Illinois	Gurnee (NSSD) Rockford
	Iowa	Lake Mills Marshalltown Sioux City
	Ohio	Newark
SOUTHWEST	Arizona	91st Avenue (Phoenix) Tolleson
	California	Victor Valley (Victorville)
	Texas	Denisen (Duck Creek, Paw Paw) Paris Sherman
NORTHWEST	Oregon	Newberg
	Washington	Metro-West Point (Seattle)

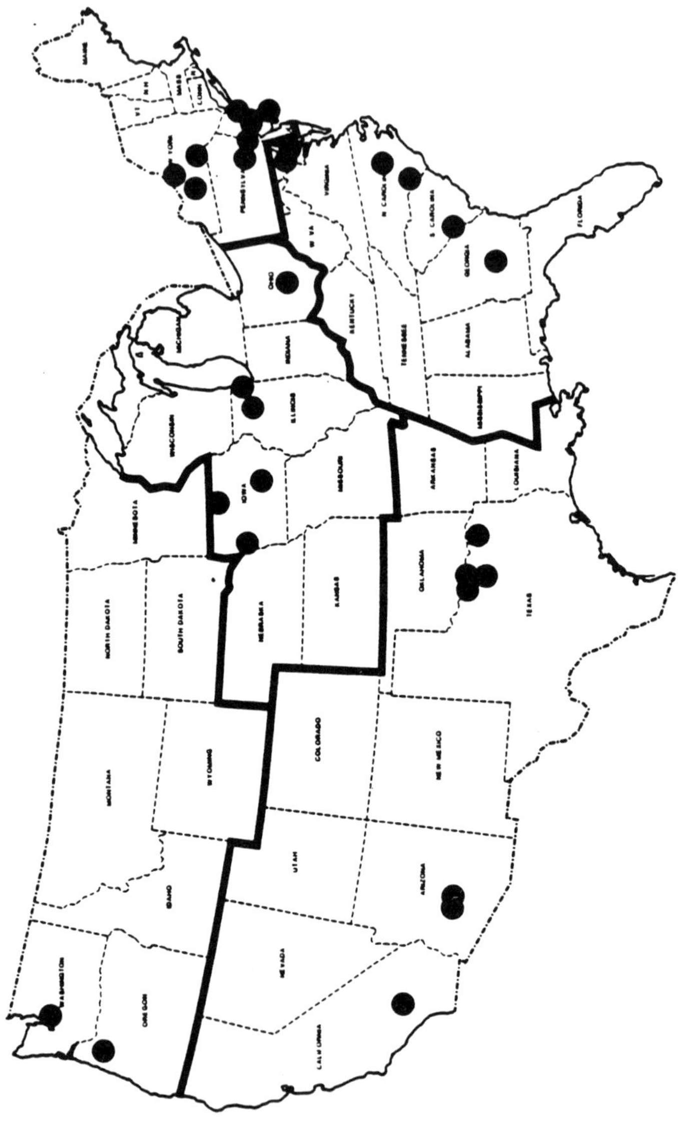

FIGURE 3
SELECTED CASE STUDIES

TABLE 10

CASE STUDY SITES-SECONDARY TREATMENT PROCESS

Activated Sludge	Fixed Film	Fixed Film and Suspended Growth	Primary
Baltimore, MD (Patapsco)	Canandaigua, NY	Baltimore, MD (Back River)	Metro, WA
Bayshore, NJ	Denison, TX (Paw Paw)	Lake Mills, IA	
Binghamton, NY	Hamilton Township, NJ	Sherman, TX	
Denison, TX (Duck Creek)	Maiden Creek, PA	Tolleson, AZ	
East Side, NY			
Gurnee, IL			
Hatfield Township, PA			
Horse Creek, SC			
Marshalltown, IA			
Newark, OH			
Newberg, OR			
Neuse River, NC			
91st Avenue, AZ			
Paris, TX			
Passaic Valley, NJ			
Raeford, NC			
Rockford, IL			
Rocky Creek, GA			
Sioux City, IA			
Victor Valley, CA			

TABLE 11

CASE STUDY SITES-TREATMENT PLANT SIZE

< 1.0 mgd	1-5 mgd	5-10 mgd	> 10 mgd
Lake Mills, IA	Denison, TX (2 plants)	Bayshore, NJ	Baltimore, MD (2 plants)
Maiden Creek, PA	East Side, NY	Canandaigua, NY	Binghamton, NY
	Hatfield Township, PA	Hamilton Township, NJ	Gurnee, IL
	Newberg, OR	Marshalltown, IA	Horse Creek, SC
	Paris, TX	Newark, OH	Neuse River, NC
	Raeford, NC	Sherman, TX	91st Avenue, AZ
	Victor Valley, CA	Tolleson, AZ	Passaic Valley, NJ
			Rockford, IL
			Rocky Creek, GA
			Seattle, WA
			Sioux City, IA

TABLE 12

CASE STUDY SITES–INDUSTRIAL CONTRIBUTORS

Food, Meat and Dairy	Metal and Electronics	Chemical and Pharmaceutical	Textile	Pulp and Paper
Canandaigua, NY	Baltimore, MD (B.R.)	Baltimore, MD (Both)	Horse Creek, SC	Binghamton, NY
Denison, TX (Both)	Gurnee, IL	Bayshore, NJ	Raeford, NC	East Side, NY
Maiden Creek, PA	Hatfield Township,	Hamilton Township, NJ		Rocky Creek, GA
Marshalltown, IA	Lake Mills, IA	Hatfield Township, PA		
Metro, WA	Marshalltown, IA	Newark, OH		
Newark, OH	Metro, WA	Passaic Valley, NJ		
Neuse River, NC	Neuse River, NC	Raeford, NC		
Paris, TX	Newberg, OR	Sioux City, IA		
Raeford, NC	91st Avenue, AZ	Victor Valley, CA		
Sherman, TX	Rockford, IL			
Tolleson, AZ	Sioux City, IA			

discussions. Most of the form was filled out during the POTW visit or shortly thereafter. A copy of the site visit interview form is included in Appendix B.

After returning to the office, the interview notes, site visit interview form and operations/laboratory data from the plant were incorporated into a written report of one to two pages, accompanied by a standardized data sheet and a treatment process schematic. The written reports were sent to the principal individuals involved in the discussions at each site for their comments, along with a followup letter of thanks. The comments and corrections to the returned draft reports were considered before making revisions to the copies included in this document.

CASE STUDY REPORTS

The twenty nine case study reports are included in Appendix C. The first page of Appendix C presents an index to the reports.

6. Results

MONITORING

A critical aspect to any successful industrial waste program is a comprehensive monitoring of industrial discharges, POTW influent and important process streams within the plant. Numerous options exist regarding such monitoring programs. The benefits derived by the municipalities in terms of understanding their influent wastewater characteristics and sources of specific contaminants are obvious. Monitoring is also performed to provide data from which to develop local limits and to later evaluate an industry's compliance with those limits. Industries can utilize monitoring to identify sidestream characteristics and incorporate the data into in-house conservation, recycle and reuse efforts. Such an approach can generate positive benefits from a program initially perceived by industry as costly and unnecessary.

POTW Influent

Most municipalities surveyed and visited on this project now have some form of influent wastewater monitoring. The most common approach is to install a composite sampler at the headworks of the plant. State of the art sampling equipment provides the POTW with three options:

- hourly sampling and collection in 24 sample containers;
- composite sampling; and
- flow-proportioned composite sampling.

Hourly samples provide a means of identifying diurnal fluctuations in wastewater characteristics. Such an approach can be costly if all 24 samples are analyzed, but is particularly useful if "midnight dumping" of prohibited substances is suspected.

Composite sampling involves the collection of a fixed volume of wastewater at regular intervals into a single, large container. A typical approach is to collect 100 ml every 15 minutes for 24 hours into a 10-liter sample bottle. This is the most common method of obtaining average daily influent samples. A better approach is to proportion the sample volume consistent with the influent volumetric discharge at the time of collection. This technique requires a feedback signal from an influent flowmeter to the sampler, but results in a sample that is consistent with the mass loadings to the POTW.

Other POTW Locations

POTW effluent is generally composite sampled and analyzed in accordance with NPDES Permit requirements. Operators may, however, select other process streams within a facility for intermittent monitoring. For example, sampling primary effluent allows for calculations of loadings to the secondary treatment system. The response of a biological process is more easily explained if one knows the specific wastewater feed characteristics, as opposed to assuming a primary clarifier performance based on influent characteristics.

An informative yet infrequently employed sampling methodology is to evaluate the strength of sidestream flows, particularly from solids processing. Recycle flows can add 50 to 100 percent of the influent solids and organics to the liquid processing trains when inefficient sludge solids capture persists. POTW design often neglects the impact of recycle streams, a problem magnified when unanticipated quantities of heavy metals and priority organics are discharged from industrial sources. While monitoring such sidestreams on a daily or weekly basis may prove impractical (and costly), periodic sampling and flow measurement permits mass balancing around solids processing units, and can provide insight into the presence of substances in the POTW effluent not necessarily present in the influent.

Recycle flows can be intermittent, or at least shift dependent, and as such are poor candidates for 24-hour composite sampling. Grab sambling is done by extracting a representative sample of sufficient quantity to perform the necessary analytical tests. Some procedures, such as the extraction methods for oil and grease, must be grab sampled to prevent deposition of the material on the container over the 24 hour composite period.

Operations Monitoring

In addition to sample collection and analysis throughout a POTW, there are numerous tools available to the plant operator to monitor the condition and performance of the facility. Suspended growth biological treatment systems generally provide more control, and therefore monitoring opportunities, than do fixed film systems. However, all POTWs have processes that can be easily checked on a daily basis which can signal the onset of an interference problem. Making use of the available tools may be the difference between total process failure and catching the problem before it fully develops.

The operational tools available fall into the following categories:

- sensory
- instrumentation/equipment
- analytical results

The sensory category refers to what operations personnel observe around the plant through the senses of sight, sound, touch and smell. Examples of what operators should typically notice as they work around a POTW are the:

- surface appearance of clarifiers;
- amount and color of foam in aeration tanks;
- presence of nuisance organisms, insects or odors near fixed film systems;
- common odors at each plant location; and
- sludge and recycle flow appearance at each processing step.

Instrumentation is designed into treatment facilities as an aid to the operations staff. Whether located at the central control panel or at the piece of equipment being monitored, digital and dial gage readouts provide instant feedback to an experienced operator concerning the conditions in the plant. Strip chart recorders maintain permanent records of the critical parameters such as raw wastewater feed to identify long-term trends and isolated indiscretions. Despite the amount of instrumentation and level of sophistication, much of the hardware is unused or simply ignored by operators because of a perceived complexity and/or unreliability. Operator distrust of the instruments would not be a problem if more time was spent in the plant, but the tendency is to remain in the control room and monitor a control panel for which the operators have little appreciation. A major thrust of the Hamilton Township, NJ interference identification program was to require that operators spend a minimum number of hours each shift "walking the grounds." Such a requirement can result in the identification of late night spill events that might otherwise go unnoticed until morning, when it may be too late for the biological processes to recover.

The use of simple portable instruments and equipment while on routine site inspection can be quite useful to the operator. The use of a device to measure the depth of sludge in clarifiers may be the best way to learn that a sludge pump did not operate as expected. Dissolved oxygen, pH, temperature and conductivity meters can be used to monitor plant performance on a routine basis with little training or time commitment required.

The final category of operations monitoring deals with the analytical testing other than what is performed on the wastewater samples, as previously discussed. Typical analyses are:

- MLSS, SVI and O_2 uptake rates of activated sludge;
- soluble BOD in each RBC stage;
- solids content of sludges throughout the plant, especially from the dewatering process;
- volatile acids-to-alkalinity ratios in anaerobic digesters;
- methane content of digester gas; and
- metals contents of dewatered sludge.

Experienced operations people can sense an upset from small fluctuations in these data. Isolation of industrial waste problems early in the spill event may prevent a catastrophic impact to the POTW.

Industrial Discharges

Sewer use ordinances and industrial waste management programs typically provide for some means of monitoring an industry's discharge to the municipal collection system. Such ordinances require measurement of both quantity and quality of the industrial or combined domestic/industrial flow. Industrial discharges are either self-monitored by industry with regular reporting requirements, or else monitoring is handled by the municipality or a contractor on behalf of the municipality. Municipal expenditures for these programs are either billed directly to the specific industries or are recovered through industrial surcharges to the normal sewer use rate.

The quantity of an industrial wastewater can be obtained from:

- metered water usage;
- metered water usage corrected for evaporative losses or product content; or
- measured sewer flow.

Industries with significant consumptive water usage have found it prudent to install flow measurement devices in their outfall sewer with continuous recorders. Pay-back periods of less than one year are not uncommon with this type of equipment. The discharge quality coupled with the flow volume will dictate the industry's sewer bill based on a formula contained within the local sewer use ordinance.

Analytical Testing

Analyses performed on a POTW influent should routinely include BOD, SS and other conventional pollutants, (such as NH_3 and P) contained on the NPDES Permit. The testing intervals for the priority organics and metals is determined on a site-specific basis as a function of permit violations, pretreatment program requirements, process upsets, types of industrial discharges and budgetary constraints.

Industries should similarly be monitored for conventional pollutants, with the testing of other compounds determined by the nature of the specific industrial waste. In the case of categorical industries, the substances of concern and pretreatment requirements are already specified by the regulations. For noncategorical industries, the initial permit application and questionnaire responses combined with supportive analytical analyses should provide sufficient information to establish a testing program.

Sewer Use Ordinances

Many of the municipalities visited for this project calculate sewer fees based on three levels of conventional pollutant (BOD, SS, NH_3) concentrations:

- less than or equal to domestic strength;
- greater than domestic strength but less than a prohibited limit; and
- greater than a prohibited waste strength limit.

The established limits and parameters included in the ordinances varies from municipality to municipality as a function of local domestic waste strength and POTW capacity. If an industry's wastewater falls into the middle category, the discharge is permitted by the POTW, but a surcharge is applied which is partially intended to create an economic incentive to the industry for waste load reduction.

A common alternative approach to the sewer use fee structure described above is a uniform charge per pound of BOD, SS and/or NH_3 discharged to the POTW. This type of unit cost approach is probably fairest, but requires frequent sampling of all industry to accurately characterize the wastewater strength. In these cases, prohibitive limits for various pollutants (such as toxics and explosives) are still established for the protection of the POTW.

In all municipalities surveyed and visited who had an established wastewater ordinance and approved pretreatment program, the language was consistent with the General Pretreatment Regulations (40 CFR Part 403) concerning prohibited discharges and categorical standards. A key deficiency was in the development of local limits for nonconventional pollutants from noncategorical industries. Local limits are perceived by municipalities as being complicated to establish and politically sensitive to enforce.

Monitoring methods and frequency are also specified by each municipality in its ordinance. Self monitoring by industry with monthly spot checking by the municipality is the simplest and least expensive alternative. Such an approach is only successful when:

- sampling procedures are clearly outlined and followed;
- a certified laboratory performs the analytical testing;
- rigorous reporting requirements are established for the industries; and
- spot checking by the municipality, including occasional sample splitting and testing, be performed on a frequent yet random basis.

The alternative to self monitoring is for the municipality to perform all sampling and analytical services on a once-per-month or once-per-quarter basis, depending on the significance of the specific industry to the POTW. Under this scenario, split samples should be made available to the industry each time they are sampled to provide them with the opportunity to verify the test results from which their compliance status and user fees will be determined.

Regardless of the approach taken, the objective of any industrial monitoring program is to obtain representative analytical results of the wastewater flow and characteristics. An industry with highly variable quality should be sampled more frequently than one with a consistent effluent quality. Municipalities tend to be more accommodating to industries who conform to the requirements of the ordinance in a spirit of cooperation than to those industries who are secretive and unyielding. Of the POTWs visited, the most successful efforts at monitoring and pollutant load reduction occurred where open lines of communication existed between industry and municipal treatment plant operators.

INTERFERENCE

As defined in Section 1, the term interference refers to a violation of an NPDES Permit or a limitation on the ability to dispose of sludge as a direct consequence of industrial wastewater being discharged to the POTW. While the recognition of a problem is trivial, attributing that upset or violation to a specific industry or group of industries can be difficult. Plant superintendents occasionally use industry as a scapegoat for poor performance, when in fact the problems may be due to poor operations and maintenance practices. Conversely, public officials (particularly if elected) are sometimes reluctant to apply too much pressure to local industry who in many cases are significant employers and taxpayers, even if directly linked to upsets at the POTW.

Identification of Interference

Officials at POTWs experiencing process upsets and permit violations sometimes find it convenient to blame industrial discharges for the problems. However, in many cases the upsets can be traced to design, administration, operation and maintenance deficiencies at the POTW unrelated to industrial wastewater. The U.S. EPA has developed a Composite Correction Program (CCP) for use by POTWs to economically improve the performance of existing facilities. A CCP has two distinct phases. The evaluation phase attempts to assess the capability of each unit process and to identify the performance limiting factors in a POTW. The performance improvement phase is designed to systematically eliminate the limiting factors and to optimize the operation of the existing facility. The U.S. EPA has published a handbook to assist POTWs in their evaluation and correction efforts (Hegg, et al., 1984).

It is incumbent upon POTW personnel to both prove that plant upsets result from industrial discharges and to identify the specific source of that upset. This "burden of proof" provides the necessary justification for a substantial monitoring program as outlined in a previous subsection. The case study examples in Section 4 tend to fall into one of three categories regarding interferences:

1. A single major industry in town dominates the waste characteristics at a relatively small POTW.

2. One or two industries among several are primarily responsible for waste strength fluctuations in small to medium-sized POTWS.

3. Industrial wastewater from numerous sources controls the wastewater feed characteristics, with no single dominant industry.

The first category listed above is by far the easiest situation to deal with from an identification standpoint. By monitoring the industry's discharge, POTW influent and effluent and other relevant plant operations, the impact of the industrial waste on the POTW can be determined. The political ramifications of category one are more complicated, as such industries can be the largest employer and taxpayer in a town. A number of the case studies from Section 5 provide examples for this category:

Bayshore, NJ
Canandaigua, NY
Denison, TX
Horse Creek Plant, SC
Lake Mills, IA
Maiden Creek Plant, PA
Paris, TX
Raeford, NC
Tolleson, AZ
Victory Valley Plant, CA

Category two corresponds to a less sensitive situation politically, but is a more difficult interference to trace. The monitoring program may be sufficient if a large data base exists over a period of time. Unfortunately, when numerous industries must be tested on a frequent basis, the sampling and analysis costs are high and difficult to justify with the non-troublesome industries. An alternate approach is to look for one or two specific pollutants unique to certain industries. Examples range from odor tracing back through the collection system to toxicity analyses to the sophisticated backtracking computer program used by the City of Baltimore. Examples from this category include:

Gurnee, IL
Hamilton Township, NJ
Hatfield Township, PA
Newark, OH
Patapsco Plant, MD
Sioux City, IA

The third and final category generally applies to larger facilities which are less likely to be susceptible to any particular industrial effluent. As pointed out in Section 3, the telephone survey indicated that most larger POTWs are infrequently affected by industrial wastes and rarely experience interference. Some exceptions to this rule were, however, identified and are presented in the case studies. Notable among them are:

Back River Plant, MD
Passaic Valley, NJ
Raleigh, NC

Types of Interference

Interference can generally be divided into two distinct types: chronic and isolated. The chronic interference refers to a more or less consistent pattern of permit violation or sludge disposal problem characterized by regular pollutant concentration levels. Chronic interferences result from both continuous and intermittent discharges. Continuous discharges of toxics can lead to inhibition of biological process efficiency without a noticeable upset. In contrast, intermittent discharges manifest themselves in a variety of ways in the treatment plant processes as observed by the operations monitoring program. Changes in DO levels, MLSS, SVI, reactor temperature, etc. are indications of process changes potentially resulting from industrial wastes. Chronic, intermittent problems tend to produce similar impacts on the POTW from incident to incident.

Isolated industrial discharge problems tend to produce a more drastic effect on the POTW, such as complete loss of unit processes. Since biological populations are not acclimated to either the specific compounds or concentration levels observed in an isolated discharge, the impact on biological processes is severe and rapid, with long recovery periods generally required. Such interferences commonly affect the effluent quality rather than the stabilized and dewatered sludge characteristics, although loss of an anaerobic digestor due to slug loads of heavy metals is not unusual.

The distinction between an isolated and an intermittent discharge is not well defined. One approach is to consider whether the biological population remains partially acclimated to the contaminant between incidents. An isolated discharge would be to a nonacclimated bacterial culture, while an intermittent waste feed would be discharged to at least a partially acclimated population. The important difference between the two is the response and rate of recovery of the biological system. Establishing a maximum recurrence interval for an intermittent discharge is site specific as a function of the type and strength of the industrial waste, the type of treatment system affected and the regularity of the discharge.

Waste haulers and industrial spill incidents makes planning for isolated interference problems difficult. One approach taken by some municipalities is to disallow waste haulers from discharging to their POTW. In cases where waste haulers are allowed, permit requirements and testing procedures have tightened up considerably over the past few years.

Another approach at the POTW is to have off-line tankage available to re-route the influent flow and contain the spill contents. Treatment plants that are currently undergoing renovation (such as Newark, Ohio or Hatfield Township, Pennsylvania) are considering the usefulness of existing tankage for this purpose. In order to recognize that such a spill is imminent, however, the POTW must depend upon the industry for immediate notification of a problem. In-line monitoring is useful for some problems such as explosivity, pH or temperature, but many of the toxic compounds would not be detected by instrumentation.

Interfering Compounds

POTW interference can be caused by a wide variety of chemical, biological and physical factors. Of the chemical factors, the types and concentrations of industrial wastewater constituents which cause interference are highly variable.

The studies reported in the literature discussing chemical interference/inhibition range from research done in the laboratory to studies of actual treatment plant operations. There has been a substantial body of work published and many researchers have devoted a great deal of effort to these types of studies. However, to draw definitive, broadly based conclusions on pollutant concentrations which cause interference/inhibition of biological systems is at best unsound and at worst, impossible.

An attempt has been made to synthesize the available information on the types and concentrations of pollutants and compounds which inhibit biological treatment systems and present it in a manner that is of use in determining likely POTW interfering levels. Tables 13-15 present ranges of concentrations for metal, organic, and conventional and inorganic pollutants which inhibit acclimated biological processes. The references used in developing these tables are:

- Revised Pretreatment Guidelines (JRB Associates, 1981a);
- Literature Study of the Biodegradability of Chemicals in Water - Volume 1 (Geating, 1981);
- Biodegradation and Treatability of Specific Pollutants (Barth and Bunch, 1979);
- the literature review presented in Section 3 of this report; and
- data obtained from the case studies.

The wide range of concentrations presented in Tables 13-15 is a result of the significant amount of data in the literature that are apparently contradictory. In general, it can be said that this situation is a result of the method of measuring inhibition and the different conditions surrounding the biological process. Unfortunately, some study conditions are seldom documented in the literature. The most important conditions that affect biological inhibition are: biomass characteristics, pH, temperature, synergism (or antagonism) and acclimation. There are various ways of measuring inhibition and the fact that different researchers use different methods results in a range of published "inhibiting concentrations", even for nearly identical study conditions. The two most typical methods of determining activated sludge inhibition are by measuring 1) decreases in COD or BOD removal or 2) decreases in oxygen utilization rates as compared to controls. Threshold inhibition levels as measured by these two methods are usually defined differently by individual researchers, but are most typically set at the 10-50 percent range. Anaerobic treatment inhibition is typically defined as increased volatile acid levels or decreased methane generation, but once again the threshold levels are variously defined. Nitrification inhibition is specified as a certain decrease in the degree of ammonia conversion.

For most studies biomass characteristics are not typified, except as it relates to acclimation. The diverse biomass population is likely to be very different from one reported study to the next and this will have a significant impact on the inhibitory concentration levels of pollutants. pH and temperature generally are reported study conditions in the literature. Once again the actual conditions vary dramatically from study to study, with the result being that inconsistent values are reported. pH plays a particularly important role in metal-caused inhibition. pH affects the solubility of metal ions, and it is the soluble metal species that is toxic to microorganisms. Synergism, or the increase in the inhibitory effect of one substance by the presence of another, is most important when considering combinations of metals. Toxic organics do not exhibit this effect as often as metals. On the other hand, some compounds are antagonistic towards each other, decreasing the inhibitory effect of either compound alone. Good examples are chelating agents, such as EDTA, which are antagonistic toward metal ions and reduce their toxic effects.

Acclimation, which is a biological phenomenon, is an important factor which must be considered to understand the inhibitory effect of pollutants. Biomass characteristics (relative populations) will change with time so that a given pollutant will become less toxic to the biomass as a whole and may in fact become a significant substrate. The values reported in Tables 13-15 are in general for acclimated biological systems. Some studies have been done and reported in the literature on the effects of pollutants on non-acclimated biological systems, as well as what could be termed shock loading or short-term non-acclimated studies. These studies were not included in developing Tables 13-15 because inconsistencies between studies were even more pronounced than for acclimated studies. Non-acclimated systems can in general withstand much lower pollutant concentrations before being inhibited and from this standpoint, the lower values in Tables 13-15 could possibly be used to estimate the upper limits of inhibition for unacclimated systems. Russell, et al. (1983) present a good summary of the non-acclimated studies performed primarily during the 1960s and early 1970s.

Table 13 presents a range of metal pollutant concentrations inhibiting acclimated biological processes. The literature is more abundant and consistent in reporting inhibitory concentrations of metals than for organic compounds. The most significant factors affecting the range of reported values in Table 13 are pH and the method of reporting inhibition, both of which were discussed previously.

Table 14 presents ranges of concentrations for toxic organic compounds which inhibit acclimated biological systems. A dearth of information exists on the effects of toxic organics on fixed film processes and so this column was eliminated from the table. Scattered data exists for other organic compounds, often only available from one study. To present information in a condensed, useable and consistent manner, organic compounds were classified into eight broad categories and the information on inhibitory pollutant concentrations was synthesized within that structure. The premise was that compounds of similar structure or characteristics would inhibit biological processes at similar concentration levels. This approach was found to be fairly consistent to the levels shown in Table 14. However it is extremely important to note that the categories in Table 14 are very broad and the concentration ranges presented are simply typical reported values for some compounds and should not be construed as defining a definitive inhibition range for all compounds that might fall into the broad classifications. Appendix A lists the compounds that fall into these broad classifications. Some compounds fit more than one classification, but an attempt has been made to place these compounds into the most appropriate category.

Additional explanation for several of the compound types is in order. It appeared from the literature that the following generalizations could also be made:

- multi-chlorinated aromatics inhibited activated sludge at much lower concentrations than did other aromatics, with threshold levels generally less than 1 mg/l; and

- for oxygenated compounds, alcohols were generally not inhibitory to activated sludge at levels below 1,000 mg/l and acids were inhibitory to anaerobic processes at levels less than 10 mg/l.

TABLE 13
METAL CONCENTRATIONS INHIBITING
ACCLIMATED BIOLOGICAL PROCESSES
(in mg/l)

Metal	Biological Process			
	Activated Sludge	Nitrification	Fixed Film	Anaerobic
Arsenic	0.04 - 0.4	-	290	0.1 - 1.0
Boron	0.05 - 10	-	-	-
Cadmium	0.5 - 10	5-9	5-20	0.02 - 1.0
Calcium	2,500	-	-	-
Chromium	0.1 - 20	0.25 - 1.0	50	1.5 - 50
Copper	0.1 - 1.0	0.05 - 0.5	25 - 50	1.0 - 100
Cyanide	0.05 - 20	0.3 - 20	-	0.10 - 1.0
Iron	500			-
Lead	0.10 - 10	0.5 - 1.7	-	50 - 70
Manganese	10	-	-	-
Magnesium	-	50	-	1,000
Mercury	0.1 - 5.0	2 - 12.5	-	-
Nickel	1 - 5	1 - 5	-	2 - 200
Silver	0.03 - 0.05	0.25	-	-
Sodium	-	-	-	3,500
Tin	-	-	-	9
Vanadium	20	-	-	-
Zinc	0.30 - 20	0.01 - 1	-	5 - 10

TABLE 14
ORGANIC COMPOUND CONCENTRATIONS INHIBITING ACCLIMATED BIOLOGICAL PROCESSES
(in mg/l)

	Biological Process		
Compound Type	**Activated Sludge**	**Nitrification**	**Anaerobic**
Agricultural Chemicals			
Common Pesticides	0.05 - 0.10	-	-
Lindane	5 - 10	-	-
Aromatics	50 - 150	-	100 - 870
Halogenated Aliphatics	150 - 250	< 0.1 - 10	1 - 20
Nitrogen Compounds	1 - 500	0.1 - 100	5 - 500
Oxygenated Compounds	120 - 500	-	50 - 1,000
Phenols	100 - 1,000	1 - 10	100 - 200
Chlorophenols	5 - 100	-	100
Nitrophenols	50 - 200	-	100
Methylphenols		5 - 50	
Phthalates	-	-	-
Polynuclear Aromatic Hydrocarbons	500 - 2,500*	-	-

*Unacclimated

TABLE 15
CONVENTIONAL POLLUTANT AND INORGANIC COMPOUND CONCENTRATIONS INHIBITNG ACCLIMATED BIOLOGICAL PROCESSES
(in mg/l, except pH)

	Biological Process		
Compound Type	**Activated Sludge**	**Nitrification**	**Anaerobic**
Alkalinity	1,600		
Ammonia	480		1,500 - 8,000
Chloride		180	
Iodine	10		
Iron Salts	100		500
pH	> (11-12)	< (7.0 - 8.5)	
Sulfate			500 - 1,000
Sulfide	25 - 50		50 - 100
Surfactants	100 - 500		

Table 15 presents information on conventional pollutants and inorganic compounds inhibitory to acclimated biological systems. Information on inhibitory concentrations is limited, possibly due in part to the perception that conventional and inorganic pollutants are not inhibitory to biological systems. Conventional organic pollutants (BOD, COD, etc.) and ammonia (for nitrification) are not presented.

Toxicity Testing

All states now operate with water quality regulations that prohibit the discharge of toxic substances in toxic amounts into natural watercourses. Such regulations are applicable to the point source effluents from POTWs, which has prompted EPA and the state agencies to encourage, and in some of the more recent permits, require municipalities to perform biomonitoring on their effluent. What this type of effluent toxicity testing actually measures is the "pass through" pollution from a POTW, and is therefore not relevant to this study. However, toxicity testing of industrial discharges, particularly from new sources, might prove useful in evaluating the impact of that wastewater on the POTW.

Biomonitoring procedures utilize certain species of fish, such as fathead minnows or rainbow trout, to indicate the level of toxicity of a particular wastewater. The rate of survival in varying concentrations of pollutants over a 24 to 96-hour test period can be measured (expressed as an LC_{50}, or median lethal concentration), and comparisons made between different chemicals or wastewaters. While biomonitoring of this type might well rank industrial discharges in terms of toxicity, the results would do little to predict the impact of such discharges on a POTW.

Recently, a number of acute toxicity test procedures have been developed which have applications to industrial wastewater evaluation and control. The most straightforward procedure is to add varying concentrations of an industrial wastewater to a BOD bottle containing an active biological culture from the secondary treatment system. A DO meter equipped with a BOD probe can be used to measure the oxygen uptake rate after the sample is saturated with oxygen. If the industrial wastewater is toxic, increased doses will result in reduced oxygen utilization due to the inhibition of the bacterial oxidation. A similar approach using respirometers allows for the use of larger reactors (up to 10-liters), continuous oxygen feed and strip-chart recording of the uptake rate with time.

Recent variations of the respirometry approach utilize special cultures of microorganisms, instead of the POTW bacteria, as more precise predictors of toxic effects. One manufacturer uses specially prepared and packaged bacterial cultures in conjunction with a DO meter to plot families of inhibition curves and to develop an LC_{50} value analogous to those obtained by bioassay testing. A second technique uses photo-luminescent marine microorganisms, whose light output decreases proportionally to the level of toxic shock when fed varying concentrations of industrial wastewater. This approach has been used extensively in Baltimore, Maryland and Chattanooga, Tennessee to evaluate the toxicity of influent wastewaters to the POTW.

A comparison between the fish bioassay technique and the use of photoluminescent organisms (Cawley, 1981) indicated that despite some limitations, the latter procedure offers a number of advantages over the former in terms of speed (15 minutes versus 24 to 96 hours) and cost-effectiveness. The author does not suggest that the fish bioassay procedures be abandoned, but does indicate that a correlation between the two techniques would be an important breakthrough in the years ahead.

Rapid toxicity testing procedures will be valuable tools for identification of interference sources as they gain acceptance by municipalities. A toxic impact can be traced upstream through a collection system very rapidly when the test procedure takes less than 30 minutes. Such an approach to interference tracing is only useful if the troublesome industry discharges toxicants. Municipalities must continue to rely on more conventional monitoring practices for upsets resulting from non-toxic contamination.

MITIGATION

The success of any effort to mitigate interference is dependent to a great extent on the characteristics of the pollutant(s) causing the interference, the characteristics of the treatment plant (size, biological process, etc.) and the type of mitigation attempted. It is important to note that while some mitigation efforts can probably be utilized with success on a wide variety of interference cases (pretreatment for instance), other measures may be more specific to a particular pollutant or set of operating conditions. The bottom line is that any POTW experiencing interference has a unique set of circumstances and conditions that may require unique mitigation efforts. The purpose for describing successful interference mitigation efforts is that they may in some way be adapted to other POTWs with similar characteristics.

The information on mitigation efforts from the previous three chapters has been synthesized and categorized under three major headings:

- Treatment Plant Control
- Pretreatment and Source Control
- Legal and Enforcement Remedies

Treatment Plant Control refers to mitigation efforts that are implemented at the POTW and are as simple as biological process control, or as extensive as new or modified treatment processes. Pretreatment and Source Control describes some of the elements of a successful pretreatment or source control program for mitigating interferences. Legal and enforcement remedies include fines, litigation and sewer disconnection.

Treatment Plant Control

Mitigation of the effects of interfering industrial pollutants is generally first attempted by modifications of some type at the POTW. In the short term (and possibly long term) these measures, typically plant operation modifications, can prevent an interference or lessen its impact. Other long-term efforts to mitigate interferences at POTWs include capital improvements or modifications of a permanent nature.

Plant Operations Modifications --
The first attempt to mitigate the deleterious effects of industrial pollutants at a POTW usually consists of modifications to plant operations. These measures might be in response to a spill of an industrial contaminant, or to a chronic industrial waste problem, but typically can be categorized into two basic types: biological process control and treatment step modifications.

Biological Process Control -- Many investigator's work in the literature review reported successful prevention or reduction of interferences by adjusting activated sludge process parameters such as the mean cell residence time (MCRT), mixed liquor suspended solids (MLSS) and food to microorganism ratio (F/M). This same conclusion was borne out in discussions with many operating personnel during the case study site visits reported in Chapter 4. Process control of other biological systems like trickling filters is not as easily accomplished, but parameters such as recirculation ratio can be modified.

The following changes to activated sludge process parameters have been observed to mitigate the effects of industrial pollutants on a biological system. Although they are interrelated, they are addressed separately because any one may be typically utilized as a process control tool at a POTW.

1. **Increase the Mean Cell Residence Time.** Increasing the MCRT (sludge age) has been shown to have the effect of reducing the toxicity of all forms of industrial contaminants. Biological systems with higher MCRTs have been observed to acclimate faster to a foreign pollutant than systems with low MCRTs. By increasing the MCRT at the first sign of a possible toxic upset, (by decreasing the solids wasting rate) the effect of any toxicant will be less than if no action is taken.

2. **Increase the Mixed Liquor Suspended Solids.** High mixed liquor suspended solids (MLSS) concentrations have been shown to offset some of the effects of industrial pollutants. A high MLSS provides the best conditions for bio-absorption and acclimation to a toxic substrate. Increasing the sludge return rate to the aeration basin at the first indication of toxic upset, while at the same time bypassing any remaining toxic influent around the aeration basins, will lessen the impact of a short term upset and cause quicker biomass acclimation to a long term problem.

3. **Decrease the Food to Microorganism Ratio.** This parameter is directly related to both the MCRT and the MLSS. It has been observed that decreasing the F/M causes improved biodegradation of toxic contaminants, and expedites biomass acclimation.

The process control modifications just described appear to apply as well to activated sludge systems attempting to achieve nitrification as they do to strictly carbonaceous removal systems. In general, increasing the MCRT and MLSS, and decreasing the F/M is beneficial for treating any type of industrial contaminant, whether it is a toxic metal, toxic organic or high-strength conventional pollutant.

Fixed film biological processes do not lend themselves to the same type of process control measures as suspended growth systems, but several techniques have been employed with some success. Varying the amount and point of recirculation in a trickling filter can modify the effect of industrial pollutants. Recirculating secondary clarifier effluent is a means of achieving the greatest diluting effect, which may be desirable for high-strength organic waste or toxics. Should excessive biomass sloughing be a problem due to toxic pollutants, returning secondary clarifier underflow may help in maintaining a proper biomass population.

Treatment step operations modifications -- A further means of mitigating the effects of industrial pollutants on POTWs is through modifying the operation of existing treatment steps. Activated sludge systems are often designed to operate in several different "modes" (i.e. step aeration, contact stabilization, etc.) by providing the appropriate physical layout. Some modes of operation have been shown to be more successful than others at mitigating the effects of industrial contaminants, particularly those dosed in highly variable concentrations. It has been shown at the laboratory and plant-scale that extended aeration and step aeration (step feed) are generally more stable than complete mix and conventional activated sludge. Some contradiction exists as to whether conventional plug flow or complete mix activated sludge treatment is more resistant to industrial contaminants, but it appears that complete mix generally provides more consistent treatment, particularly under shock loading conditions. The contact stabilization mode has been reported as being less successful at treating industrial pollutants than other modes, particularly when the organic matter is predominantly soluble and waste strength fluctuations are common.

A successful means of mitigating the effects of industrial contaminants on any biological treatment process is through the use of staged treatment. Many treatment systems have realized improved conventional and industrial pollutant removal when switching from parallel treatment to series treatment. For example, two aeration basins operating in series are generally more successful at mitigating the effects of industrial contaminants than the same two basins operating in parallel. The same principles have been observed to apply equally to fixed film processes and fixed film/suspended growth combinations.

A typical response of a fixed film process to industrial waste stressing is excessive biomass growth, resulting in clogged media and reduced treatment efficiency. Treatment is generally improved if the biomass population (thickness) can be reduced, and several modifications have been successful in achieving this end. By increasing or altering shearing forces, biomass sloughing increases. This can be accomplished by altering the direction of flow through RBCs and submerged fixed film basins, or by increasing or altering the aeration pattern (if any) in the basins. A second means of inducing increased biomass sloughing is through chemical addition, but this approach is potentially dangerous and should only be attempted under the guidance of professionals skilled in the use of such chemicals.

Treatment Plant Modifications --
The most permanent type of industrial pollutant mitigation efforts come in the form of physical addition to or modification of the treatment system. Successful modification of treatment plants for industrial waste effects mitigation have included the addition of new plant facilities such as flow equalization and physical/chemical treatment steps, the addition of facilities for adding chemicals to existing treatment processes, and the modification of existing biological systems (i.e. converting to oxygen activated sludge and replacing old rock trickling filter media).

Adding flow equalization prior to biological treatment units has the effect of dampening any slug or diurnal loads of non-compatible or high-strength industrial contaminants entering a treatment plant. Pollutants that periodically enter a POTW in inhibiting concentrations can be diluted by flow equalization to non-inhibitory levels and thus, not adversely impact the biological system.

Some POTWs use a variation of this principle with success, especially when toxic metal pollutants are involved. pH and conductivity of influent wastewater are measured and recorded continuously in the influent. When the pH drops or conductivity rises drastically, possibly indicating an increased heavy metal level, the influent flow is diverted to a holding basin until such time that the metal concentrations in the influent return to normal. At that time, the diverted wastewater can be bled back to the influent wastestream in a manner such that metal concentrations are diluted and do not inhibit the biological system. This type of technique may become more useful in the future as continuously recording specific ion electrodes are developed for more pollutants.

Other treatment steps that might be added depend on the interfering industrial pollutant. The addition of flotation/skimming tanks are beneficial for removing pollutants like oils, greases or other water immiscible compounds. Separate settling basins may be beneficial in some cases for chemical treatment to precipitate metals or cause coagulation of unsettleable solids.

The addition of chemicals or nutrients to the wastewater stream in existing treatment steps has been shown in many instances to mitigate some industrial pollutant's effects. The following chemicals or additives have been shown to improve industrial wastestream treatability or biological process stability for one reason or another:

- chlorine
- phosphorus
- nitrogen
- lime (pH control)
- polymers
- alum and ferric chloride
- powdered activated carbon

Chlorine has been shown to be successful in controlling bulking sludge caused by industrial pollutants from such industries as textiles, and wood and paper products. Points of chlorine addition vary, but best results generally occur when chlorine is added to the aeration basin effluent or RAS.

Phosphorus addition and, to a lesser extent, sulfur and nitrogen addition occasionally improves biological treatment and sludge settleability of industrial wastewater with high carbonaceous content. In general, better treatment and settleability is attributed to correcting a nutrient deficient condition resulting from a high industrial/domestic wastewater ratio.

Lime is sometimes successful at mitigating the effects of some heavy metals on activated sludge systems. Addition of lime before primary treatment has the effect of raising the pH which improves precipitation of heavy metals in primary clarifiers. Optimum pH ranges exist for metal insolubilities, but these ranges are affected by many factors and are therefore system dependent.

Polymers and inorganic coagulants such as alum and ferric chloride are introduced to POTW wastestreams in part to help mitigate the effects of industrial pollutants. Added prior to primary treatment, the coagulants improve primary sedimentation and may increase the removal of toxic pollutants before they reach the aeration basin. Added after the aeration basin, the coagulant aids can assist in controlling bulking sludge and reducing effluent suspended solids. Jar testing is an important part of any chemical addition program as the best means of determining optimum dosages.

The addition of powdered activated carbon to an activated sludge unit has been successful at reducing the effect of toxic organic chemicals. By providing adsorption sites, the organic pollutants are removed by the activated carbon. The activated carbon also improves sludge settleability by providing dense floc nuclei. A patented process exists employing this treatment concept.

Another type of treatment plant modification that has seen some success at mitigating industrial interference is the replacement of an existing activated sludge unit with oxygen activated sludge. Pure oxygen activated sludge has been reported to be a more biologically stable process than conventional air facilities when responding to toxic or high-strength organic loadings. In addition, sludge settleability is reported to be improved. Increasing the efficiency of oxygen transfer in an aeration basin will help mitigate the effects of high-strength conventional pollutants. Retrofitting existing coarse bubble or turbine aeration units with fine bubble units may work to provide the added treatment capacity necessary to treat a high-strength waste.

Pretreatment and Source Control

Pretreatment and source control of interfering industrial pollutants is the most direct and efficient way of mitigating the effects of industrial pollutants because the cause of the interference never reaches the POTW. This reasoning is the impetus for the U.S. EPA promulgated General Pretreatment Regulations which specify the guidelines under which municipalities must develop pretreatment programs. It is not the intent of this report to discuss pretreatment guidelines, complete program development or details of industrial treatment processes. Rather, this discussion is intended to document elements important to bringing about pollutant source control, whether as part of a municipal/industrial cooperative agreement or a fully approved pretreatment program. The discussions are divided into three major headings:

- Industrial Waste Discharge Agreement Development
- Industrial Pretreatment Works
- Industrial Waste Discharge Agreement Monitoring and Enforcement

Industrial Waste Discharge Agreement Development -- With the exception of the categorical industries who are required to construct applicable pretreatment facilities, industries are not likely to voluntarily reduce the quantity or improve the quality of their wastewater. It is incumbent upon the municipalities to convince their industries of the importance of an effective industrial waste management program to the proper operation of a POTW. Judging from comments made during the 29 case study site visits, the single most important element necessary to realize successful implementation of source control (pretreatment) is cooperation between industry and the POTW. It was frequently stated by POTW officials that without the cooperation of industry, their source control programs would not be as successful, nor would it have been as quickly and easily implemented.

Although cooperation appears to be the key to successfully implementing source control (and thus mitigating interference at the POTW), a number of specific program elements have proven to be important:

- involve industry from the inception;
- treat industry fairly and equitably; and
- set local discharge limits.

It is important to solicit the input of industry when initiating or revising source control measures. Communicating the hows and whys of a source control program is tied to the emphasis on trust and cooperation. In some cases industries do not realize they are capable of impacting a POTW, particularly if they discharge high-strength conventional pollutants which they perceive to be compatible.

Industrial involvement in all phases of a pretreatment program is essential. In the Paris, Texas case study described in Section 4, the municipality firmly upheld the regulations, yet demonstrated a reasonable approach to their program by establishing fair milestone dates in the industrial compliance schedules. An important element is that all industries were treated in an equitable manner within a specified size classification. That is, no industry was singled out as being the "biggest problem" or "worst offender". If an industry perceives a POTW to be singling them out, they are very likely not to cooperate. It is important to include all industries in a source control program whether they are presently a problem or not.

The final point about developing and setting local limits is also illustrated by the Paris example. Industry will respond most favorably to discharge limits when the allocation of a treatment plants capacity for a particular pollutant is done in a fair and equitable manner, based on a rational wastestream monitoring program. If pollutant discharge limits for non-categorical industries are arbitrarily set, resistance to these limits are more likely than if they are developed rationally based on the treatment capacity of the POTW, the impact on the receiving stream or sludge quality considerations. Setting local limits for categorical industries may also be perceived better by industries than arbitrarily applying federal standards, despite the fact that local limits would possibly be more stringent. A rational, methodical approach to setting industrial waste discharge limits will be most acceptable to industry.

Industrial Pretreatment Works --
A source control agreement sets the framework for the mitigation of industrial interference, but it is the industrial process modifications or added industrial treatment steps which result that actually effect the change. There exists a wide variety of treatment processes applicable to industrial pretreatment, depending on the wastestream pollutants, the volume of the wastestream and the extent to which the waste must be treated. The application of specific treatment streams will not be discussed here, as it is not in the scope of work. However, typical municipal treatment processes can be applied to industrial wastestreams, and the comments in that portion of this section apply to industrial plants as well.

In many cases where industries have been forced to alter waste discharges, it has been found that wastewater flow equalization, pH neutralization or simple process modifications have been all that are necessary to meet discharge limits and eliminate interferences. Process modifications or wastestream recovery processes (such as for metals) have in some cases ended up saving industries money in addition to reducing pollutant loads.

Industrial Waste Agreement Monitoring and Enforcement --
With a source control program developed, monitoring of industries and enforcement of the industrial waste discharge permits (or sewer use ordinance) is designed to prevent and control industries from causing interferences. The extent to which this works is greatly dependent on the way in which the source control program was developed -- many of the elements of successful source control development are equally as important in monitoring. Details on wastestream monitoring, detection of interferences and tracking the sources of an interference back through a sewer system are found elsewhere in this document. The following points are monitoring and enforcement program elements that have been found to be important to the success of some programs:

- Continued industry/POTW communication
- Frequent industrial discharge sampling and analysis
- Development of an effective spill prevention program
- Consistent and appropriate response to industry discharge violations
- Listing of industrial discharge violators in a local newspaper or publication

Communication and the exchange of information must continue between the POTW and industries even after the pretreatment program has been developed. These open lines of communication can often divert an impending problem or lessen its effect. As an example, many of the 29 case study POTWs have received calls from industries when an accidental pollutant spill has occurred, thus warning operators of an impending upset condition and allowing them to possibly take precautionary measures.

It is in the best interests of any medium to large municipality to develop an accidental spill prevention program (ASPP). The EPA has funded a Guidance Manual for POTWs on the subject. The purpose of an ASPP is to provide "...a set of procedures and a regulatory structure that will minimize the chance that accidental spills of toxic materials will damage a municipality's collection system or treatment plant" (SAIC, 1986). The principal elements of an effective municipal ASPP are:

- identification of potential sources and types of spill materials;
- adequate regulatory control;
- POTW review of industrial user spill prevention programs;
- complete emergency response procedures; and
- documentation of the development strategy.

Industrial users should be primarily concerned with spill prevention, containment and cleanup procedures in their ASPP.

Consistency and fairness are stressed throughout these discussions on source control measures and they are particulary important for the response to industrial discharge violators. A consistent approach to violation punishment, tailored to the specific industries in a system, will set the stage for fewer complaints and less litigation because of violations.

Finally, as was mentioned earlier, public pressure goes a long way toward inducing industries to comply with discharge permits. Publishing the names of industrial waste violators in local newspapers is specified by EPA to be part of a permitted pretreatment program, but can also be an element of any source control program. For some industries, the possibility of adverse public reaction arising from their names being published as a "polluter" was the single most effective deterrent to discharge permit violation. This is particularly true of retail sales industries, where the company name is listed on product labels to be selected or passed over by the consuming public.

Legal and Enforcement Remedies

The final methods of mitigating POTW interference are those formal remedies used when less formal attempts to solve the problems in a cooperative fashion have failed. Often only the threat of their use is needed for other actions to resolve an issue. As defined here, the three types of measures available to most POTWs to attempt to stop an interference are: fine the nuisance industry, sue the nuisance industry or disconnect them from the sewer system altogether. These three measures are in some way tied to a sewer use ordinance, industrial waste permit, or some other sort of agreement or law regulating discharges to public sewers.

Fines --
After noncompliance warnings and efforts to cause industries to pretreat their wastes properly have failed, the enforcement option most POTWs first turn to is fines (provided regulations exist for this measure). Fines can be structured many ways, but to be effective must be linked to the severity of the interference

caused, or in proportion to the quantity of interfering pollutant. Fines are typically not intended to be a means of actually correcting an interference problem,. but rather a measure to economically pressure an industry into providing acceptable pretreatment. The intent is clearly not to put industry out of business, however a POTW needs to have methods available to protect itself from uncooperative industrial polluters.

Litigation --
POTW-initiated litigation (including public hearings) can be a first step in causing an industry to comply with a sewer use ordinance or other regulations, or it can be a further attempt to cause compliance after fines have failed to bring about the desired result. Litigation is sometimes a necessary step before proceeding to sewer disconnection, or it may be a way of enlisting public pressure to bring about a change. Litigation brings media attention and public pressure and has caused more than one industry to implement or improve pretreatment when pressure from a sewer authority has failed. In some cases (such as Bayshore, New Jersey), litigation is instituted to collect unpaid fines, which can be a sizeable sum of money if the delinquent industry is a significant contributor.

Sewer disconnection --
If litigation does not precede a sewer disconnection, disconnecting an industrial sewer line is likely to bring about litigation. The most drastic of all measures, disconnection will eliminate an interfering pollutant, but will significantly reduce the likelihood of a negotiated settlement being reached. Sewer disconnection should be judiciously considered before being implemented, but is an option available to POTWs under extreme circumstances.

References

Abeliovich, A. and Azov, Y. (1976) "Toxicity of Ammonia to Algae in Sewage Oxidation Ponds." Appl. & Environ. Microbiol. 31:801-806.

Andersen, D. R. (1981) "Pharmaceutical Wastewater Treatment: A Case Study." Proc. 35th Ind. Waste Conf. Ann Arbor Science, Ann Arbor, Michigan. 456-462.

Argaman, Y., Hucks, C. E. and Shelby, S. E., Jr. (1984) "The Effects of Organotin on the Activated Sludge Process." Water Res. 18:535-542.

Austin, S., Yunt, F. and Wuerdeman, D. (1981) "Parallel Evaluation of Air and Oxygen-Activated Sludge." EPA-600/2-81-155. U.S. EPA, Cincinnati, Ohio. 43pp.

Avendt., R. J. and Avendt, J. B. (1983) "Municipal Activated Sludge Treatment of Organometallic Pesticide Residues." Proc. 15th Mid-Atlantic Ind. Waste Conf. Butterworth Publ., Boston, MA. 619-628.

Bagby, M. M. and Sherrard, J. H. (1981) "Combined Effects of Cadmium and Nickel on the Activated Sludge Process." Journal WPCF. 53:1609-1619.

Bailey, S. W. and Zimomra, D. C. (1981) "Nationwide Survey of Heavy Metals in Municipal Sludge." Proc. 13th Mid-Atlantic Ind. Waste Conf. Ann Arbor Publishing, Ann Arbor, Michigan. 15-25.

Baird, R., Carmona, L. and Jenkins, R. L. (1977) "Behavior of Benzidine and Other Aromatic Amines in Aerobic Wastewater Treatment." Journal WPCF. 49:1609-1615.

Bard, C. C., Murphy, J. J., Stone, D. L. and Terhaar, C. J. (1976) "Silver in Photoprocessing Effluents." Journal WPCF. 48:389-394.

Barth, E. F., and Bunch, R. L. (1979) "Biodegradation and Treatability of Specific Pollutants." EPA-600/9-79-034. U.S. EPA, Cincinnati, Ohio. 60 pp.

Bauer, G. L., Hardie, M. G. and Vollstedt, T. J. (1981) "Biophysical Treatment of Coke Plant Wastewaters." Proc. 35th Ind. Waste Conf. Ann Arbor Publishing, Ann Arbor, Michigan. 332-342.

Baxter, R. A., Gilbert, P. E., Lidgett, R. A., Mainprize, J. H. and Vodden, H. A. (1975) "The Degradation of Poylchlorinated Biphenyls by Micro-Organisms." The Science of the Total Envir. 4:53-61.

Baxter, J. C., Aguiler, M. and Brown, K. (1983) "Heavy Metals and Persistent Organics at a Sewage Sludge Disposal Site." J. Environ. Qual. 12:311-316

Beckett, P. H. T., Phil, D. and Davis, R. D. (1982) "Heavy Metals in Sludge - Are Their Toxic Effects Additive?" Water Pollution Control 1982. Water Research Center, Stevenage, Herts. 112-119.

Beckwith, E., Greaves, K., Meyer, M.A., Vasudevan, C., Aulenbach, D.B. and Clesceri, N.L (1983) "Removal of Mercury in POTW Using Aluminum Salts for Phosphorus Removal." Proc. 15th Mid-Atlantic Ind. Waste Conf. Butterworth Publishers, Boston, Massachusetts. 293-303.

Bedard, R.G. (1976) "Biodegradability of Organic Compounds." M.S.. Thesis. Univ. of Connecticut. Storrs, Connecticut. 84 pp.

Beg, S. A., Siddiqi, R. H. and Ilias, S. (1981) "Effect of Toxicants on Biological Nitrification for Treatment of Ferttilizer Industry Wastewater." Proc. 35th Ind. Waste Conf. Ann Arbor Publishing, Ann Arbor, Michigan. 826-834.

Beg, S. A., Siddiqi, R. H. and Ilias, S. (1982) "Inhibition of Nitrification by Arsenic, Chromium and Fluoride." Journal WPCF. 54:482-488.

Beg, S. A., and Atiqullah, M. (1983) "Interactions of Noncompetitive Inhibitors on the Nitrification Process." Journal WPCF. 55:1080-1086.

Belly, R. T., Lauff, J. J. and Goodhue, C. T. (1975) "Degradation of Ethylenediaminetetracetic Acid by Microbial Populations from an Aerated Lagoon." Appl. Microb. 29:787-794.

Benjamin, M. M. and Leckie, J. O. (1981) "Competitive Adsorption of Cd, Cu, Zn, and Pb on Amorphous Iron Oxyhydroxide." J. Colloid and Interface Sci. 83:410-419.

Bhattacharyya, D., Jumawan, A.B. and Grieves, R.B. (1979) "Charged Membrane Ultrafiltration of Heavy Metals from Nonferrous Metal." Journal WPCF. 51:176-186.

Bhattacharyya, D., Kermode, R.I. and Dickinson, R.L. (1983) "Coal Gasification Process Wastewater Reusability: Separation of Organics by Membranes." Environ. Prog. 2:38-46.

Biesinger, M. G. and Jenkins, D. (1981) "Brewery Wastewater Treatment Without Activated Sludge Bulking Problems." Proc. 35th Industrial Waste Conf. Ann Arbor Science, Ann Arbor, Michigan. 596-609.

Bieszkiewicz, E. and Hoszowski, A. (1978) "Effect of Copper and Tri- and Hexavalent Chromium on the Work of an Activated Sludge." Acta Microbiol. Polonica. 27:147-153.

Bieszkiewicz, E., Hoi, D. V. and Matusiak, K. (1979) "Effect of Methyl Alcohol and Ethylene Glycol on the Work of Activated Sludge." Acta Microbiol. Polonica. 28:255-260.

Bishop, D. F. (1982) "The Role of Municipal Wastewater Treatment in Control of Toxics." EPA-600/D-82-360. U.S. EPA, Cincinnati, Ohio. 38 pp.

Blevins, W. T. (1982) "Factors Affecting Floc Formation and Bulking in the Activated Sludge Process for Treatment of Textile Waste Waters." Office of Water Research and Technology, U.S. Dept of Interior, Washington, D.C. 26 pp.

Blumenschein, C. D. and Helwick, R. (1983) "Removal of Chlorinated Organics Utilizing Rotating Biological Contactors." Proc. 44th Annual Mtg., Int'l Water Conf. IWC-83-2:10-15.

Braam, F. and Klapwijk, A. (1981) "Effect of Copper on Nitrification in Activated Sludge." Water Res. 15:1093-1098.

Bracewell, L. W., Jenkins, D. and Cameron, W. (1980) "Treatment of Phenol-Formaldehyde Resin Wastewater Using Rotating Biological Contactors." Proc. First Nat'l. Symposium on RBC Tech. Univ. of Pittsburgh. 733-758.

Broecker, B. and Zahn, R. (1977) "The Performance of Activated Sludge Plants Compared with the Results of Various Bacterial Toxicity Tests - A Study with 3,5-Dichlorophenol." Water Res. 11:165-172.

Brown, J. A. and Weintraub. M. (1982) "Biooxidation of Paint Process Wastewater." Journal WPCF. 54:1127-1130.

Butler, L. and Nandan, S. (1981) "Destructive Oxidation of Phenols and Sulfides using Hydrogen Peroxide." Water 1980: AIChE Symposium Series. 77:108-111.

Button, M. P. and Gaudy, A. F., Jr. (1982) "A Process for Removal of Heavy Metals from Secondary Sludge." Proc. 36th Ind. Waste Conf. Ann Arbor Publishing, Ann Arbor, Michigan. 509-518.

Cain, C. B., Russell, L. L., and Keramida, V. (1983) "Development of Reasonable Limitations on Discharge of Industrial Pollutants to Municipal Sewers." Proc. 56th Annual Conf., WPCF.

Capestany, G. J., McDaniels, J. and Opgrande, J. L. (1977) "The Influence of Sulfate on Biological Treatment of Phenolbenzaldehyde Wastes." Journal WPCF. 49:256-261.

Casey, J. D. and Wu, Y. C. (1978) "Removal of Copper and Cadmium by Metabolically Controlled Activated Sludge." Proc. 32nd Ind. Waste Conf. Ann Arbor Publishing, Ann Arbor, Michigan. 141-152.

Cawley, W. A. Jr. (1981) "Comparison of Bacterial Luminescence Bioassay and Fish Bioassay for Complex Wastewater." M.S. Thesis, University of Cincinnati, Cincinnati, Ohio. 75 pp.

Cenci, G. and Morozzi, G. (1977) "Evaluation of the Toxic Effect of Cd^{2+} and $Cd(CN)_4^{2-}$ Ions on the Growth of Mixed Microbial Population of Activated Sludges. The Science of Total Environ. 7:131-143.

Chanda, M., O'Driscoll, K. F. and Rampel, G. L. (1983) "Cyanide Detoxification by Selective Ion Exchange with Protonated Poly (4-Vinyl Pyridine)." J. Chem. Tech. Biotechnol. 33A:97-108.

Chang, A. C., Page, A. L. and Bingham, F. T. (1981) "Re-utilization of Municipal Wastewater Sludges - Metals and Nitrate." Journal WPCF. 53:237-245.

Chang, S. Y., Huang, J. C., Liu, Y.C. (1985) "Effects of Cd and Cu on a Biofilm Treatment System." Proc. 39th Ind. Waste Conf. Butterworth Pub., Boston, Massachusetts. 305-313.

Chavalitnitikul, C. and Brunker, R.L. (1981) "The Removal of Heavy Metals from Sewage Influent Waters by Foam Flotation." Proc. 13th Mid-Atlantic Ind. Waste Conf. Ann Arbor Publishing, Ann Arbor, Michigan. 72-86.

Chen, Y.S.E., Peluso, R.A. and Mureebe, A.K. (1984) "Mercury Waste Treatment" Proc. Ind. Wastes Symposia, 57th Annual WPCF Conf. 242-256.

Chesler, P. G. and Eskelund, G. R. (1980) "RBC for Munitions Wastewater Treatment." Proc. First Nat'l. Symposium on RBC Tech. Univ. of Pittsburgh. 711-723.

Choate, W. T., Houldsworth, D. and Butler, G. A. (1983) "Membrane-Enhanced Anaerobic Digesters." Proc. 37th Ind. Waste Conf. Ann Arbor Publishing, Ann Arbor, Michigan. 661-666.

Chou, W. L., Speece, R. E., Siddiqi, R. H. and McKeon, K. (1978) "The Effect of Petrochemical Structure on Methane Fermentation Toxicity." Prog. Water Tech. 10:545-558.

Chow, S. T. and Ng, T. L. (1983) "The Biodegradation of N-Methyl-2-Pyrrolidone in Water by Sewage Bacteria." Technical Note. Water Res. 17:117-118.

Christensen, E. R. and Delwiche, J. T. (1982) "Removal of Heavy Metals from Electroplating Rinsewaters by Precipitation, Flocculation and Ultrafiltration." Water Res. 16:729-737.

Clevenger, T. E., Hemphill, D. D., Roberts, K. and Mullins, W. A. (1983) "Chemical Composition and Possible Mutagenicity of Municipal Sludges." Journal WPCF. 55:1470-1475.

Coker, E. G. and Matthews, P. J. (1983) "Metals in Sewage Sludge and Their Potential Effects in Agriculture." Wat. Sci. Tech. 15:209-225.

Cook, K. A. (1979) "Degradation of the Non-Ionic Surfactant Dobanol 45-7 By Activated Sludge." Water Res. 13:259-266.

Cornwell, D. A. and Westerhoff, G. F. (1980) "Extract Heavy Metals Via Liquid Ion-Exchange." Water and Wastes Engr. 17:36-42.

Coulter, R. R. (1984) "Pilot Plant Study of Landfill Leachate Treatability by Rotating Biological Contactors." Proc. 2nd Int'l Conf. on Fixed-Film Biological Processes. Univ. of Pittsburgh. 777-813.

Cox, D. P. (1978) "The Biodegradation of Polyethylene Glycols." Adv. Appl. Microbiol. 23:173-194.

Culp, Wesner, Culp (1978) "Field Manual for Performance Evaluation and Troubleshooting at Municipal Wastewater Treatment Facilities." EPA-430-9-78-001. U.S. EPA, Washington, DC. 387 pp.

Cummins, M. D. (1981) "Effect of Sanitary Landfill Leachate on the Activated Sludge Process." EPA-600/9-81-002a. U.S. EPA, Cincinnati, Ohio. 170-178.

Dagon, T. J. (1973) "Biological Treatment of Photo Processing Effluents." Journal WPCF. 45:2123-2135.

Dechev, G. B. and Matveeva, E. G. (1977) "Study of Oxygen Uptake Variations of Activated Sludge Depending on Phenol Concentration." Biologie Biophysique. 30:595-598.

DeLuca, S.J., Chao, A.C. and Smallwood, C. Jr. (1983) "Removal of Organic Priority Pollutants by Oxidation-Coagulation." Journal Environ. Engr. Div., ASCE. 109:36-46.

DeWalle, F. B., Kalman, D. A., Dills, R., Norman, D., Chian, E. S. K., Giabbai, M. and Ghosal, M. (1982) "Presence fo Phenolic Compounds in Sewage, Effluent and Sludge from Municipal Sewage Treatment Plants." Water Science Tech. 14:143-150.

DiGeronimo, M. J., Nikaido, M. and Alexander, M. (1979) "Utilization of Chlorobenzoates by Microbial Populations in Sewage." Appl. and Environ. Microbiol. 37:619-625.

Dohanyos, M., Madera, V. and Sedlacek, M. (1978) "Removal of Organic Dyes by Activated Sludge." Prog. Water Tech. 10:559-575.

Dunn, G. F., Jr. and Hutton, D. G. (1983) "The Combined Powdered Activated Carbon-Activated Sludge (PACT) Process for Toxics Control." Water Resources Symposium No. 10. Univ. of Texas, Austin, Texas. 53-76.

E. C. Jordan Co. (1982) "Fate of Priority Pollutants in Publicly Owned Treatment Works: 30 Day Study." EPA-440/1-82/302. U.S. EPA, Washington, DC. 263 pp.

E. C. Jordan Co. (1984) "Combined Sewer Overflow Toxic Pollutant Study." EPA-440/1-84-304. U.S. EPA, Washington, DC. 212 pp.

Edgehill, R. U. and Finn, R. K. (1983) "Activated Sludge Treatment of Synthetic Wastewater Containing Pentachlorophenol." Biotech and Bioengr. 25:2165-2176.

Eis, B. J., Ferguson, J. F. and Benjamin, M. M. (1983) "The Fate and Effect of Bisulfate in Anaerobic Treatment." Journal WPCF. 55:1355-1365.

El-Dib, M. A. and Badawy, M. I. (1979) "Adsorption of Soluble Aromatic Hydrocarbons on Granular Activated Carbon." Water Res. 13:255-258.

Elliott, W. R., Riding, J. T. and Sherrard, J. H. (1978) "Maximizing Phosphorus Removal in Activated Sludge." Wat. & Sew. Works. March, 1978:88-92.

Environmental Resources Training Center, Southern Illinois University (1979) "A Course on Troubleshooting O&M Problems in Wastewater Treatment Facilities." U.S. EPA, Cincinnati, Ohio.

Farmwald, J. A. and MacNaughton, M. G. (1981) "Effects of Hydrazine on the Activated Sludge Process" Journal WPCF. 53:565-575.

Fenger, B. H., Mandrup, M., Rohde, G. and Sorensen, J. C. K. (1979) "Degradation of a Cationic Surfactant in Activated Sludge Pilot Plants." Water Res. 7:1195-1208.

Ferguson, J. F., Keay, G. F. P., Merrill, M. S. and Benedict, A. H. (1979) "Powdered Activated Carbon in Contact Stabilization Activated Sludge." Journal WPCF. 51:2314-2323.

Fox, C. R. (1979) "Removing Toxic Organics from Wastewater." Chem Engr. Prog. 75:70-77.

Fox, L. L. and Merrick, N. J. (1983) "Controlling Residual Polychlorinated Biphenyls in Wastewater Treatment Through Conventional Means." Proc. 37th Industrial Waste Conf. Ann Arbor Science, Ann Arbor, Michigan. 413-423.

Fricke, C., Clarkson, C., Lomnitz, E. and O'Farrell, T. (1985) "Comparing Priority Pollutants in Municipal Sludges." Biocycle. Jan./Feb. 1985:35-37.

Fry, F. F., Smith, T. G. and Sherrard, J. H. (1982) "Start-up and Shock Loading Characteristics of a Rotating Biological Contactor Package Plant." Proc. First Int'l Conf. on Fixed-Film Biological Processes. Univ. of Pittsburgh. 542-568.

Furukawa, K., Tonomura, K. and Kamibayashi, A. (1978) "Effect of Chlorine Substitution on the Biodegradability of Polychlorinated Biphenyls." Appl and Environ. Microbiol. 35:223-227.

Gaffney, P.E. (1976) "Carpet and Rug Industry Case Study I: Water and Wastewater Treatment Plant Operation." Journal WPCF. 48:2590-2598.

Ganczarczyk, J. J. (1979) "Second-Stage Activated Sludge Treatment of Coke-Plant Effluents." Water Res. 13:337-342.

Gaudy, A. F., Jr., Gaudy, E. T., Feng, Y. J. and Brueggemann, G. (1982 a) "Treatment of Cyanide Waste by the Extended Aeration Process." Journal WPCF. 54:153-164.

Gaudy, A. F. Jr., Kincannon, D. F. and Manickam, T.S. (1982 b) "Treatment Compatibility of Municipal Waste and Biologically Hazardous Industrial Compounds." EPA-600/2-82-075a. U.S. EPA, Ada, Oklahoma. 203 pp.

Geating, J. (1981) "Literature Study of the Biodegradability of Chemicals in Water (Vols. 1 and 2)" U.S. EPA, Cincinnati, Ohio. 241 pp.

George, T. K. and Gaudy, A. F., Jr. (1973a) "Transient Reponse of Continuously Cultured Heterogeneous Populations to Changes in Temperatures." Appl. Microbiol. 26:796-803.

George, T. K. and Gaudy, A. F., Jr. (1973b) "Response of Completely Mixed Systems to pH Shock." Biotech & Bioengr. 15:933-949.

Gerardi, M. H. (1981) "Sludge Settling Hampered by Industrial Discharge." Public Works. June, 1981:95-97.

Gerber, V. Y., Gorobets, L. D. and Ioakimis, E. G. (1979) "Effects of Urea and Dichloromethane on Biochemical Treatment of Refinery Wastewater" in Chemistry and Technology of Fuels and Oils. Plenum Publishing. 269-272.

Gerike, P., Fischer, W. K. and Jasiak, W. (1978) "Surfactant Quaternary Ammonium Salts in Aerobic Sewage Digestion." Water Res. 12:1117-1122.

Gledhill, W. E. (1975) "Biodegradation of 3,4,4' - Trichlorocarbanilide (TCC) in Sewage and Activated Sludge." Water Res. 9:649-654.

Gray, A. C. Jr., Paul, P. E. and Roberts, H. D. (1979) "Evaluation of Operation and Maintenance Factors Limiting Biological Wastewater Treatment Plant Performance." EPA-600/2-79-078. U.S. EPA, Cincinnati, Ohio. 149 pp.

Greenfield, J. H. and Neufeld, R. D. (1982) "Quantification of the Influence of Steel Industry Trace Organic Substances on Biological Nitrification." Proc. 36th Ind. Waste Conf. Ann Arbor Publishing, Ann Arbor, Michigan. 772-783.

Gutierrez, A. G., McIntyre, A. E., Perry, R. and Lester, J. N. (1984) "Behaviour of Persistent Organochlorine Micropollutants During Primary Sedimentation of Wastewater." The Science of the Total Envir. 39:27-47.

Hahn, W. H., Barnhart, E. L. and Meighan, R. B. (1977) "The Biodegradability of Synthetic Size Material Used in Textile Processing." Proc. 30th Ind. Waste Conf. Ann Arbor Publishing, Ann Arbor, Michigan. 530-539.

Haines, J. R. and Alexander, M. (1975) "Microbial Degradation of Polyethylene Glycols." Appl. Microbiol. 29:621-625.

Hammer, M. J. (1983) "Rotating Biological Contactors Treating Combined Domestic and Cheese-Processing Wastewaters." Proc. 37th Ind. Waste Conf. Ann Arbor Science, Ann Arbor, Michigan. 29-38.

Hamza, A. A., El-Sharkawi, F. M. and Younis, M. A. (1982) "Biofiltration of Tannery Wastewater." Proc. First Intl. Conf. on Fixed-Film Biological Processes. Univ. of Pittsburgh. 1093-1112.

Hannah, S. A., Austern, B. M., Eralp, A. E. and Wise, R. H. (1985) "Comparative Removal of Toxic Pollutants By Six Wastewater Treatment Processes." submitted to Journal WPCF for publication.

Hathaway, S.W. (1980) "Sources of Toxic Compounds in Household Wastewater." U.S. EPA, Cincinnati, Ohio. 83 pp.

Haugh, R., Niku, S., Schroeder, E. and Tchobanoglous, G. (1981) "Performance of Trickling Filter Plants: Reliability, Stability and Variability." Project Summary. EPA-600/S2-81-228, U.S. EPA, Cincinnati, Ohio. 13 pp.

Healy, J. B., Jr. and Young, L. Y. (1978) "Catechol and Phenol Degradation by a Methanogenic Population of Bacteria." Appl. and Envir. Microb. 35:216-218.

Healy, J. B., Jr. and Young, L. Y. (1979) "Anaerobic Biodegradation of Eleven Aromatic Compounds to Methane." Appl and Envir. Microb. 38:84-89.

Heddle, J. F. (1979) "Activated Sludge Treatment of Slaughterhouse Wastes with Protein Recovery." Water Res. 13:581-584.

Hegg, B. A., Rakness, K. L., Schultz, J. R. and DeMers, L. D. (1980) "Evaluation of Operation and Maintenance Factors Limiting Municipal Wastewater Treatment Plant Performance: Phase II." EPA-600/2-80-129. U.S. EPA, Cincinnati, Ohio. 169 pp.

Hegg, B.A., Schultz, J.R., and Rakness, K.L. (1984) "Improving POTW Performance Using the Composite Correction Program Approach." EPA-625/6-84-008. U.S. EPA, Cincinnati, Ohio. 258 pp.

Henney, R. C., Fralish, M. C. and Lacina, W. V. (1980) "Shock Load of Chromium (VI)." Journal WPCF. 52:2755-2760.

Herbst, E., Scheunert, I., Klein, W. and Korte, F. (1977) "Fate of PCB's-^{14}C In Sewage Treatment-Laboratory Experiments with Activated Sludge." Chemosphere. 11:725-730.

Hill, D. O. and Gelman. S. R. (1978) "Effects of Chloride on Nitrification Rates in Activated Sludge Systems." Proc. 32nd Ind. Wastes Conf. Ann Arbor Publishing, Ann Arbor, Michigan. 206-213.

Hill, N. P., McIntyre, A. E., Perry, R. and Lester, J. N. (1985) "Behavior of Chlorophenoxy Herbicides During Primary Sedimentation." Journal WPCF. 57:60-67.

Holladay, D. W., Hancher, C. W., Scott, C. D. and Chilcote, D. D. (1978) "Biodegradation of Phenolic Waste Liquors in Stirred-Tank, Packed-Bed, and Fluidized-Bed Bioreactors." Journal WPCF. 50:2573-2589.

Horning, R. H. (1978) "Textile Dyeing Wastewaters: Characterization and Treatment." EPA-600/2-78-098. U.S. EPA, Research Triangle Park, North Carolina. 312 pp.

Hrubec, J., van Kreijl, C. F., Morra, C. F. H. and Slooff, W. (1983) "Treatment of Municipal Wastewater by Reverse Osmosis and Activated-Carbon-Removal of Organic Micropollutants and Reduction of Toxicity." The Science of the Total Environ. 27:71-88.

Huang, C. P. and Wirth, P. K. (1981) "Treatability of Cadmium (II) Plating Wastewater by Aluminosilicate Adsorption." Proc. 13th Mid-Atlantic Ind. Wastes Conf. Ann Arbor Publishing, Ann Arbor, Michigan. 87-94.

Huang, J. Y. C. and Skeikhdeslami, B. (1983) "Metal Inhibition on Nitrification." Proc. 37th Ind. Waste Conf. Ann Arbor Publishing, Ann Arbor, Michigan. 85-93.

Hung, Y. T., Fossum, G. O., Paulson, L. E. and Willson, W. G. (1981) "Assessment of Activated Sludge Process in Treating Solvent-Extracted Coal Gasification Wastewaters." Proc. 13th Mid-Atlantic Ind. Waste Conf. Ann Arbor Publishing, Ann Arbor, Michigan. 228-238.

Hutton, D. G. and Temple, S. (1979) "Priority Pollutant Removal: Comparison of Dupont PACT Process and Activated Sludge." Chemical and Petrochemical Waste Treatment Session. Proc. Ind. Waste Symposia, 52nd WPCF Conf. 1-19.

Iannone, J., Pai, M., and Papamichael, F. (1984) "Organic Priority Pollutants in New York City Wastewater: Their Sources and Impacts." Proc. Ind. Wastes Symposia, 57th Annual Conf., WPCF. 392-405.

Imai, H., Endoh, K., Kobayashi, C. (1979) "Activated Sludge During Acclimation to Saline Water." J. Ferment. Technol. 57:454-459.

Ishikawa, T., Ose, Y. and Sato, T. (1979) "Removal of Organic Acids by Activated Sludges." Water Res. 13:681-685.

Ito, K. and Matsuo, T. (1980) "The Effect of Organic Loading on Nitrification in RBC Wastewater Treatment Processes." Proc. First Nat'l Symposium on RBC Tech. Univ. of Pittsburgh. 1165-1175.

Janeczek, J. Jr. and Lamb, J. C. III (1983) "Treatability of a Coal Gasification Wastewater Using the Powdered Activated Carbon/Activated Sludge Process." Proc. 37th Ind. Waste Conf. Ann Arbor Publishing, Ann Arbor, Michigan. 497-505.

Jank, B. E., Guo, H. M. and Cairns, V. W. (1974) "Activated Sludge Treatment of Airport Wastewater Containing Aircraft De-Icing Fluids." Water Res. 8:875-880.

Jeffus, H. M. (1981) "Problems with Metals in the Residue from Combined Municipal/Industrial Waste Treatment." Proc. Conf. on Combined Mun./Ind. Wastewater Treat. EPA-600/9-81-021. U.S. EPA, Ada, Oklahoma. 544-550.

Jenke, D. R., Pagenkopf, G. K. and Diebold, F. E. (1983) "Chemical Changes in Concentrated, Acidic, Metal-Bearing Wastewaters When Treated with Lime." Environ. Sci. Technol. 17:217-223.

Jenkins, R. L., Scheybeler, B. J., Smith, M. L., Baird, R., Lo, M. P. and Haug, R. T. (1981) "Metals Removal and Recovery from Municipal Sludge." Journal WPCF. 53:25-32.

Joel, A. R. and Grady, C. P. L., Jr. (1977) "Inhibition of Nitrification - Effects of Aniline After Biodegradation." Journal WPCF. 49:778-788.

Johnson, L. D. and Young, J. C. (1983) "Inhibition of Anaerobic Digestion by Organic Priority Pollutants." Journal WPCF. 55:1441-1449.

Jones, D. D., Speake, J. L., White, J. and Gauthier, J. J. (1984) "Biological Treatment of High-Strength Coke-Plant Wastewater." Proc. 38th Ind. Waste Conf. Butterworth Pub., Boston, MA. 561-570.

JRB Associates (1981a) "304(g) Guidance Document: Revised Pretreatment Guidelines (Vols. I and II)." Internal Report. U.S. EPA, Cincinnati, Ohio.

JRB Associates (1981b) "Assessment of the Impacts of Industrial Discharges on Publicly Owned Treatment Works." Report submitted to the Office of Water Enforcement, U.S. EPA, Washington, D.C.

JRB Associates (1982-1984) "POTW Inspection Reports(s)." A Series of Internal Reports (Contract No. 68-01-6514). U.S. EPA, Washington, DC.

Kang, S. J., Bulkkey, J. W. and Spangler, J. L. (1981) "Fate of Heavy Metals and Tolerance Limits in POTW." Proc. 1981 Environ. Engr. Div. Conf., ASCE. 400-407.

Kang, S. J. and Borchardt, J. A. (1982) "Inhibition of Nitrification by Chromium in a Biodisc System." Proc. First Int'l. Conf. on Fixed-Film Biological Processes. Univ. of Pittsburgh. 990-1006.

Kao, J. F., Hsieh, L. P., Cheng, S. S. and Huang, C. P. (1982) "Effect of EDTA on Cadmium in Activated Sludge Systems." Journal WPCF. 54:1118-1126.

Kao, J. F. and Kang, S. F. (1984) "A Study of Soda Pulp Wastewater Treatment by Submerged Biofilter." Proc 2nd Int'l. Conf. on Fixed-Film Biological Processes. Univ. of Pittsburgh. 1681-1692.

Kashiwaya, M. and Yoshimoto, K. (1980) "Tannery Wastewater Treatment by the Oxygen Activated Sludge Process." Journal WPCF. 52:999-1007.

Kennedy, K. J. and vandenBerg, L. (1982) "Effects of Temperature and Overloading on the Performance of Anaerobic Fixed-Film Reactors." Proc. 36th Ind. Waste Conf. Ann Arbor Publishing, Ann Arbor, Michigan. 678-685.

Khan, K. A., Suidan, M. T. and Cross, W. H. (1981) "Anaerobic Activated Carbon Filter for the Treatment of Phenol-Bearing Wastewater." Journal WPCF. 53:1519-1532.

Khararjian, H. A., Smith, J. W. and Ledoux, G. A. (1979) "Treatment of Phenolic Wastewater." Proc. 11th Mid. Atlantic Ind. Waste Conf. Ann Arbor Publishing, Ann Arbor, Michigan, 189-195.

Kim, J. W. and Armstrong, N. A. (1981) "A Comprehensive Study on the Biological Treatabilities of Phenol and Methanol-II: The Effects of Temperature, pH, Salinity and Nutrients." Water Res. 15:1233-1247.

Kim, B. M. and Amodeo, P. A. (1983) "Calcium Sulfide Process for Treatment of Metal-Containing Wastes." Environ. Prog. 2:175-180.

Kincannon, D. F., Gaudy, A. F. Jr., and Manickam, T.S. (1981) "Treatment of Municipal Wastewaters Containing Biologically Hazardous Industrial Compounds by Conventional Activated Sludge and Extended Aeration." Proc. Conf. on Combined Mun./Ind. Wastewater Treat. EPA-600/9-81-021. U.S. EPA, Ada, Oklahoma. 60-78.

Kincannon, D. F., Esfandi, A. and Manickam, T. S. (1982) "Compatibility of Semiconductor Industry Wastewater with Municipal Activated Sludge Systems." Proc. 36th Ind. Waste Conf. Ann Arbor Publishing, Ann Arbor, Michigan. 533-539.

Kincannon, D. F., Stover, E. L., Nichols, V. and Medley, D. (1983) "Removal Mechanisms for Toxic Priority Pollutants." Journal WPCF. 55:157-163.

Klecka, G. M. (1982) "Fate and Effects of Methylene Chloride in Activated Sludge." Appl and Environ. Microbiol. 44:701-707.

Klein, S. A. (1974) "NTA Removal in Septic Tank and Oxidation Pond Systems." Journal WPCF. 46:78-88.

Klemetson, S.L. and Scharbow, M.D. (1979) "Filtration of Phenolic Compounds in Coal Gasification Wastewater." Journal WPCF. 51:2752-2763.

Kobylinski, E. A. and Bell, B. A. (1983) "Light Metal Cation Inhibition in Anaerobic Digestion." Proc. 1983 Nat'l. Conf. on Env. Engr., ASCE. 399-402.

Koch, C. M., Stroka, J. G., Perna, R. K. and Forester, R. E. (1982) "Impact of Pretreatment on Sludge Content of Heavy Metals." Journal WPCF. 54:339-343.

Kunz, R. G., Giannelli, J. F. and Stensel, H.D. (1976) "Vanadium Removal From Industrial Wastewaters." Journal WPCF. 48:762-770.

Landon-Arnold, S. and Chan, D. B. (1982) "Application of Rotating Biological Contactor (RBC) Process for Treatment of Wastewater Containing a Firefighting Agent (AFFF). Proc. First Int'l. Conf. on Fixed-Film Biological Processes. Univ. of Pittsburgh. 927-943.

Landon-Arnold, S. and Chan, D. B. (1984) "Microbial Treatability of Aqueous Film Forming (AFFF) with a Rotating Biological Contactor." Proc. 2nd Int'l. Conf. on Fixed-Film Biological Processes. 1296-1314.

Lau, C. M. (1978) "Staging Aeration for High-Efficiency Treatment of Aromatic Acids Plant Wastewater." Proc. 32nd Ind. Waste Conf. Ann Arbor Publishing, Ann Arbor, Michigan. 63-74.

Lawson, C. T. and Siegrist, S. A. (1981) "Removal Mechanisms for Selected Priority Pollutants in Activated Sludge Systems." Proc. of the 1981 National Conf. on Environ. Engr., ASCE. 356-363.

Leipzig, N. A. and Hockenbury, M. R. (1980) "Powdered Activated Carbon/Activated Sludge Treatment of Chemical Production Wastewaters." Proc. 34th Ind. Waste conf. Ann Arbor Publishing, Ann Arbor, Michigan. 195-205.

Lense, F. T., Mileski, S. E. and Ellis, C. W. (1978) "Effects of Liquid Detergent Plant Effluent on the Rotating Biological Contactor." EPA-600/2-78-129. U.S. EPA, Cinncinnati, Ohio. 57 pp.

Lester, J. N. (1983) "Significance and Behviour of Heavy Metals in Wastewater Treatment Processes: I. Sewage Treatment and Effluent Discharge." The Science of the Total Environment. 30:1-44.

Levins, P., Adams, J., Brenner, P., Coons, S., Harris, G., Jones, C., Thrun, K. and Wechsler, A. (1981) "Sources of Toxic Pollutants Found in Influents to Sewage Treatment Plants: Vol. VI. Integrated Interpretation." EPA-440/4-81-008. U.S. EPA, Washington, DC. 125 pp.

Li, C. T., Chen, H. T. and Wu, Y. C. (1982) "Treatment of Starch Industrial Waste by RBCs." Proc. First Int'l. Conf. on Fixed-Film Biological Processes. Univ. of Pittsburgh. 960-989.

Li, C. T. Hwu, N. T. and Whang, J. S. (1984) "Treatment of Slaughterhouse Wastewater by Biofiltration Tower." Proc. 2nd Intl. Conf. on Fixed-Film Biological Processes. Univ. of Pittsburgh. 1336-1359.

Liu, D. (1980) "Enhancement of PCBs Biodegradation by Sodium Ligninsulfonate." Water Res. 14:1467-1475.

Lordi, D. T., Lue-Hing, C. and Whitebloom, S. W. (1980) "Cyanide Problems in Muicipal Wastewater Treatment Plants." Journal WPCF. 52:597-609.

Lowry, J. D., and Chwirka, J. (1983) "Papermill Wastewater Treatment." Envir. Progress. 2:158-166.

Luthy, R. G., Sekel, D. J. and Tallon, J. T. (1980) "Biological Treatment of Synthetic Fuel Wastewater." J. Environ. Engr. Div., ASCE. 106:609-629.

Luthy, R. G. and Jones, L. D. (1980) "Biological Oxidation of Coke Plant Effluent." J. Environ. Engr. Div., ASCE. 106:847-851.

Luthy, R. G. (1981) "Treatment of Coal Coking and Coal Gasification Wastewaters." Journal WPCF. 53:325-339.

Lytle, P. E. (1984) "Treatment of Photofinishing Effluents Using Rotating Biological Contactors (RBCs)." Journal of Imaging Tech. 10:221-226.

Manickan, T. S. and Gaudy, A. F., Jr. (1983) "Comparison of Activated Sludge Response to Quantitative, Hydraulic and Combined Shock for the Same Increases in Mass Loading." Proc. 37th Ind. Waste Conf. Ann Arbor Publishing, Ann Arbor, Michigan. 601-618.

Matsui, S., Murakami, T., Sasaki, T., Hirose, Y. and Iguma, Y. (1975) "Activated Sludge Degradability of Organic Substances in the Wastewater of the Kashima Petroleum and Petrochemical Industrial Complex in Japan." Prog. Water Tech. 7:645-659.

Matsumoto, J. and Noike, T. (1978) "Effects of Heavy Metals on Anaerobic Sludge Digestion: I. Studies by the Batch Digestion Experiment." Tech. Reports. Tohoka Univ. 43:173-189.

McCarty, P.L. and Reinhard, M. (1980) "Trace Organics Removal by Advanced Wastewater Treatment." Journal WPCF. 52:1907-1922.

McIntyre, A. E., Lester, J. N. and Perry, R. (1981a) "The Influence of Chemical Conditioning and Dewatering on the Distribution of Polychlorinated Biphenyls and Organochlorine Insecticides in Sewage Sludges." Envir. Pollut. (Series B). 2:309-320.

McIntyre, A. E., Perry, R. and Lester, J. N. (1981b) "The Behaviour of Polychlorinated Biphenyls and Organochlorine Insecticides in Primary Mechanical Wastewater Treatment." Envir. Poll. (Series B). 2:223-233.

McIntyre, G., Rodriguez, J.J., Thackston, E.L. and Wilson, D.J. (1983) "Inexpensive Heavy Metal Removal by Foam Flotation." Journal WPCF. 55:1144-1149.

McKinney, R. E. (1977) "Performance Evaluation of an Existing Lagoon System at Eudora, Kansas." EPA-600/2-77-167. U.S. EPA, Cincinnati, Ohio. 240 pp.

McManus, A. M. C., Werthman, P. H. and Westendorf, J. R. (1985) "Granular Activated Carbon Removal of Priority Pollutants in a Combined Municipal/Industrial Wastewater." Proc. 39th Ind. Waste Conf. Butterworth Pub., Boston, Massachusetts. 719-735.

Medley, D.R. and Stover, E.L. (1983) "Effects of Ozone on the Biodegradability of Biorefractory Pollutants." Journal WPCF. 55:489-494.

Medwith, B. W. and Lefelhocz, J. F. (1982) "Single-Stage Biological Treatment of Coke-Plant Wastewaters with a Hybrid Suspended Growth-Fixed Film Reactor." Proc. 36th Ind. Waste Conf. Ann Arbor Publishing, Ann Arbor, Michigan. 68-76.

Miller, S., Abeliovich, A. and Belfort, G. (1977) "Effects of High Organic Loading on Mixed Photosynthetic Wastewater Treatment." Journal WPCF. 49:436-440.

Mirzadeh, A., Maeda, Y. and Fazeli, A. (1977) "Effects of Sodum Bentonite and Ferric Chloride on Activated Sludge Treatment of Wastewater." J. Ferment. Technol. 55:258-264.

Monnig, E. C., Little, L. W. and Zweidinger, R. (1981) "Investigations on the Suitability of Various Pesticide Manufacturing Wastewater for Discharge to Municipal Waste Treatment Facilities." Proc. Conf. on Combined Mun./Ind. Wastewater Treat. EPA-600/9-81-021. U.S. EPA, Ada, Oklahoma. 259-271.

Moore, L. and Barth, E. F. (1976) "Degradation of NTA Acid During Anaerobic Digestion." Journal WPCF. 48:2406-2409.

Moos, L. P., Kirsch, E. J., Wukasch, R. F. and Grady, C. P. L., Jr. (1983) "Pentachlorophenol Biodegradation - I:Aerobic." Water Res. 17:1575-1584.

Morimoto, M. K. and Nambu, S. (1976) "The Response of Activated Sludge to a Polychlorinated Biphenyl (KC-500)." Water Res. 10:157-163.

Morozzi, G. and Cenci, G. (1978) "Comparison of the Toxicity of Some Metals and Their Tetracyanide Complexes on the Respiration of Non Acclimated Activated Sludges." Zbl. Bakt. Hyg. 167:478-488.

Mosey, F. E. and Hughes, D. A. (1975) "The Toxicity of Heavy Metal Ions to Anaerobic Digestion." Water Pollution Control 1975. Water Research Center, Stevenage, Herts. 18-39.

Muttamora, S. and Islam, S. (1983) "Effect of Chromium on Activated Sludge Process Performance." Proc. 15th Mid-Atlantic Ind. Waste Conf. Butterworth Pub., Boston, MA. 144-154.

Nall, A.E. (1980) "Economically Remove Toxics." Water and Wastes Engr. Feb. 1980:43, 48, 52-53.

Nay, M. W. Jr., Randall, C. W. and King, P. H. (1974) "Biological Treatability of Trinitrotoluene Manufacturing Wastewater." Journal WPCF. 46:485-497.

Naylor, L. M. and Loehr, R. C. (1982a) "Priority Pollutants in Municipal Sewage Sludge." Biocycle. July/Aug. 1982:18-22.

Naylor, L. M. and Loehr. R. C. (1982b) "Priority Pollutants in Municipal Sewage Sludge - Part II." Biocycle. Nov./Dec. 1982:37-42.

Neel, J. K., Vennes, J. W., Fossum, G. O. and Orthmeyer, F. B. (1976) "Anaerobic and Aerobic Treatment of Combined Potato Processing and Municipal Wastes." EPA-600/2-76-236. U.S. EPA, Cincinnati, Ohio. 142 pp.

Neiheisel, T. W., Horning, W. B., Petrasek, A. C., Asberry, V. R., Jones, D. A., Marcum, R. L. and Hall, C. T. (1982) "Effects on Toxicity of Volatile Priority Pollutants Added to a Conventional Wastewater Treatment System." Internal Report. U.S. EPA, Cncinnati, Ohio. 9 pp.

Nelson, P. O., Chung, A. K. and Hudson, M. C. (1981) "Factors Affecting the Fate of Heavy Metals in the Activated Sludge Process." Journal WPCF. 53:1323-1333.

Neufeld, R. D. (1976) "Heavy Metals-Induced Deflocculation of Activated Sludge." Journal WPCF. 48:1940-1947.

Neufeld, R. D. and Valiknae, T. (1979) "Inhibition of Phenol Biodegradation by Thiocyanate." Journal WPCF. 51:2283-2291.

Neufeld, R. D., Hill, A. J. and Adekoya, D. O. (1980a) "Phenol and Free Ammonia Inhibition to Nitrosomonas Activity." Water Res. 14:1695-1703.

Neufeld, R. D., Mack, J. D. and Strakey, J. P. (1980b) "Anaerobic Phenol Biokinetics." Journal WPCF. 52:2367-2377.

Neufeld, R. D., Mattson, L. and Lubon, P. (1981) "Thiocyanate Bio-Oxidation Kinetics." J. Environ. Engr. Div., ASCE. 107:1035-1049.

Newbry, B. W., Macaulay, M. N., Musterman, J. L. and Davison, W. E., Jr. (1982) "Troubleshooting an Existing RBC Facility." Proc. First Int'l Conf. on Fixed-Film Biological Processes. Univ. of Pittsburgh. 1710-1730.

Nielson, J. S. and Hrudey, S. E. (1983) "Metal Loadings and Removal at a Municipal Activated Sludge Plant." Water Res. 17:1041-1052.

Niku, S., Schroeder, E.D., Tchobanoglous, G. and Samaniego, F. J. (1981) "Performance of Activated Sludge Processes: Reliability, Stability and Variability." Project Summary. EPA-600/S2-81-227. U.S. EPA, Cincinnati, Ohio. 11 pp.

Novak, J. T. and Kraus, D. L. (1973) "Degradation of Long Chain Fatty Acids by Activated Sludge." Water Res. 7:843-851.

Nutt, S.G. and Zaidi, S.A (1984) "Treatment of Cyanide-Containing Wastewaters by the Copper-Catalyzed SO_2/Air Oxidation Process." Proc. 38th Ind. Wastes Conf. Ann Arbor Publishing, Ann Arbor, Michigan. 357-368.

Nyer, E. K. and Bourgeois, H. J., Jr. (1981) "Operational Troubleshooting in Industrial Biological Treatment Systems." Proc. 35th Ind. Waste Conf. Ann Arbor Publishing, Ann Arbor, Michigan. 849-854.

Olthof, M. and Oleszkiewicz, J. (1983) "Benzol Plant Wastewater Treatment in a Packed-Bed Reactor." Proc. 37th Ind. Waste Conf. Ann Arbor Publishing, Ann Arbor, Michigan. 519-525.

Ongerth, J. E. and DeWalle, F. B. (1980) "Pretreatment of Industrial Discharges To Publicly Owned Treatment Works." Journal WPCF. 52:2246-2256.

Osantowski, R. and Hendriks, R.V. (1982) "Physical/Chemical and Biological Treatment of Coke-Plant Wastewater." Proc. 36th Ind. Waste Conf. Ann Arbor Publishing, Ann Arbor, Michigan. 168-176.

Otake, T., Tone, S., Kono, K. and Nakao, K. (1979) "Photo-Oxidation of Phenols with Ozone." J. Chem. Engr. Japan. 12:289-295.

Ouyang, C. F. (1984) "Effect of Influent Conditions on Nutrient Removal in RBC System." Proc. 2nd Int'l Conf. on Fixed-Film Biological Processes. Univ. of Pittsburgh. 683-708.

Painter, H. A. and King, E. F. (1978) "The Effect of Phosphate and Temperature on Growth of Activated Sludge and on Biodegradation of Surfactants." Water Res. 12:909-915.

Pajak, A. P., Martin, E. J., Brinsko, G. A., and Erny, F. J. (1977) "Effect of Hazardous Material Spills on Biological Treatment Processes." EPA-600/2-77/239. U.S. EPA, Cincinnati, Ohio. 204 pp.

Panzer, C. C. (1982) "Biological Nitrogen Control - A Comparison of Methods." J. Am. Leather Chem. Assoc. 77:149-160.

Parkin, G. F. and Speece, R. E. (1982) "Modeling Toxicity in Methane Fermentation Systems." Journal Env. Engr. Div., ASCE. 108:515-531.

Parkin, G. F. and Miller, S. W. (1983) "Response of Methane Fermentation to Continuous Addition of Selected Industrial Toxicants." Proc. 37th Ind. Waste Conf. Ann Arbor Publishing, Ann Arbor, Michigan. 729-743.

Parkin, G. F., Speece, R. E., Yang, C. H. J. and Kocher, W. M. (1983) "Response of Methane Fermentation Systems to Industrial Toxicants." Journal WPCF. 55:44-53.

Patterson, J. W., Kodukula, P. and Aratani, T. (1983) "Removal of Metals in Combined Treatment Systems." EPA-600/2-83-051. U.S. EPA, Ada, Oklahoma. 274 pp.

Pearson, F., Shiun-Chung, C. and Gautier, M. (1980) "Toxic Inhibition of Anaerobic Biodegradation." Journal WPCF. 52:472-482.

Pelosi, P. and McCarthy, J. (1982a) "Preventing Fouling of Ion-Exchange Resins - I." Chem. Engr. Aug. 9, 1982:75-78.

Pelosi, P. and McCarthy, J. (1982b) "Preventing Fouling of Ion-Exchange Resins - II." Chem. Engr. Sept. 6, 1982:125-128.

Petrasek, A. C. and Kugelman, I. J. (1983) "Metals Removal and Partitioning in Conventional Wastewater Treatment Plants." Journal WPCF. 55:1183-1190.

Petrasek, A. C., Kugelman, I. J., Austern, B. M., Pressley, T. A., Winslow, L. A. and Wise, R. H. (1983) "Fate of Toxic Organic Compounds in Wastewater Treatment Plants." Journal WPCF. 55:1286-1296.

Petrasek, A. C. (1981a) "Inhibition, Removal and Partitioning Interactions Between Lead and the Activated Sludge Process." Internal Report. U.S. EPA, Cincinnati, Ohio 62 pp.

Petrasek, A. C. (1981b) "Removal and Partitioning Interactions and the Inhibitory Effects of Mercury on the Activated Sludge Process." Internal Report. U.S. EPA, Cincinnati, Ohio. 58 pp.

Petrasek, A. C. (1981c) "Inhibition of the Activated Sludge Process by Cadmium." Internal Report. U.S. EPA, Cincinnati, Ohio. 39 pp.

Pitkat, C. A. and Berndt, C. L. (1981) "Textile Waste Treatment of a Municipal PACT Facility." Proc 35th Ind. Waste Conf. Ann Arbor Publishing, Ann Arbor, Michigan. 178-185.

Pitter, P. (1976) "Determination of Biological Degradability of Organic Substances." Water Res. 10:231-235.

Porter, J. J. and Snider, E. H. (1976) "Long-Term Biodegradability of Textile Chemicals." Journal WPCF. 48:2198-2210.

Radick, K. A. (1984) "Start-up Problems at a Dairy Waste Treatment Plant." Proc. Ind. Wastes Symposia, 57th Annual WPCF Conf. 63-77.

Raef, S. F., Characklis, W. G., Kessick, M. A. and Ward, C. H. (1977) "Fate of Cyanide and Related Compounds in Aerobic Microbial Systems - II. Microbial Degradation." Water Res. 11:485-492.

Reinhold, M. and Mallevialle, M. (1975) "Biodegradation by the Activated Sludge System of Steam Process Water in the Timber Industry." Water Res. 9:87-93.

Remacle, J. and Hauba, C. (1983) "The Removal of Heavy Metals from Industrial Effluents in a Biological Fluidized Bed." Envir. Tech. Letters. 4:53-58.

Renn, C. E. (1974) "Biodegradation of NTA Detergents in a Wastewater Treatment System." Journal WPCF. 46:2363-2371.

Rice, R.G. (1981) "Ozone for the Treatment of Hazardous Materials". Water 1980: AIChE Symposium Series. 77:79-93.

Richards, A. D., Fricke, A. M. and Scott, J. E. (1983) "Phenol Degradation in a Three-Phase, Fluidized-Bed Bioreactor." ORNL/MIT-361. Oak Ridge National Lab, Oak Ridge, Tennessee. 57 pp.

Ripley, L. E., Kmet, N. M. Boyle, W. C. and Converse, J. C. (1985) "The Effects of Ammonia Nitrogen on the Anaerobic Digestion of Poultry Manure." Proc. 39th Ind. Waste Conf. Butterworth Pub., Boston, Massachusetts. 73-79.

Rivera, A. L. (1983) "Heavy Metal Removal in a Packed-Bed, Anaerobic Upflow (ANFLOW) Bioreactor." Journal WPCF. 55:1450-1456.

Robins, J.H. and Green, A.C. (1974) "Development of On-Shore Treatment System for Sewage from Watercraft Waste Retention System." EPA 670/2-74-056. U.S. EPA, Cincinnati, Ohio. 114 pp.

Rooney, M. C. and Wu, M. H. (1982) "Joint Treatment of Meat-Packing and Municipal Wastewater by Full-Scale AWT Facilities." Proc. 36th Ind. Waste Conf. Ann Arbor Publishing, Ann Arbor, Michigan. 301-309.

Rossin, A. C., Sterritt, R. M. and Lester, J. N. (1982) "Removal of Heavy Metals in Activated Sludge." Water, Air and Soil Poll. 17:185-198.

Rozich, A. F. and Gaudy, A. F., Jr. (1985) "Response of Phenol-Acclimated Activated Sludge Process to Quantitive Shock Loading." Journal WPCF. 57:795-804.

Russell, L.L., Cain, C.B. and Jenkins, D.I. (1983) "Impact of Priority Pollutants on Publicly Owned Treatment Works Processes: A Literature Review." Proc. 37th Ind. Waste Conf. Ann Arbor Publishing, Ann Arbor, Michigan. 871-883.

Rusten, B. and Thorvaldsen, G. (1983) "Treatment of Food Industry Effluents - Activated Sludge vs. Aerated Submerged Biological Filters." Environ. Letters. 4:441-450.

Saeger, V. W., Kaley, R. G., Hicks, O., Tucker, E. S. and Mieure, J. P. (1976) "Activated Sludge Degradation of Adipic Acid Esters." Appl. and Environ. Microbiol. 31:746-749.

Sampayo, F. F. and Hollopeter, D. C. (1979) "The Influence of Industrial Waste on Nitrification." Proc. 33rd Ind. Waste Conf. Ann Arbor Publishing, Ann Arbor, Michigan. 142-154.

Schwartz, M. (1984) "Biological Treatment of Formaldehyde Wastes in a Mixture of Wastewater From Methanol, Acetylene, Vinyl Acetate and Acetic Acid Production." Proc. Ind. Wastes Symposia, 57th Annual WPCF Conf. 26-41.

Science Applications International Corp. (1986) "Guidance Manual for the Development of an Accidental Spill Prevention Program." U.S. EPA, Regional X, Seattle, Washington.

Selna, M. W. and Schroeder, E. D. (1978) "Response of Activated Sludge Processes to Organic Transients-Kinetics." Journal WPCF. 50:944-957.

Shaul, G. M., Barnett, M. W. and Dostal, K. A. (1983) "Treatment of Dye and Pigment Processing Wastewaters by the Activated Sludge Process." Proc. 37th Ind. Waste Conf. Ann Arbor Publishing, Ann Arbor, Michigan. 677-689.

Silva, S.J. (1981) "EPA Moving to Control Industrial Toxic Pollutants with New NPDES Permits." Civil Engr. 51:76.

Skrinde, J. R. and Bhagat, S. K. (1982) "Industrial Wastes as Carbon Sources in Biological Denitrification." Journal WPCF. 54:370-377.

Slonim, Z., Lien, L. T., Eckenfelder, W. W. and Roth, J. A. (1984) "Removal of 4,6-Dinitro-o-Cresol: Pilot-Scale Anaerobic-Aerobic System." Proc. 2nd Int'l Conf. on Fixed-Film Biological Processes. Univ. of Pittsburgh. 1095-1118.

Smith, L. L. (1980) "Evaluation of an Anaerobic Rotating Surface System for Treatment of a Munition Wastewater Containing Organic and Inorganic Nitrates." Proc. 34th Ind. Waste Conf. Ann Arbor Publishing, Ann Arbor, Michigan. 628-634.

Smith, J. W., Khararjian, H. and Harvell, G. (1981) "Performance of Three Types of Activated Sludge Processes Under Variable Organic and Pesticide Loadings." Proc. 13th Mid-Atlantic Ind. Waste Conf. Ann Arbor Publishing, Ann Arbor, Michigan. 250-263.

Smith, L. L. and Greene, W. C. (1982) "Operation of a RBC Facility for the Treatment of Munition Manufacturing Plant Wastewater." Proc. First Int'l Conf. on Fixed-Film Biological Processes. Univ. of Pittsburgh. 944-959.

Smith, J. W. and Moore, L. W. (1984) "Biodetoxification of Hazardous Wastewaters with Activated Rotating Biological Contactors." Proc. 16th Mid-Atlantic Ind. Waste Conf. Technomic Publishing, Lancaster, Pennsylvania. 15-28.

Snyder, D.J. III, ed. (1978) "Digest of the Clean Water Act of 1977." Roy F. Weston, Inc., West Chester, Pennsylvania. 17 pp.

Soderberg, R. W. and Bockrath, R. E. (1985) "Treatability of Diverse Waste Streams in the PACT Activated Carbon-Biological Process." Proc. 39th Ind. Waste Conf. Butterworth Pub., Boston, Massachusetts. 121-127.

Speece, R. E., Parkin, G. F. and Gallagher, D. (1983) "Nickel Stimulation of Anaerobic Digestion." Water Res. 17:677-683.

Stafford, D. A. (1974) "The Effect of Phenols and Heterocyclic Bases on Nitrification in Activated Sludges." J. Appl. Bact. 37:75-82.

Sterritt, R. M. and Lester, J. N. (1981) "The Influence of Sludge Age on Heavy Metal Removal in the Activated Sludge Process." Water Res. 15:59-65.

Stover, E. L. and Kincannon, D. F. (1983) "Biological Treatability of Specific Organic Compounds Found in Chemical Industry Wastewaters." Journal WPCF. 55:97-109.

Stover, E. L., Gonzalez, R. and Gomathinayagem, G. (1984) "Anaerobic Fixed-Film Biological Treatment Kinetics of Fuel Alcohol Production Wastewaters." Proc. 2nd Int'l Conf. on Fixed-Film Biological Processes. Univ. of Pittsburgh. 1625-1646.

Stover, E. L. and Rakness, K. L. (1984) "Process Evaluation at Fixed Film Biological Treatment Plants-Two Case Studies." Proc. 2nd Intl. Conf. on Fixed-Film Biological Processes. Univ. of Pittsburgh. 814-830.

Strachan, S. D., Nelson, D. W. and Sommers, L. E. (1983) "Sewage Sludge Components Extractable with Nonaqueous Solvents." J. Environ. Qual. 12:69-74.

Stracke, R. J. and Baumann, E. R. (1978) "Biological Treatment of a Toxic Industrial Waste-Performance of an Activated Sludge and Trickling Filter Pilot Plant." Proc. 30th Ind. Waste Conf. Ann Arbor Science, Ann Arbor, Michigan. 1131-1160.

Stratta, J. M., Long, D. A. and Doherty, M. C. (1982) "Improvement of Nitrification in Rotating Biological Contactors by Means of Alkaline Chemical Addition." Proc. First Int'l Conf. on Fixed-Film Biological Processes. Univ. of Pittsburgh. 758-801.

Suidan, M. T., Cross, W. H., Fong, M. and Calvert, J. W. (1981) "Anaerobic Carbon Filter for Degradation of Phenols." Journal Envir. Engr. Div., ASCE. 107:563-579.

Sujarittanonta, S. and Sherrard, J. H. (1981) "Activated Sludge Nickel Toxicity Studies." Journal WPCF. 53:1314-1322.

Sukes, G.L., Pordon, R.G. and Gupta, K. (1984) "The Destruction of Cyanide in Wastewater with Ozone." Proc. Ind. Wastes Symposia, 57th Annual WPCF Conf. 154-170.

Sullivan, D. E. (1983) "Biodegradation of a Cationic Surfactant in Activated Sludge." Water Res. 17:1145-1151.

Sundstrom, D. W., Klei, H. E., Tsui, T. and Nayar, S. (1979) "Response of Biological Reactors to the Addition of Powdered Activated Carbon." Water Res. 13:1225-1231.

Surampalli, R. Y., TeKippe, R. J. and Baumann. E. R. (1984) "The Value of Supplemental Air in Improving RBC Performance." Proc. 2nd Int'l Conf. on Fixed-Film Biological Processes. Univ. of Pittsburgh. 944-964.

Sykes, R. M., Rubin, A. J., Roth, S. A. and Chang, M. C. (1979) "Treatability of a Nonionic Surfactant by Activated Sludge." Journal WPCF. 51:71-77.

Tabak, H. H., Quave, S. A., Mashni, C. I. and Barth, E. F. (1981) "Biodegradability Studies with Organic Priority Pollutant Compounds." Journal WPCF. 53:1503-1518.

Thabaraj, G. J.. and Gaudy, A. F., Jr. (1969) "Effect of Dissolved Oxygen Concentration on the Metabolic Response of Completely Mixed Activated Sludge." Journal WPCF. 41:R322-R335.

Theis, T. L. and Padgett, L. E. (1983) "Factors Affecting the Release of Trace Metals from Municipal Sludge Ashes." Journal WPCF. 55:1271-1279.

Tokuz, R. Y. and Eckenfelder, W. W., Jr. (1979) "The Effect of Inorganic Salts on the Activated Sludge Process Performance." Water Res. 13:99-104.

Tolaney, M. (1977) "Treatment of High Strength Citrus Wastewater with High Purity Oxygen Activated Sludge Process." Proc. 30th Ind. Wastes Conf. Ann Arbor Publishing, Ann Arbor, Michigan. 171-183.

Troxler, R. W. and Hopkins, K.S. (1982) "Case Histories: Carpet Manufacturing Wastewater Treatment in Municipal Plants." Proc. 36th Ind. Waste Conf. Ann Arbor Publishing, Ann Arbor, Michigan. 755-765.

Turner, C. D., Wernberg, K., Strain, J. H. and Gallagher, J. R. (1984) "Treatment of Coal Gasification Wastewater Using Rotating Biological Contactors." Proc. 2nd Int'l Conf. on Fixed-Film Biological Processes. Univ. of Pittsburgh. 1257-1276.

U.S. EPA (1977a) "Federal Guidelines: State and Local Pretreatment Programs (Vols. I, II, and III)." MCD-43, EPA-430/9-76/017. U.S. EPA, Washington, DC.

U.S. EPA (1977b) "Results of Investigations, T. E. Maxson WTP and Significant Industrial Contributors, Memphis, Tennessee." EPA-904/9-77-005. U.S. EPA, Athens, Georgia. 246 pp.

U.S. EPA (1981) "Treatability Manual (Vols. I through V)." EPA-600/2-82/001. U.S. EPA, Washington, D.C.

U.S. EPA (1985) "Pretreatment Implementation Review Task Force: Final Report to the Administrator." U.S. EPA, Washington, DC. 75 pp.

Vaicum, L. and Eminovici, A. (1974) "The Effect of Trinitro-Phenol and -Hexachlorocyclohexane on the Biochemical Characteristics of Activated Sludge." Water Res. 8:1007-1012,

Volesky, B., Samak, Q. and Waller, P. (1977) "The Effect of Metallic and Ionic Species on the Performance of a Biological Effluent Treatment System." Proc. 12th Canadian Symp. Water Pollution Research (Canada). 191-212.

Vuccta, J., Anderson, J. R., TeKippe, R. J., Calkins, R. J. and Bishop, W. J. (1979) "Bench-Scale Testing for Residual Waste Treatment." Journal WPCF. 51:2366-2383.

Wagner, F. (1984) "Studies on the Causes and Prevention of Bulking Sludge in Germany." Water Sci. Tech. 16:1-14.

Watson, G. K. and Jones, N. (1977) "The Biodegradation of Polyethylene Glycols by Sewage Bacteria." Water Res. 11:95-100.

Watt, J. C. and Cahill, C. J. (1980) "Wastewater Treatability Studies for a Grassroots Chemical Complex Using Bench Scale Rotating Biological Contactors." Proc. First Nat'l. Symposium on RBC Tech. Univ. of Pittsburgh. 661-690.

Webber, W. G., Kemp, D. W. and Rice, S. E. (1977) "Study of the Effect of Boron Toxicity on an Activated Sludge System." Proc. 31st Ind. Waste Conf. Ann Arbor Publishing, Ann Arbor, Michigan. 743-752.

Webber, M. D., Monteith, H. D. and Corneau, D. G. M. (1983) "Assessment of Heavy Metals and PCB's at Sludge Application Sites." Journal WPCF. 55:187-195.

Weber, A. S. and Sherrand, J. H. (1980) "Effects of Cadmium on the Completely Mixed Activated Sludge Process." Journal WPCF. 52:2378-2388.

Weber, W. J., Corfis, N. H. and Jones, B. E. (1983) "Removal of Priority Pollutants in Integrated Activated Sludge-Activated Carbon Treatment Systems." Journal WPCF. 55:369-376.

Werthman, P. H., Matthews, R. R., Martens, R. R. and McManus, K. R. (1982) "Residuals Management at a Combined Industrial/Municipal Wastewater Treatment Plant." Proc. 14th Mid-Atlantic Ind. Waste Conf. Ann Arbor Publishing, Ann Arbor, Michigan. 1-14.

Wilson, A. W. (1981) "Case Study of a Potato Chip Producer Discharging to a Small Municipal Treatment System." Proc. Conf. on Combined Mun./Ind. Wastewater Treat. EPA-600/9-81-021. U.S. EPA, Ada, Oklahoma. 329-352.

Wilson, T. E., Lukasik, G. and Ogle, D. (1980) "Treatment of a Filamentous Industrial Waste in a Municipal Step Aeration Plant." Prog. Wat. Tech. 12:189-199.

Wood, K. N., Hunt, O. R. and Anderson, J. J. (1983) "Carbon Treatment of Pesticide Wastewaters in Light of Best Available Technology Effluent Limitation Guidelines." Proc 37th Ind. Wastes Conf. Ann Arbor Publishing, Ann Arbor, Michigan. 451-463.

Wozniak, D. J. and Huang, J. Y. C. (1982) "Variables Affecting Metal Removal from Sludge." Journal WPCF. 54:1574-1580.

Wozniak, D. J. and Huang, J. Y. C. (1982) "Variables Affecting Metal Removal from Sludge." Journal WPCF. 54:1574-1580.

Wu, Y. C., Kennedy, J. C., Gaudy, A. F., Jr. and Smith, E. D. (1982) "Treatment of High-Strength Organic Wastes by Submerged Media Anaerobic Reactors: State- of-the-Art Review." Proc. First Int'l Conf. on Fixed-Film Biological Processes. Univ. of Pittsburgh. 1212-1238.

Yall, I. and Sinclair, N.A. (1971) "Mechanisms of Biological Luxury Phosphate Uptake." U.S. EPA, Project No. 17010 DDQ, Washington, D.C., 77 pp.

Yost, K.J., Wukasch, R.F., Adams, T.G. and Michalczyk, B. (1981) "Heavy Metal Sources and Flows in a Municipal Sewage System: Literature Survey and Field Investigation of the Kokomo, Indiana Sewage System," EPA-600/2-81-244. U.S. EPA, Cincinnati, Ohio. 282 pp.

Yu, T. S., Jiang, M. L. and Denny, R. G. (1984) "Evaluation of Nitrification Performance by Rotating Biological Contactors." Proc. 2nd Int'l. Conf. on Fixed-Film Biological Processes. 1405-1426.

Yurovskaya, E. M. (1982) "Microbial Nitrification in Coking Plant Wastewater." Sov. J. Water Chem. Technol. 4:106-111.

Zogorski, J. S. (1984) "Comparison of RBC Treatment Performance Between the Conventional (Hydraulic) and an Alternative (Organic) Flow Configuration. Proc. 2nd Int'l Conf. on Fixed-Film Biological Processes. Univ. of Pittsburgh. 906-927.

Zwikl, J. R., Buchko, N. S. and Junkins, D. R. (1982) "Physical/Chemical Treatment of Coke Plant Wastewaters." Environ. Prog. 1:244-251.

Appendix A: Interfering Substances

CONVENTIONAL AND INORGANIC

- Alkalinity
- Ammonia
- Biochemical Oxygen Demand
- Chemical Oxygen Demand
- Chloride
- Fats, Oil and Grease
- Iodine
- Iron Salts
- Nutrients
- pH
- Sulfate
- Sulfide
- Surfactants
- Suspended Solids

METALS

- Arsenic
- Barium
- Beryllium
- Boron
- Cadmium
- Calcium
- Chromium
- Cobalt
- Copper
- Cyanide
- Iron
- Lead
- Magnesium
- Manganese
- Mercury
- Molybdenum
- Nickel
- Selenium
- Silver
- Sodium
- Tin
- Vanadium
- Zinc

AGRICULTURAL CHEMICALS

- Aldrin/Dieldrin
- Chlordane
- Chlorophenoxy Herbicides
- DDT
- Endrin
- Heptachlor
- Lindane
- Malathion
- Organometallic Pesticides
- PCBs
- Toxaphene

AROMATICS

- Benzene
- Chlorobenzene
- Dichlorobenzene
- Hexachlorobenzene
- Nitrobenzene
- Toluene
- Xylene

HALOGENATED ALIPHATICS

Carbon Tetrachloride
Chloroform
Dichloroethane
Dichloroethylene
Dichloropropane
Hexachlorobenzene
Hexachlorobutadiene
Hexachlorocyclohexane
Hexachloroethane
Methylene Chloride
Tetrachlorodibenzodioxins
Tetrachlorodibenzofurans
Tetrachloroethane
Tetrachloroethylene
Trichloroethane
Trichloroethylene
Vinyl Chloride

NITROGEN COMPOUNDS

Acetanilide
Acetonitrile
Acrylonitrile
Aniline
Benzidine
Benzonitrile
Chloroaniline
Dichlorobenzidine
Dimethylnitrosamine
Dyes
EDTA
Ethylpyridine
Fluorenamine
Hydrazine
Nitrosodiphenylamine
Pyridine
Trisodium Nitrilotriacetate
Urea

OXYGENATED COMPOUNDS (Acids, Alcohols, Aldehydes, Esters, Ethers, Ketones)

Acetone
Acrolein
Adipic Acid Esters
Allyl Alcohol
Benzoic Acid
Boric Acid
Butanol
Butyl Benzoate
Chlorobenzoate
Chloroethyl Ether
Cinnamic Acid
Crotonol
Cyclohexanecarboxylic Acid
Diethylene Glycol
Ethoxy Ethanol
Ethyl Acetate
Ethylene Glycol
Formaldehyde
Formic Acid
Heptanol
Hexanol
Isophorone
Linoleic Acid
Malonic Acid
Methanol
Methylethyl Ketone
Methylisobutyl Ketone
Octanol
Polyethylene Glycols
Polyvinyl Alcohols
Protocatechuic Acid
Syringic Acid

PHENOLS

Catechol
Chlorophenol
Cresol
Dichlorophenol
Dinitrophenol
Nitrophenol
Pentachlorophenol
Phenol
Trichlorophenol
Trinitrophenol
Vanillin

PHTHALATES

Dimethylphthalate
Dioctylphthalate
Ethylhexylphthalate

POLYNUCLEAR AROMATIC HYDROCARBONS

Anthracene
Benzo (a) Anthracene
Chloronaphthalenes
di-Isopropylnaphthalene
Naphthalene
Pyrene

Appendix B: Project Forms

INTERFERENCES AT POTW's
LITERATURE REVIEW

Authors: ____________________

Title: ____________________

Citation: ____________________

JMM Reviewer: ____________________ JMM Code: ____________________

- On a scale of 1 (lowest) to 10 (highest), rate this reference in terms of its:

  applicability to this EPA project ____ technical quality ____

- The focus of this reference was:

  contaminant treatability in a POTW unit process ☐
  impact of contaminant on POTW process operation ☐
  other (specify) ____________________

- The conclusions of the study were based on:

  bench scale testing ☐ treatment plant records/testing ☐ City/Plant Name: ____________
  pilot scale testing ☐ literature ☐

- In the space provided below, list the treatment processes (PROCESS) analyzed in the study, the industrial waste or contaminants (CAUSE) identified, any immediate or secondary effects (EFFECT) on the treatment process, and actions useful in mitigating the negative effects (RESPONSE). Include quantitative information on loadings, concentrations, etc. If a biological treatment process is considered, indicate whether the bacterial populations were acclimated to the contaminants. Use the back of this sheet if additional space is required.

PROCESS	CAUSE	EFFECT	RESPONSE

- What other variables were considered in this work, and how did they impact the process or contaminant? Be specific.

- Any other relevant information from this reference?

INTERFERENCES AT POTWs
STATE AGENCY TELEPHONE SURVEY

State:	________________	**Date:**	________________
Agency:	________________		
Office/Section:	________________	**Agency Contact:**	________________
Phone No.:	( ) ______________	**JMM Staff Members:**	________________

EXPLANATION: We have been contracted by EPA to prepare a guidance manual for operators pertaining to interferences at POTWs. One of our project tasks is to survey all state regulatory agencies in order to uncover specific operations problems created by the introduction of contaminants from industrial sources.

QUESTIONS	ANSWER	COMMENTS
• How many municipal treatment plants do you have subject to state or EPA regulations?		
• Do you compile data on NPDES Permit violations other than Quarterly Non-compliance Reports sent to EPA?		
• To what extent are these violations related to industrial discharges?		
• Have there been problems in POTW operations associated with industrial discharges? Are summary data available?		
• Does documentation exist on specific case histories?		
• Are there other contacts I can make within your state that might prove helpful?		

END OF CALL

Further Action Required?

INTERFERENCES AT POTWs

TREATMENT FACILITY SITE VISIT

JMM Staff Member: ______________

Date: ______________

1. **General Information**

   A. Plant Name: ______________

   B. Address: ______________

   ____

   C. Telephone No.: ______________

   D. Key Personnel:

Name	Title
______	______
______	______
______	______
______	______

   E. Capacity: ______ Ave. Flow: ______ Pop. Served: ______

   F. NPDES Permit:

   Permit No. ______________ Date Issued: ______________

   Receiving Water ______________

   Discharge Limits BOD_5 = ______ mg/l

   TSS = ______ mg/l

   NH_3 = ______ mg/l (summer)

   ______ mg/l (winter)

   Other ______________

   ____

2. **Plant History**

   A. Date of initial construction: ______

   B. Significant upgrades:

Year	Description
______	______
______	______
______	______

3. Flow Diagram

Sketch or attach a flow diagram of the wastewater treatment facility, indicating the number and size of each major process component.

4. Plant Operations - Liquid Processing

A. Preliminary:

Racks/Screens ______________________

Shredding ______________________

Grit Removal ______________________

Flow Measurement ______________________

Other ______________________

B. Primary Treatment:

Scraping/Skimming ______________________

Sludge Pumping ______________________

Detention Time ____________ Overflow rate ____________

C. Secondary Biological Treatment:

	Fixed Film	Suspended Growth
Process	____________	____________
Aeration	____________	____________
Organic Loading	____________	____________
Recirculation	____________	____________
Detention Time	____________	____________
D.O.	____________	____________
MLSS		____________
F/M		____________
MCRT		____________
SVI		____________

D. Secondary Clarification:

Scraping/Skimming ______________________

Sludge Pumping ______________________

Detention Time ____________ Overflow rate ____________

E. Disinfection:

Chemical Feed ____________ Residual ____________

Detention Time ____________ Coliforms ____________

F. Other: ______________________

____

____

5. **Plant Operations - Sludge Processing**

A. Thickening:

Process ________________________________
Solids Loading ________________________________
Chemical Feed ________________________________
Return Flow ________________________________
Other ________________________________

B. Stabilization:

Process ________________________________
Detention Time ________________________________
Chemical Feed ________________________________
Withdrawal ________________________________
Return Flow ________________________________
Other ________________________________

C. Dewatering:

Process ________________________________
Solids Loading ________________________________
Chemical Feed ________________________________
Return Flow ________________________________
Other ________________________________

D. Disposal: ________________________________

E. Other: ________________________________

6. Performance Data

Indicate in the space provided (or attach) any relevant performance data for typical and upset conditions. Include design data if available.

A. Treatment Plant Influent:

B. Primary Clarifier Effluent:

C. Treatment Plant Effluent:

D. Sludge:

7. Industrial Discharges

A. Pretreatment Program:

Effective Date ________ Local Limits ____________

Total No. Industries ________ ____________

Categorical Industries ________ ____________

B. Significant Contributors:

Industry	Flow	Pollutants (Concentrations)
________	________	____________
		____________
		____________
________	________	____________
		____________
		____________
________	________	____________
		____________
		____________
________	________	____________
		____________
		____________
________	________	____________
		____________
		____________
________	________	____________
		____________
		____________

C. Waste Haulers:

Number ________ Approximate Flow ____________

Types of Waste ______________________________

____

8. Interferences

A. Compliance Record: ____________________

____

____

____

____

B. Chronic Problems: (include rate of pollutant introduction)

Cause	Effect	Detection/Mitigation

8. Interferences (cont.)

C. Isolated Problems:

Cause	Effect	Detection/Mitigation

D. Comments:

9. JMM Evaluation

Category	Rating*	Comments
A. Management:	____	____
		____
B. Operations:	____	____
		____
C. Maintenance:	____	____
		____
D. Laboratory:	____	____
		____
E. Pretreatment Program:	____	____
		____

* Rating Code: 1 = Excellent, to 5 = Poor

10. Miscellaneous Comments

Appendix C: Case Study Reports

BAYSHORE REGIONAL SEWERAGE AUTHORITY
Union Beach, New Jersey

The Bayshore Regional Sewerage Authority (BRSA) operates an activated sludge treatment facility whose performance is largely dictated by a single industrial waste discharger. Three manufacturers of flavors and fragrances (one of whom is a perfume retailer) represent the total industrial wastewater flow of 325,000 gpd, or less than 5 percent of the POTW total. All three industries discharge high concentrations of conventional pollutants and routinely violate the maximum allowable monthly concentration limits for BOD (500), COD (1500) and TSS (500) as specified in their industrial waste permits. Two of the three manufacturers contribute less than 0.5 percent of the POTW flow, hence their impact is minimal. However, one building of the largest industry produces in excess of 200,000 gpd of wastewater with the following characteristics (in mg/l):

	1984			October 1985		
Parameter	**Ave.**	**Monthly High**	**Monthly Low**	**Ave.**	**Daily High**	**Daily Low**
BOD	1004	2054	245	2624	5250	522
COD	3238	4998	1440	7084	11380	2520
TSS	776	1835	94	1113	1698	672

The large variation in wastewater quality indicates that an activated sludge pretreatment system located at the industry at times produces a suitable effluent, but is obviously not sufficient to meet the fluctuating demands of their process wastes.

The potential impact of such an industrial discharge is evident when analyzing Figure C-1. The bar graph represents the percentage of total BOD being contributed by the industry on a daily basis. The upper plot on the line graph corresponds to the mass BOD loading, with the industry's contribution plotted beneath. This graph clearly demonstrates that the effluent from this single industry has increased the BRSA plant loading above the design limit of 15,000 pounds of BOD per day.

The BRSA has been particularly aggressive in their dealings with the industry in question. Since the manufacturer is not a retailer, adverse publicity has little effect, particularly since the industry is the largest employer in town. Consequently, the BRSA has taken a two-pronged approach:

- notification of violation and intent to pursue fines with a subsequent discontinuation of service if noncompliance persists after 30 days, and
- legal action to recover $1.25 million in back surcharge payments and costs.

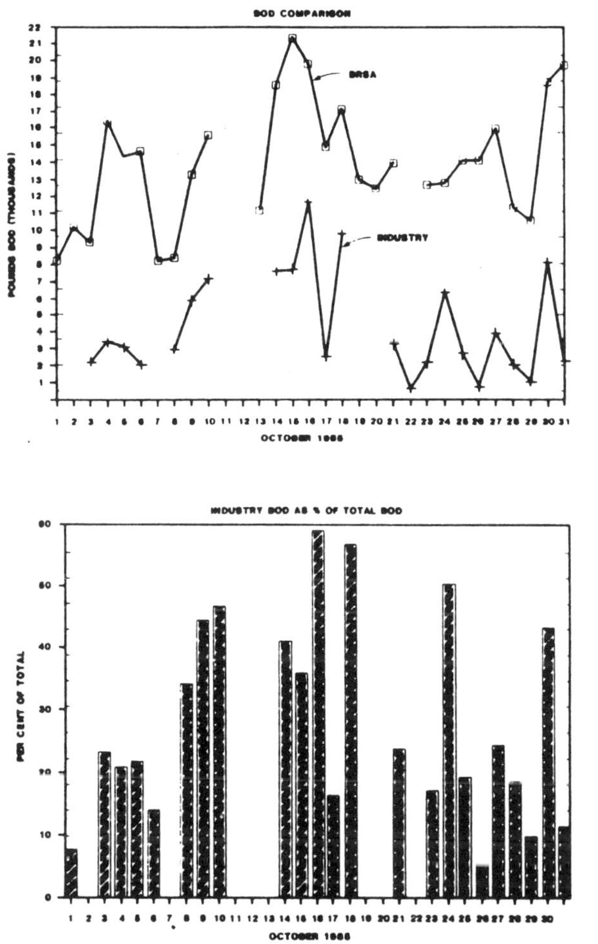

Figure C-1
Impact of Industrial Waste Discharge on POTW Loadings
October 1985

A similar approach proved successful during the 1960's, when the Keansburg, NJ water treatment plant was pumping the contents of their backwash water storage tank into the sewer system approximately twice per year. In the absence of an industrial wastewater permitting system, the BRSA's only recourse was to take the Keansburg authority to court and have them disconnected until a backwash recycle system could be installed at the water treatment plant.

BAYSHORE REGIONAL SEWERAGE AUTHORITY
Union Beach, New Jersey

Design Flow:	**8.0**	Location:	**Eastern shore**
Secondary Treatment:	**Activated Sludge (Modified Contact Stabilization)**	Population Served:	**80,000**

INFLUENT WASTEWATER

	Typical (Upset)
Ave. Flow, mgd	6.6
% Industrial	5
BOD_5, mg/l	220 (380)
SS, mg/l	250 (400)

SIGNIFICANT INDUSTRIES

Industry	Flowrate (1000 gpd)	Problem Pollutants
Flavors & Fragrances (3 industries)	325	BOD, TSS, COD

PLANT LOADING

Primary Clarifiers	Typical (Upset)
Overflow Rate, gal/sf/day	825
Detention Time, hours	1.75
Effluent BOD_5, mg/l	150 (250)
Effluent SS, mg/l	100 (200)

Aeration Basins	Typical (Upset)
F/M, lbs BOD_5/lbs MLSS/day	0.65 (1.25)
MCRT, days	8-10
MLSS, mg/l	2000-2500
Return Flow, %	25
Detention Time, hours	
Contact	3
Reaeration	12

Secondary Clarifiers	Typical (Upset)
Overflow Rate, gal/sf/day	540
Detention Time, hours	3.35
SVI, ml/gm	125 (500)

PLANT PERFORMANCE

	Permit Limit	Typical (Upset)
BOD_5, mg/l	30	35 (400)
SS, mg/l	30	27 (80)
D.O. mg/l	5	2-5

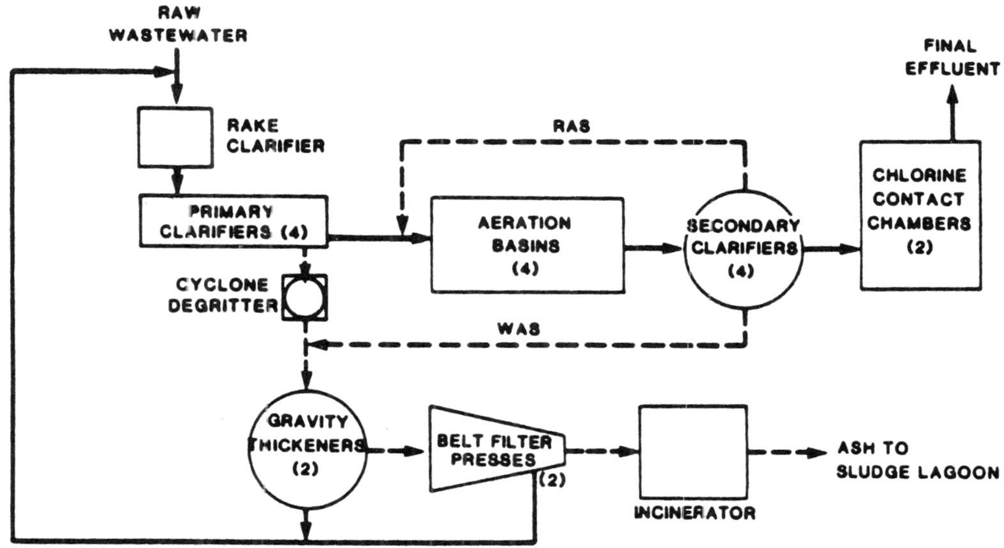

HAMILTON TOWNSHIP WASTEWATER TREATMENT PLANT
Trenton, New Jersey

The Hamilton Township Wastewater Treatment Plant (HTWTP) is an unusual facility in that plant upgrades over the past 30 years have been constructed as parallel flow processes rather than as replacements for older, outdated technology. Although this results in a complicated plant schematic (see below), parallel flow paths do provide operational flexibility and an opportunity to study the impact of a combined industrial/domestic wastewater on different fixed-film biological treatment processes. The HTWTP has had a difficult time meeting its permit limit for BOD over the past few years, and is currently under a Consent Order and Agreement and Compliance Schedule from the State Department of Environmental Protection.

Despite being at just over 50 percent of the plant's hydraulic capacity, Hamilton Township has experienced organic overloads, resulting in at least partial failure of 15 of the 48 RBC units. With the advent of an Industrial Waste Monitoring Program as part of a Sewers and Sewage Disposal Ordinance, the reasons for such overloading became apparent. Although the industrial waste program is still in its' infancy, observations and analytical data have identified a pharmaceuticals manufacturer as a significant and potentially harmful discharger to the POTW.

Dating back to the summer of 1984, high concentrations of volatile organics were being discharged to the POTW on a once or twice-per-week basis. A monitoring program at the HTWTP uncovered an increase in influent BOD from 150 to 350-500 mg/l and high atmospheric levels of organic constituents with this discharge pattern. The specific industry was identified when a high influent pH reading lead Hamilton Township personnel to the pharmaceuticals manufacturer in March, 1985. Sampling conducted at that time detected significant levels of ethyl benzene, toluene and xylene in the industry's effluent. These findings precipitated an extensive testing program by the Township, with an independent engineering study conducted by the industry. The results indicated a correlation between the pharmaceutical discharges and high influent soluble BOD at the POTW. Analyses conducted on the industry's flow streams resulted in the following calculated average effluent concentrations:

Parameter	Concentration (mg/l)
Arsenic	2.6
Phenols	25.7
Total Toxic Volatile Organics (TTVO)	1.3
BOD	21,800
TSS	557
TDS	65,800

Based on an average flow of 15,000 gpd, these wastewater characteristics should not be harmful to an 8.5 mgd facility if discharged on a steady basis. It is the intermittent discharge of this wastewater which has contributed to the overloading of the biological population of the POTW.

During a three week shutdown of the industry in July of 1985, the HTWTP recovered to the point of meeting their permit limits. Consequently, the Township only permitted the industry access to the sewer system after the installation of metering pumps to equalize flows. This requirement initially improved POTW performance during the Fall of 1985, but a gradual deterioration in effluent quality (indicating possible toxicity effects) lead the Township to terminate service to the industry in late-November.

While the most recent action is being challenged, the industry is constructing an anaerobic pretreatment facility on site which should reduce the financial impact of a surcharge to be instituted with the next version of the industrial waste management program.

A number of operations and personnel changes have been instituted at the HTWTP to help mitigate the impact of the industrial discharges. These changes include:

- installation of aeration equipment in the influent channels to the RBCs to increase the first stage DO to 2-3 mg/l;
- extensive use of sludge judges and visual monitoring to augment reliance on control room instrumentation;
- performance of bioassay testing by an independent contractor to assess toxicity effects;
- purchase of a toxicity tester to be used in calculation of local limits for toxic contaminants; and
- hiring of four more people plus the purchase of a vehicle for an extensive industrial sampling program.

At this time, only a few industries have been sampled to any degree. One of the electroplaters in town was discovered with up to 60 mg/l of cadmium and 160 mg/l of chromium in their wastewater. Although pretreatment has not been installed, conservation efforts on the part of the industry has reduced their discharge from 14,000 to under 10,000 gpd.

HAMILTON TOWNSHIP WASTEWATER TREATMENT PLANT
Trenton, New Jersey

Design Flow:	**16 mgd**	Location:	**Central Western Border**
Secondary Treatment:	**Trickling Filter and RBC**	Population Served:	**87,000**

INFLUENT WASTEWATER

	Typical (Upset)
Ave. Flow, mgd	8.5
% Industrial	10 (est)
BOD_5, mg/l	240 (500)
SS, mg/l	160 (400)

SIGNIFICANT INDUSTRIES

Industry	Flowrate (1000 gpd)	Problem Pollutants
Pharmaceutical	15	BOD, phenol, ethyl benzene, toluene, xylene
Electroplaters (2)	160	Cd, Cr, Zn, Ni

PLANT LOADING

Primary Clarifiers	Typical (Upset)
Overflow Rate, gal/sf/day	830, 260, 320
Detention Time, hours	1.8, 4.8, 5.6

Trickling Filters	Typical (Upset)
Plant Flow (mgd)	2.5, 1.0
Hydraulic Loading, gal/sf/day	100, 210
Organic Loading, lbs BOD/1,000 cf/day	15, 16 (30)
Return Flow, %	20,100

Secondary Clarifiers	Typical (Upset)
Overflow Rate, gal/sf/day	520, 260, 265
Detention Time, hours	2.8, 4.8, 6.8

RBCs	Typical (Upset)
Plant Flow (mgd)	5.0
First Stage Organic Loading, lbs BOD/1,000 sf/day	
- Total	5.3 (10.8)
- Soluble	3.5 (6.7)

PLANT PERFORMANCE

	Permit Limit	Typical (Upset)
BOD_5, mg/l	30	45 (100)
SS, mg/l	30	20 (50)
NH_3, mg/l (Effective 6/86)	10	20 (30)

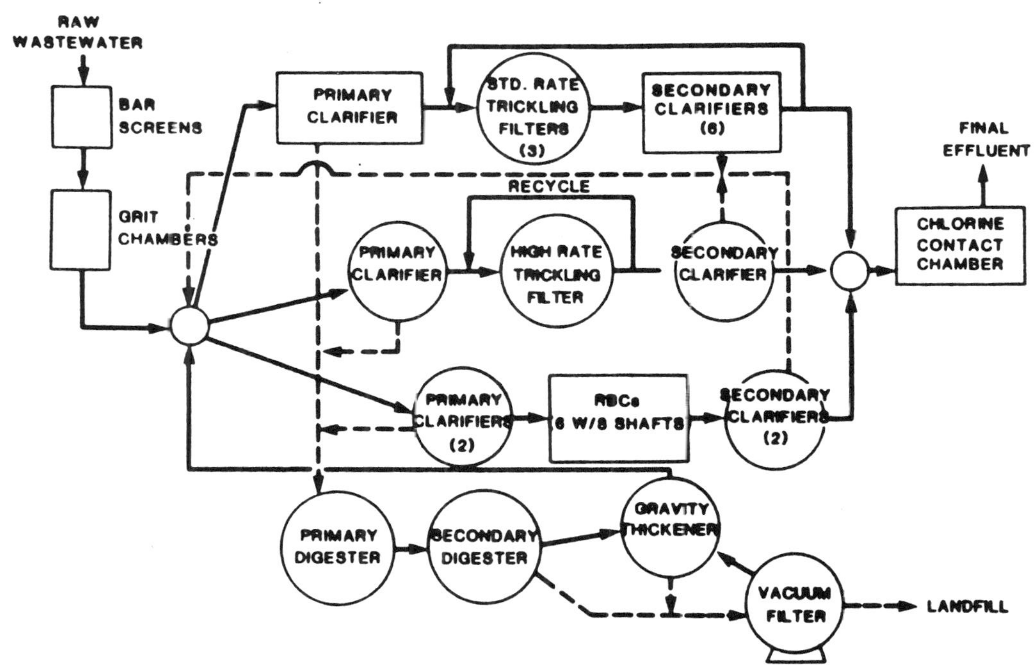

PASSAIC VALLEY WASTEWATER TREATMENT PLANT
Newark, New Jersey

Coping with industrial waste discharges to a 300 mgd POTW in a highly industrialized area is a challenging task. The Passaic Valley Sewerage Commissioners (PCSC) maintain an industrial waste control staff to monitor nearly 400 industries that contribute 20 percent of the wastewater volume and 50 percent of the waste strength. The PVSC performed their first Industrial Waste Survey for database development in 1972, and adopted a set of Rules and Regulations (including local limits) in 1976. By 1982, a comprehensive system consistent with the Federal Clean Water Act of 1977 had been adopted, which established uniform user fees for mass and volumetric loadings in the Passaic Valley plant.

The influent wastewater to the POTW is considered a high-strength waste, with typical BOD and TSS values of 290 and 450 mg/l, respectively. Despite the strength of the influent, the plant is close to meeting the 30/30 NPDES discharge limits, even though the primary clarifiers are not scheduled to go on-line until later this year (1986). The PVSC believes that the addition of primary treatment coupled with the economic incentives for pretreatment created by the user charge system will reduce the effluent to consistently below the limits.

The individual constituents of concern to the PVSC fall into three general categories:

- metals
- flammables
- fibers

The sources of heavy metals are chemical manufacturers, platers and tanneries. One of the smaller (30,000 gpd) chemical companies had been identified as a signficant contributor (120 lbs/day) of mercury to the POTW. Although the mercury level of 50 ug/l at the influent was not inhibitory to the activated sludge, the concentration of mercury in the sludge limited the municipality's disposal options. It is anticipated that ocean disposal of sludge will not be permitted much longer, which will require the PVSC to incinerate. The Federal Air Pollution Standards limit the mercury discharge to 3,200 g/day, which translates into a local limit of 0.4 lbs/day in the wastewater from the industry in question. The chemical company responded by isolating the relevant process streams and utilizing a batch recovery system for the mercury, reducing the discharge from 120 down to 5 lbs/day. When ocean disposal is formally eliminated as a disposal option, the company can employ carbon treatment for removal of the remaining mercury.

The oxidation of trivalent chromium to the hexavalent form in a POTW sludge incinerator is a problem caused by the chromium-laden discharge from various industrial users. An additional problem caused by the tanning industrial category is the clogging of local sewers that results from hides being inadvertently discharged from the companies. Similar clogging problems existed at the pretreatment plant due to the balled-up fibers from the pulp and paper manufacturers which close off sludge return lines, orifices and nozzles. This condition improved substantially when the moving-bridge primary clarifiers were placed in service in December, 1985.

The Passaic Valley plant had a unique problem with high concentrations of flammable materials in the influent wastewater. The lower explosive limit (LEL) is defined as the "lowest concentration of a combustible substance in air through which a flame, once ignited, will continue to propogate". When a wastewater approaches 50 percent of the LEL, it is important that it not be discharged into the sewer collection system. The pure oxygen process has a control built into the system which vents all oxygen away from the activated sludge treatment process when high LEL is detected. Since the venting of the oxygen reduces the treatment efficiency and can result in a permit violation, such discharges are not only health hazards, but interferences as well.

The PVSC instituted a three-part program in October of 1984 to mitigate the problems of flammables:

- required industries using or manufacturing solvents which come in contact with discharged wastewater to install LEL detection instruments, and to provide pretreatment to isolate the flammables if high LELs were detected;

- surveyed other industries which used solvents but had no such discharge to determine if a potential existed, requiring necessary control mechanisms; and

- monitored the collection system more closely for illegal dumping of such chemicals.

Representatives of Passaic Valley made it clear that a cooperative attitude on the part of industry was an important factor in successful mitigation of interference problems. In fact, it was the local pharmaceutical manfacturer that conducted the research resulting in the type of LEL instrument recommended by the Advisory Committee when the LEL regulation was adopted.

PASSAIC VALLEY WASTEWATER TREATMENT PLANT
Newark, New Jersey

Design Flow:	330 mgd	Location:	Adjacent to Newark Bay
Secondary Treatment:	Activated Sludge (Pure Oxygen)	Population Served:	1.5 Million

INFLUENT WASTEWATER

	Typical (Upset)
Ave. Flow, mgd	250
% Industrial	19
BOD_5, mg/l	290 (500)
SS, mg/l	450 (750)

SIGNIFICANT INDUSTRIES

Industry	Flowrate (mgd)	Problem Pollutants
Pulp and Paper (2)	10.3	Fibers
Pharmaceuticals	6	Xylene, Toluene, Hexane
Tanneries (3)	1.5	Cr
Chemicals (3)	0.5	Cd, Cr, Hg, Pb

PLANT LOADING

Primary Clarifiers	Typical (Upset)
Overflow Rate, gal/sf/day	1,100
Detention Time, hours	2.0
Effluent BOD_5, mg/l	225
Effluent SS, mg/l	125

Aeration Basins	Typical (Upset)
F/M, lbs BOD_5/lbs MLSS/day	0.6
MCRT, days	5
MLSS, mg/l	2,800
Detention Time, hours	1.6
Return Flow, %	35
D.O. Level, mg/l	4-12

Secondary Clarifiers	Typical (Upset)
Overflow Rate, gal/sf/day	480
Detention Time, hours	5
SVI, ml/gm	65

PLANT PERFORMANCE

	Permit Limit	Typical (Upset)
BOD_5, mg/l	30	25 (40)
SS, mg/l	30	25 (60)

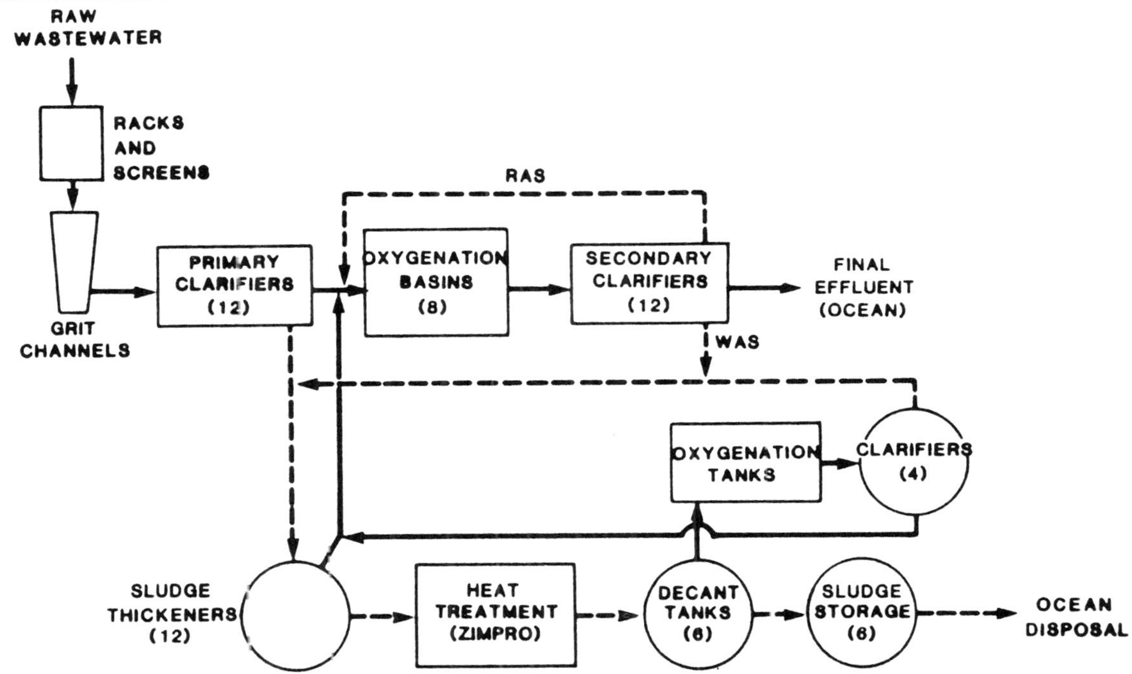

BINGHAMTON-JOHNSON CITY JOINT SEWAGE TREATMENT PLANT
Binghamton, New York

In 1981, the Binghamton-Johnson City Joint Sewage Treatment Plant had to terminate landspreading of sludge because the sludge did not meet the cadmium criteria established by the New York State Department of Environmental Conservation (NYDEC) for such disposal. A paper coating industry in Binghamton was the only identified industry that used and discharged high concentrations of cadmium. The industry cooperated with the Joint Sewage Authority and reduced their cadmium discharge levels by installing a pretreatment system that utilizes ammonia stripping followed by metal precipitation. During the 1975-1979 period, cadmium levels in the sludge from the Joint Sewage Treatment Plant were in the range of 100 to 150 mg/kg. In 1982, the level was reduced to 53 mg/kg, and presently the level is at 15 mg/kg. The cadmium limit imposed by the NYDEC for land spreading of the sludge is 25 mg/kg. The treatment plant is presently landfilling sludge but intends to install sludge composting equipment in the future.

There are approximately twelve significant industries contributing flow to the Joint Sewage Treatment Plant. There are several electroplaters who discharge metals, (other than cadmium), to the plant. Most of the electroplaters have installed pretreatment equipment in order to comply with industry categorical standards. The Binghamton-Johnson City Joint Sewage Board adopted rules and regulations effective March 1, 1985 that gave the board direct control over industrial wastewaters discharged to the sewage treatment plant. In the near future the Joint Sewer Board will issue industrial discharge permits in order to enforce local limits which will be set to ensure compliance with state water quality standards, (which are presently being revised), and sludge disposal criteria. At the present time, chromium, zinc and nickel concentration levels in the sludge occasionally approach or exceed the NYDEC landspreading limits. The imposition of local industrial waste discharge limits are expected to result in reduction in metal levels in the sludge to acceptable levels.

The Joint Sewage Treatment Plant has had consistent operational problems associated with biosolids losses from the secondary clarifiers. The poor operation of the clarifiers has been due to severe short circuiting as well as hydraulic surges from infiltration/inflow and stormwater flows. The Joint Sewage Board is presently undertaking corrective action to alleviate the problem. The secondary clarifiers are being modified by the addition of new influent baffling and the relocation of the effluent weirs. The reduction of infiltration/inflow and stormwater flows into the plant will be reduced by extensive sewer system rehabilitation.

BINGHAMTON - JOHNSON CITY JOINT
SEWAGE TREATMENT PLANT
BINGHAMTON, NEW YORK

Design Flow:	**18 mgd**	Location:	**Central New York**
Secondary Treatment:	**Activated Sludge (Completely Mixed or Contact Stabilization)**	Population Served:	**120,000**

INFLUENT WASTEWATER

	Typical (Upset)
Ave. Flow, mgd	20
% Industrial	1
BOD_5, mg/l	210
SS, mg/l	210

SIGNIFICANT INDUSTRIES

Industry	Flowrate (1000 gpd)	Problem Pollutants
Paper Coating	250	Cadmium
Electroplaters	35	Zinc, Nickel, Chromium

PLANT LOADING

Primary Clarifiers	Typical (Upset)
Overflow Rate, gal/sf/day	1,000
Detention Time, hours	2
Effluent BOD_5, mg/l	125
Effluent SS, mg/l	95

Secondary Clarifiers	Typical (Upset)
Overflow Rate, gal/sf/day	850
Detention Time, hours	2.5
SVI, ml/gm	170

Aeration Basins	Typical (Upset)
MCRT, days	2-5
MLSS, mg/l	2,500
Detention Time, hours	2
D.O. Level, mg/l	2

PLANT PERFORMANCE

	Interim Permit Limit: Summer	Interim Permit Limit: Remainder of Year	Typical (Upset)
BOD_5, mg/l	45	70	55
SS, mg/l	80	100	90

SLUDGE METAL CONCENTRATION

	Landspreading Limit	1982 Level	Present Level
Cadmium, mg/kg	25	53	15
Chromium, mg/kg	1,000	889	840
Nickel, mg/kg	200	219	319

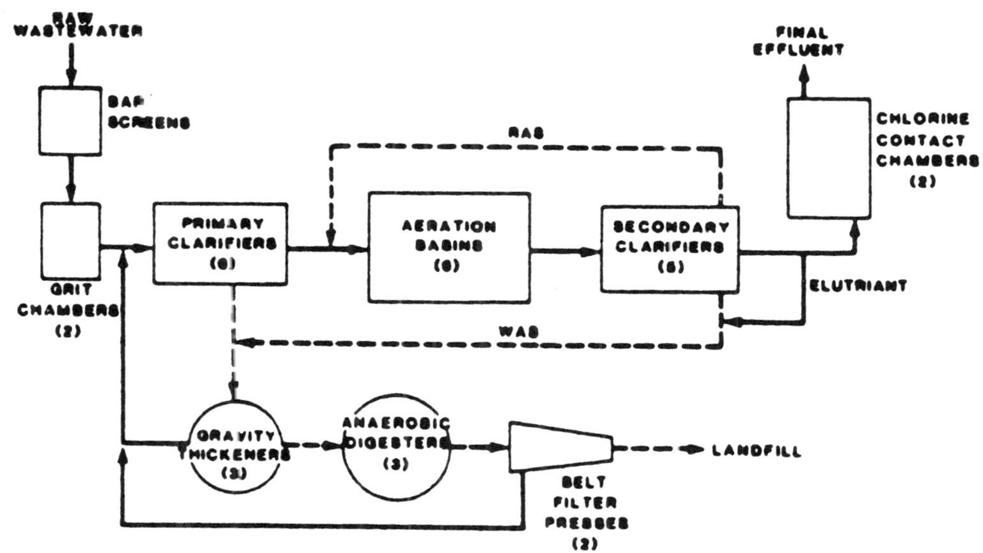

CITY OF CANANDAIGUA WASTEWATER TREATMENT PLANT
Canandaigua, New York

During its first year of operation, the City of Canandaigua Wastewater Treatment Plant failed to meet its NPDES Permit limits for BOD and Ultimate Oxygen Demand (UOD)* approximately half of the time. Violations were due to high influent organic loads from a winery in the city. In 1982, the winery discharged an average flow of 100,000 gpd having a BOD concentration of 3,500 mg/l. This was in violation of the discharge limitations that were in effect for the winery under the sewer use ordinance. The pretreatment limits for the winery had been set at the following concentrations:

COD	600 mg/l
BOD	300 mg/l
Suspended Solids	350 mg/l
Total Kjeldahl Nitrogen	55 mg/l
Phosphorus	10 mg/l

The City of Canandaigua initiated court action against the winery in early 1983 for violations of the sewer use ordinance. Subsequently, the city and winery agreed out of court on a compliance schedule for the winery.

The limits in the sewer use ordinance will be integrated into the industrial discharge permit that will be issued to the winery in the near future. The winery expanded its pretreatment facility and in 1984 its discharge had an average BOD concentration of 400 mg/l. The City of Canandaigua Wastewater Treatment Plant operation has improved dramatically as its effluent met its discharge permit requirements for all of 1984.

The City of Canandaigua Wastewater Treatment Plant uses air driven rotating biological contactors (RBCs) for carbonaceous organics removal followed by nitrification. In 1982, the RBCs were not effectively removing CBOD or ammonia from the wastestream because of excessive organic overloading from the winery. In addition, the organic overload caused excessive growth on the RBCs resulting in inadequate rotation of the units. During 1982, up to half of the RBCs were not operational.

The winery, which is the only major industry in the city, had increased production and was overloading its existing pretreatment facility in 1982. As a result of the agreement in early 1983 that set a compliance schedule, an expansion of the winery's pretreatment facility was on line by late 1983. Both the original and expanded pretreatment systems utilize an extended aeration activated sludge process for organic concentration reduction. The winery's expanded pretreatment facility has experienced problems with filamentous

* UOD = (1.5 x $CBOD_5$) + (4.5 x TKN), where $CBOD_5$ is the five day carbonaceous biochemical demand and TKN is total nitrogen.

organism growth in its sludge. The winery is attempting to control this problem through chlorination. The City of Canandaigua Wastewater Treatment Plant has not experienced problems with sludge bulking in their secondary clarifiers. Generally RBC sludge does not tend to bulk and additionally the use of ferric chloride for phosphorus removal prior to secondary clarification further enhances sludge settleability. Bulking of the sludge has occasionally occurred, however, in the city's sludge gravity thickeners. The winery periodically exceeds its organic discharge limits due to pretreatment plant upsets. When this occurs simultaneously with inflows of leachate and septic tank waste from truckers, a slight reduction in gas production from the city's anaerobic digestors has been noted.

CANANDAIGUA WASTEWATER TREATMENT PLANT
CANANDAIGUA, NEW YORK

Design Flow:	**6.5 mgd**	Location:	**Central New York State**
Secondary Treatment:	**Rotating Biological Contactors**	Population Served:	**25,000**

INFLUENT WASTEWATER

	Typical (Upset)
Ave. Flow, mgd	3.0
% Industrial	3
BOD_5, mg/l	180 (360)
SS, mg/l	230

SIGNIFICANT INDUSTRIES

Industry	Flowrate (1000 gpd)	Problem Pollutants
Winery	100	BOD, COD

PLANT LOADING

Primary Clarifiers	Typical (Upset)
Overflow Rate, gal/sf/day	800
Detention Time, hours	2.7
Effluent BOD_5, mg/l	120 (230)
Effluent SS, mg/l	80

Secondary Clarifiers	Typical (Upset)
Overflow Rate, gal/sf/day	600
Detention Time, hours	3

Rotating Biological Contactors	Typical (Upset)
Organic Loading, lbs BOD_5/1,000 sf/d	1.5 (2.9)
Hydraulic Loading, gal/sf/day	
Carbonaceous Contactors	2.5
Nitrification Contactors	1.6

PLANT PERFORMANCE

	Permit Limit	Typical (Upset)
BOD_5, mg/l	30	11 (45)
SS, mg/l	30	11
UOD, mg/l		
Summer and Fall	50.7	31 (76)
Spring	97.0	22 (148)
Winter	134.0	35 (73)

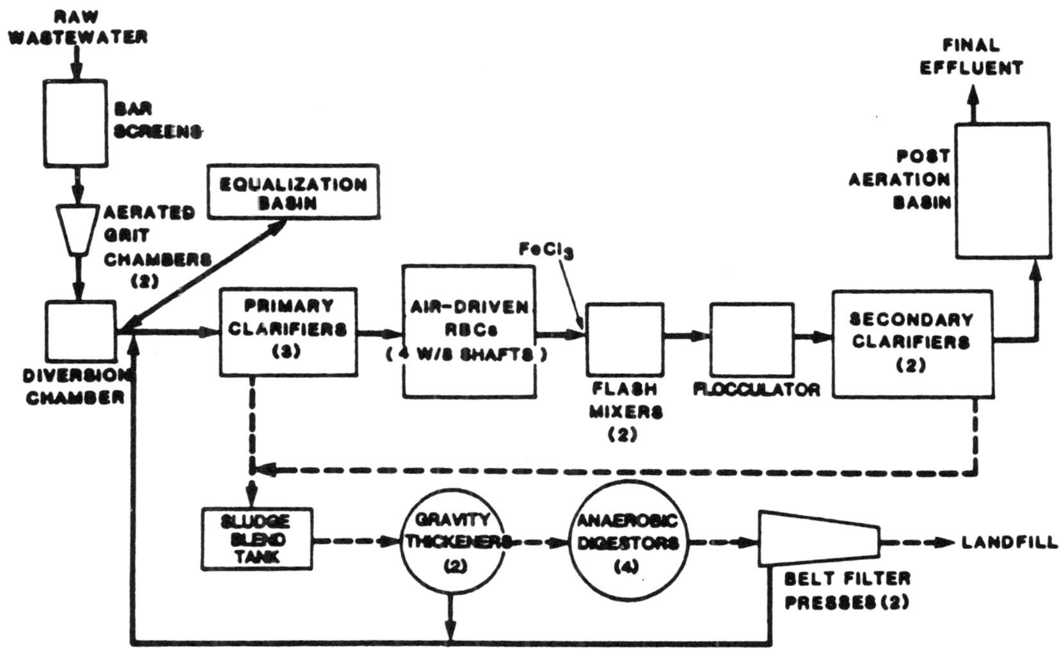

EAST SIDE SEWAGE TREATMENT PLANT
Oswego, New York

The City of Oswego, East Side Treatment Plant has experienced significant noncompliance problems associated with the loss of solids from their secondary clarifiers. Half of the plant's hydraulic flow is from a paper mill which is the only major industry in the city. From 1981 to 1983, the noncompliance problems at the plant were attributed to severe hydraulic and organic load peaks from the paper mill as well as operational difficulties such as frequent breakdowns of the return sludge pump drives. It is not known whether filamentous growth in the sludge occurred at that time. In 1983 the paper mill voluntarily reduced the hydraulic and organic peaks to the plant. Solids losses from the secondary clarifier still remained a problem. During 1984, the plant frequently exceeded their NPDES discharge suspended solids by five times the limit and the BOD by three times the limit. During that period, the plant still occasionally received hydraulic peaks from the paper mill which were twice the average rate for two to three hour periods, but a substantial cause of the problem was identified as poor sludge settleability due to filamentous growth. The frequent washout of biosolids from the secondary clarifiers resulted in a low mean cell residence time and the generation of a young sludge that did not settle well. In the spring of 1985, the belt drives on the return sludge pumps which had frequently been out of service were replaced with electronic variable speed drives. This improvement allowed the plant operators to maintain better control of the solids inventory in the aeration tanks. Plant performance was still poor, however, because of sludge bulking.

Several measures have been taken at the plant in an attempt to alleviate the sludge bulking problem. The measures that were taken are:

- switching from plug flow feed to a step feed in the aeration tanks in order to achieve better dissolved oxygen distribution;

- varying process control strategies such as sludge return and wasting rates, and sludge blanket depth; and

- chlorination of the return sludge for the destruction of filamentous growth in the sludge.

The step feed operation has resulted in better dissolved oxygen distribution but did not significantly improve sludge settleability. The second two mitigation efforts were ongoing at the time of writing. A chlorination dosage of 6 lb Cl_2/1000 lb solids had been applied to the return sludge. Microscopic examination of the sludge indicated that the filaments had shrunk and the SVI level had dropped to the range of 60-80. The plant operators intend to chlorinate whenever the SVI increases to 150. It has not been determined if these mitigation measures can result in plant performance that will consistently meet the permit discharge limits.

The paper mill periodically discharges slugs of waste containing high suspended solids to the treatment plant. At these times, the sludge in the primary tanks takes on a gelatinous quality which makes sludge removal difficult. High periodic input of clay filler materials from the paper mill has resulted in poor sludge incineration with associated high fuel usage.

The City of Oswego is presently preparing an industrial discharge permit for the paper mill. The permit will restrict the monthly and daily average BOD and suspended solids levels in the influent from the paper mill as well as restrict the daily maximum hydraulic peak allowed. Under the permit provisions the paper mill will be required to submit listings of the chemicals used in their processes. The paper mill is presently voluntarily investigating the possible relationship of the chemicals used in their manufacturing processes to the occurrence of filamentous growth in the activated sludge process.

EAST SIDE SEWAGE TREATMENT PLANT
OSWEGO, NEW YORK

Design Flow:	3 mgd	Location:	Northern New York
Secondary Treatment:	Plug or Step Feed Activated Sludge	Population Served:	10,000

INFLUENT WASTEWATER

	Typical (Upset)
Ave. Flow, mgd	2.5
% Industrial	50

	Municipal	Paper Mill
BOD_5, mg/l	100	300
SS, mg/l	120	450 (1000)

SIGNIFICANT INDUSTRIES

Industry	Flowrate (1000 gpd)	Problem Pollutants
Paper Mill	1,200	SS, BOD

PLANT LOADING

Primary Clarifiers	Typical (Upset)	
Overflow Rate, gal/sf/day	600	
Detention Time, hours	2	
	Municipal	Paper Mill
Effluent BOD_5, mg/l	70	120
Effluent SS, mg/l	40	100

Secondary Clarifiers	Typical (Upset)
Overflow Rate, gal/sf/day	800
Detention Time, hours	2
SVI, ml/gm	100 (1000)

Aeration Basins	Typical (Upset)
F/M, lbs BOD_5/lbs MLSS/day	0.2
MCRT, days	7 (3)
MLSS, mg/l	2,000 (300)
Detention Time, hours	7
Return Flow, %	25 - 45
D.O. Level, mg/l	2 - 4

PLANT PERFORMANCE

	Permit Limit Summer	Permit Limit Remainder of Year	Typical (Upset)
BOD_5, mg/l	30	45	20 (120)
SS, mg/l	30	70	25 (300)

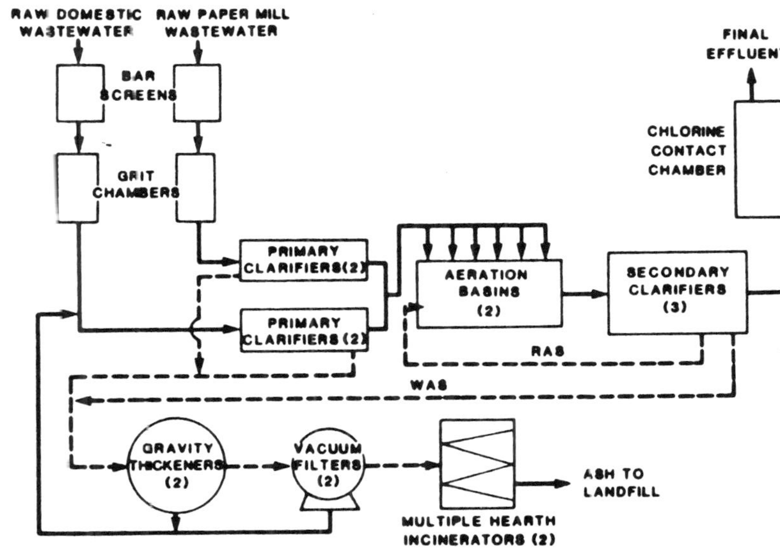

HATFIELD TOWNSHIP ADVANCED TREATMENT FACILITY
Colmar, Pennsylvania

The Hatfield Township Municipal Authority (HTMA) operates an advanced wastewater treatment facility which receives two-thirds of its domestic flow from Hatfield Township and one-third from Montgomery Township, Pennsylvania. Although less than 10 percent of the plant flow is supplied by industry, up to 60 percent of the influent waste strength can result from industrial and waste hauler sources. The HTMA was issued a Consent Order and Agreement by the Pennsylvania Department of Environmental Resources in March 1985 for non-compliance with their NPDES Permit due to hydraulic and organic overloading at the POTW. The Order gives the HTMA until May 1, 1987 to meet a new set of discharge limits, which, in addition to those shown on the attached data sheet, includes a 2/6 mg/l (summer/winter) ammonia limit. A new 6.4 mgd Shreiber process facility including nitrification/denitrification capability is currently under construction to meet the goals of the post-1987 permit.

Violations of the total phosphorus limit (2 mg/l) and occasional problems with the ammonia (32) and BOD (15, summer) limits are the primary reasons for the Consent Order. The advanced treatment facility has performed well enough in the area of suspended solids removal that pressure filters installed following the tube settlers in the flow schematic have been taken out of service. Feeding $FeSO_4$ to the return activated sludge improved the total-P removal from 43 percent to 65 percent, and is enough to consistently reduce the effluent P to below 2 mg/l.

While industial discharges can not be blamed for exceeding the phosphorus limits, high influent BOD, SS and nitrogen are directly attributable to a few of the key industries. The Industrial Wastewater Discharge Permits issued by the HTMA limit the concentrations of these compounds to the following monthly averages (in mg/l):

Parameter	Maximum	Surchargeable
BOD	2,000	195
SS	800	180
TKN + NO_2 + NO_3	160	15
Total-P	20	8
FOG	250	

The most significant violator of the discharge limits is a 120,000 gpd industrial waste pretreatment facility that uses a physical-chemical process for metals removal and pH neutralization. The average effluent from this plant was measured in 1984 as:

BOD (soluble)	3109
SS	1718
Nitrogen	586
Total-P	7.7

Additionally, numerous organic compounds have been identified in their discharge with concentrations ranging from 10 to 5,500 ug/l, but no interference due to these organics has been detected to date. Given the nature of the wastewater, it is not surprising that the HTMA has denied a permit to the industrial waste pretreater for expansion of their facility. However, the feasibility of utilizing the existing POTW to treat larger volumes of this wastewater after the new municipal facility is on-line is now being evaluated.

A second significant contributor of conventional pollutants had been a 17,000 gpd dairy, who was being surcharged for excess BOD, SS and nitrogen. In this case, the solution was to truck the whey waste rather than discharge to the sewer, resulting in a cost savings to the dairy and reduced loadings at the POTW.

A small (1,500 gpd) chemical company was being surcharged $ 10,000 per quarter for an ammonia discharge of up to 30,000 mg/l. At times, the NH_3 concentration at the POTW influent would reach 100 mg/l. In 1983, the industry installed a pretreatment system which reduced the NH_3 concentration, so that the company's quarterly surcharge now ranges from $ 1,000 to $ 1,500.

The HTMA permits 13 haulers of septage, holding tank contents and leachate to discharge 115,000 gpd into the effluent launder of the primary clarifiers. While this practice generated over $ 350,000 of income in 1984, the impact of these discharges on the treatment plant are difficult to assess given the limited sampling of these wastewaters. A single day's testing of the discharges in April 1985 produced suspended solids results ranging from 23 to 65,000 mg/l, and COD values of from 800 to 36,000 mg/l. HTMA has estimated that the hauler wastewater increases the sludge production at the POTW by 40 percent over the volume generated by the influent wastewater.

HATFIELD TOWNSHIP ADVANCED WASTE TREATMENT FACILITY
COLMAR, PENNSYLVANIA

Design Flow:	3.6 mgd	Location:	Southeastern Pennsylvania
Secondary Treatment:	Activated Sludge (Complete Mix)	Population Served:	20,000

INFLUENT WASTEWATER

	Typical (Upset)
Ave. Flow, mgd	3.4 (6)
% Industrial	10
BOD_5, mg/l	230 (350)
SS, mg/l	235 (390)
NH_3, mg/l	25 (35)

SIGNIFICANT INDUSTRIES

Industry	Flowrate (1000 gpd)	Problem Pollutants
Ind. Waste Treatment	120	BOD, SS, N, Organics
Dairy	17	BOD, SS, NH_3
Steel (Paint Shop)	25	Paint slugs
Chemical	1.5	NH_3
Waste Haulers (13)	115	COD, SS

PLANT LOADING

Primary Clarifiers	Typical (Upset)
Overflow Rate, gal/sf/day	600 (1,050)
Detention Time, hours	3 (1.7)
Effluent BOD_5, mg/l	170
Effluent SS, mg/l	205

Aeration Basins	Typical (Upset)
F/M, lbs BOD_5/lbs MLSS/day	0.33
MCRT, days	4.3
MLSS, mg/l	4,500-5,000
Detention Time, hours	4 (2.2)
Return Flow, %	50
D.O. Level, mg/l	4

Secondary Clarifiers	Typical (Upset)
Overflow Rate, gal/sf/day	720 (1,250)
Detention Time, hours	3 (1.7)
SVI, ml/gm	75
Effluent BOD_5, mg/l	50
Effluent SS, mg/l	75

PLANT PERFORMANCE

	Permit Limit	Typical (Upset)
BOD_5, mg/l	15/30	12 (20)
SS, mg/l	20	5 (10)
NH_3, mg/l	32	24 (40)
$NO_2 + NO_3$, mg/l	N/A	3 (6)
Total-P, mg/l	2	4 (20)

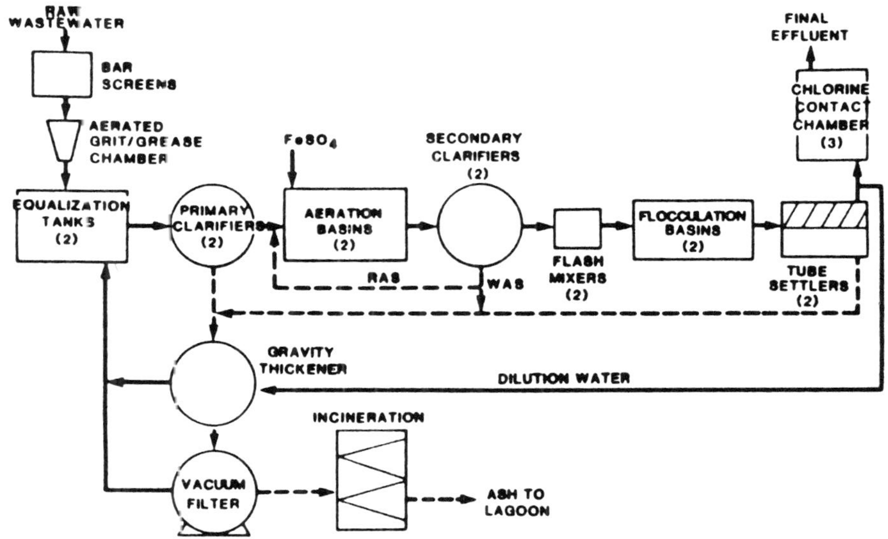

MAIDEN CREEK WASTEWATER TREATMENT PLANT
Blandon, Pennsylvania

The Maiden Creek Wastewater Treatment Plant (MCWTP) went on-line in December, 1981 as a secondary treatment facility designed to remove both carbonaceous and nitrogenous BOD. The plant uses a patented aerated submerged fixed film biological treatment system, where flat asbestos plates hanging vertically in the settled wastewater provide a growth surface for the bacteria. Each of three contact basins contains 320 plates with 200 sq. ft. of surface area. Oxygen is provided by fine bubble aeration through ceramic diffusers.

During the first six months of operation following an initial acclimation period, the MCWTP experienced gradual flow increases from 0.1 to 0.15 mgd while consistently meeting their permit limits. In August of 1981, a local mushroom processor began batch discharging high BOD wastewater to the POTW at flows sometimes exceeding 100 gpm. The hydraulic and organic shock loadings resulted in nitrifier washouts, solids carryover, reduced BOD removal efficiency and at times total biological process failure. Although the industry was not measuring their wastewater flow rates at that time, they were the only significant non-domestic contributor. After factoring out any potential infiltration/inflow from stormwater flows, the discharge pattern from the industry was obvious from an inspection of the weekly flow recordings at the POTW. Figure C-2 illustrates the dramatic effect of the industrial discharges on the MCWTP influent.

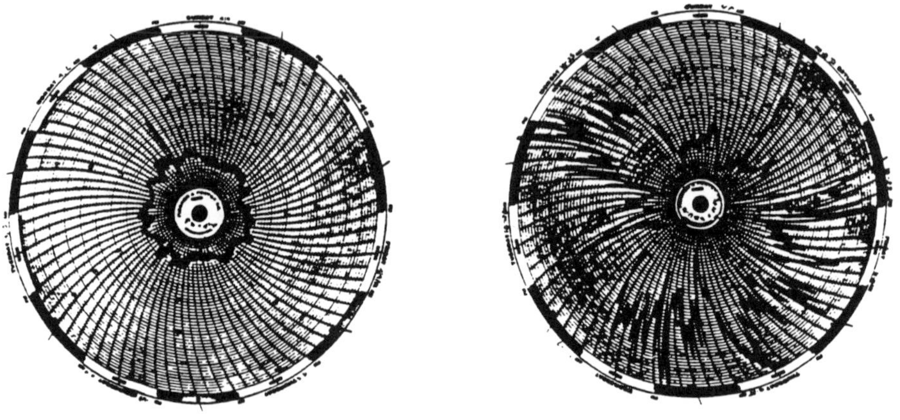

April, 1982 **October, 1982**

FIGURE C-2
WASTEWATER DISCHARGE AT INFLUENT METERING STATION (MGD)

As a result of significant time and effort on the part of Maiden Creek Township Municipal Authority two years ago, the food processor installed a physical-chemical treatment system which included surge control tanks and aeration. The system did reduce the solids load and partially mitigated the flow spike problem, although the surge tanks were not capable of providing complete equalization. Unfortunately, the great percentage of their organic waste is soluble, so the pretreatment facility is ineffective in reducing the BOD loading to the POTW. Additionally, wastewater production far exceeds the 50,000 gpd limit imposed by their permit, so occasional flow spikes are still evident. The industry has requested nearly ten times the current flow limit, necessitating the design of a full secondary system to reduce their waste strength to domestic levels. Such a system, including a 650,000 gallon aerated equalization basin, is scheduled to go on-line in mid-1986. In the interim, the municipality has required that the industry:

- control flow surges;
- meter and record their flows continuously;
- reduce the BOD in the effluent by in-house methods; and
- composite sample their discharge on a regular basis.

Failure to comply with the abovementioned program will result in a shut off by the POTW, a measure used previously in February, 1985 when the industry's wastewater was responsible for total process failure at the plant.

A number of operational changes were instituted in May of 1985 to help combat the high organic loads in the contact basins. These changes included:

- increasing the aeration by using all blowers at the plant, resulting in an increase in the first stage D.O. from 2 mg/l to 5 mg/l;

- addition of selective strains of bacteria to increase the rate of BOD removal;

- recycling the plant effluent to the head of the plant to dilute the incoming wastewater; and

- reducing the allowable flow from the food processor and closely monitoring their adherence to the limits.

Since these changes were implemented concurrently, it is impossible to isolate the individual impacts of each operations change. However, the collective result was a substantially improved compliance record. There have also been no flow spikes at the POTW since mid-December, 1985, indicating better flow control on the part of the food processor.

MAIDEN CREEK WASTEWATER TREATMENT PLANT
BLANDON, PENNSYLVANIA

Design Flow:	**0.45 mgd**	Location:	**Southeastern Pennsylvania**
Secondary Treatment:	**Aerated Submerged Fixed Film (Contact Aeration)**	Population Served:	**2,000**

INFLUENT WASTEWATER

	Typical (Upset)
Ave. Flow, mgd	0.25
% Industrial	20 (60)
BOD_5, mg/l	350 (900)
SS, mg/l	200
NH_3, mg/l	60

SIGNIFICANT INDUSTRIES

Industry	Flowrate (1000 gpd)	Problem Pollutants
Food Processor	50	BOD, Flow surges
Dental Office	negl.	Hg

PLANT LOADING

Primary Clarifiers	Typical (Upset)
Overflow Rate, gal/sf/day	350 (1,000)
Detention Time, hours	3.75 (1.25)
Effluent BOD_5, mg/l	260
Effluent SS, mg/l	100

Secondary Clarifiers	Typical (Upset)
Overflow Rate, gal/sf/day	450 (1,300)
Detention Time, hours	2.8 (1.0)

Contact Basins	Typical (Upset)
Organic Loading (lbs BOD_5/1000 sf/day)	
Total Plant	2.8
First Stage	8.4
Detention Time, hours	12
D.O. Level, mg/l	5 - 10

PLANT PERFORMANCE

	Permit Limit	Typical (Upset)
BOD_5, mg/l	30	15 (400)
SS, mg/l	30	10 (50)
NH_3, mg/l	10/20	1 (60)

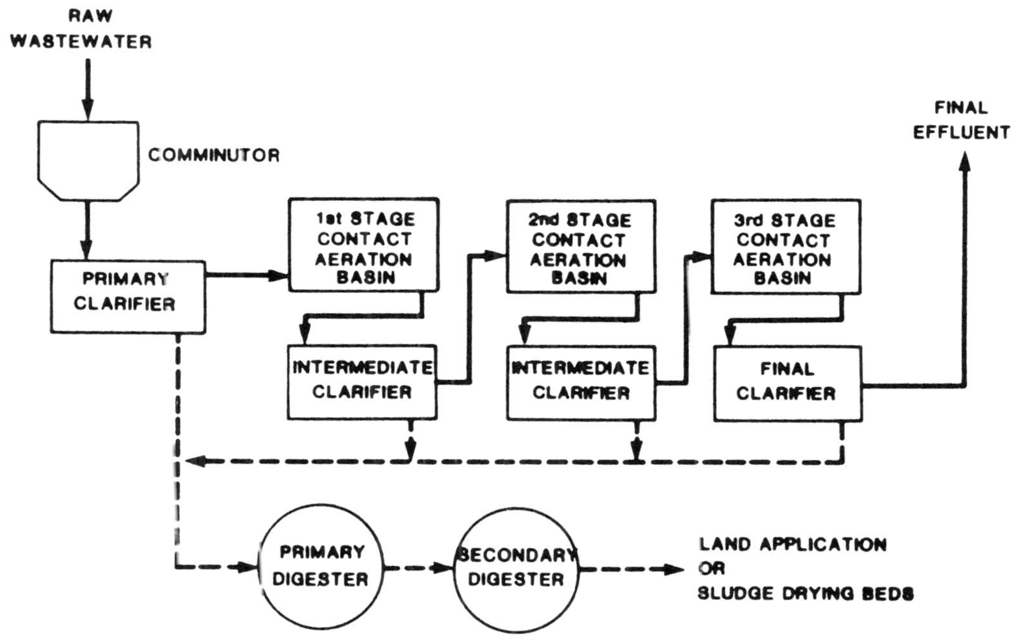

ROCKY CREEK WATER POLLUTION CONTROL PLANT
Macon, Georgia

The Rocky Creek Water Pollution Control Plant (RCTP) treats an average of 12 mgd of wastewater, nearly half of which is contributed by industrial users. 40 percent of the total plant flow and 70 to 80 percent of the organic and solids loading is contributed by one paper products manufacturer. Additional major industrial users are an animal food processor, two food processors and a wood preserving plant. The RCTP has been in substantial non-compliance of its NPDES permit since coming on line in 1975, primarily because of variable discharge of high strength organic waste. Although industrial wastes continue to make up a large portion of the organic loading to the RCTP, the plant has not experienced a NPDES permit violation in the last six months, coinciding with the development and implementation of an industrial pretreatment program.

The RCTP utilizes the extended aeration activated sludge process to treat the high-strength domestic/industrial wastewater. The large organic contribution of the paper products manufacturer is nutrient deficient and requires phosphorus and nitrogen addition for proper biological treatment. Despite the large organic contribution and poor solids settling characteristics of this industrial wastestream, it has not historically presented chronic treatment problems because of its fairly consistent strength. Interferences identified at the RCTP were primarily attributed to the other major industrial users, in particular the animal food processor. Operations at the animal food processing plant were such that periodic slugs of organic wastes were discharged to the RCTP with COD values as high as 30,000 mg/l. Typical daily average RCTP influent organic and solids loadings were 350 mg/l BOD and 230 mg/l SS, but would rise as high as 525 mg/l and 500 mg/l, respectively, during upset periods. These stress conditions resulted in treatment plant organic overload and poorly settling sludge, with effluent BOD and SS levels rising to 80 mg/l and 150 mg/l, respectively. Oils and creosote from the wood preserving plant are not typically discharged in high enough concentrations to upset the biological treatment process on their own, but occasionally contribute to the magnitude of interferences during organic overloads by decreasing sludge settleability. Chlorine addition was used with some success to improve the settleability of the activated sludge.

To control a worsening problem, a pretreatment program was developed and final industrial wastewater discharge permits were issued in October, 1985 by the Macon-Bibb County Water and Sewerage Authority. In order to meet their permit limits, nearly all major industries have hired consultants to examine their pretreatment programs or have installed or upgraded existing pretreatment plants. The paper products manufacturer upgraded its existing stabilization pond by installing a new 15 million gallon clarifier, a thickener tank and belt press. The dewatered sludge is incinerated. The animal food processor modified a sump pump station in addition to constructing a new drain catch-basin. These modifications allow for the recovery of previously wasted sugar and molasses sludges and cause wastewater organic loadings to be significantly decreased, in addition to being equalized. The wood preserving operation installed a compact clarifier that recovers floatable oils and grease as well. The industrial pretreatment improvements have been a major factor in the RCTP treatment improvements, as the plant has gone from being overloaded and upset 50 percent of the time to being upset twice a month at most. As a result, no NPDES permit violations have been experienced since September, 1985.

ROCKY CREEK WATER POLLUTION CONTROL PLANT
Macon, Georgia

Design Flow:	**14 mgd**	Location:	**Central Georgia**
Secondary Treatment:	**Extended Aeration Activated Sludge**	Population Served:	**75,000**

INFLUENT WASTEWATER

	Typical (Upset)
Ave. Flow, mgd	12
% Industrial	50
BOD_5, mg/l	350 (525)
SS, mg/l	230 (500)

SIGNIFICANT INDUSTRIES

Industry	Flowrate (1000 gpd)	Problem Pollutants
Paper products	5000	BOD, COD, SS
Animal Food processing	7	BOD, COD, SS
Food processing	130	BOD,SS
Wood preserving	9	oils, phenols

PLANT LOADING

Aeration Basins	Typical (Upset)
MCRT, days	20–40
MLSS, mg/l	3000
Detention Time, hours	28
Return Flow, %	40–60
D.O. Level, mg/l	2.0 (4.0)

Secondary Clarifiers	Typical (Upset)
Overflow Rate, gal/sf/day	350
Detention Time, hours	5.4

PLANT PERFORMANCE

	Permit Limit	Typical (Upset)
BOD_5, mg/l	30	20 (80)
SS, mg/l	75	40 (150)

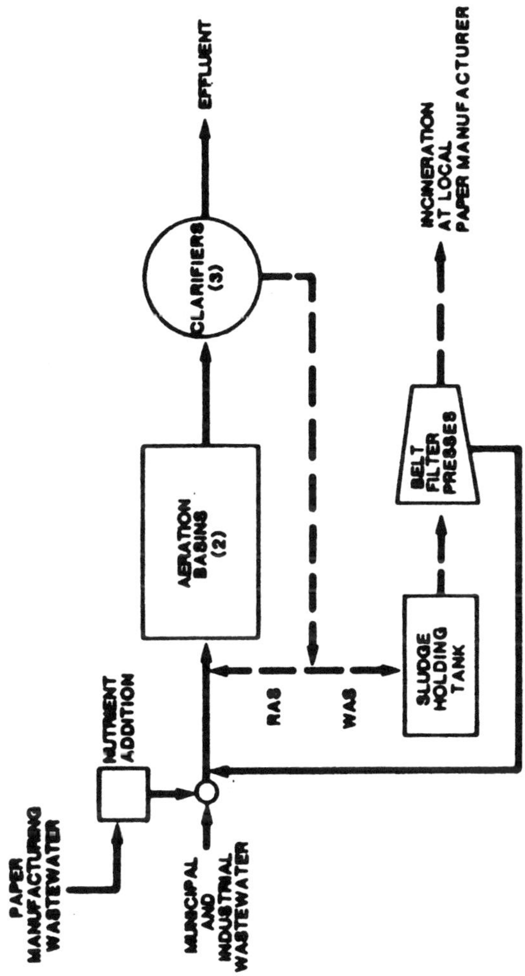
PAPER MANUFACTURING WASTEWATER
NUTRIENT ADDITION
MUNICIPAL AND INDUSTRIAL WASTEWATER
RAS
WAS
AERATION BASINS (2)
CLARIFIERS (3)
EFFLUENT
SLUDGE HOLDING TANK
BELT FILTER PRESSES
INCINERATION AT LOCAL PAPER MANUFACTURER

CITY OF BALTIMORE, MARYLAND

The City of Baltimore owns and operates two wastewater treatment facilities, Back River and Patapsco, with a combined volumetric flow rate of approximately 250 million gallons per day. The plants serve a combined population of nearly 1.3 million, residing in the City and the Counties of Baltimore, Anne Arundel and Howard. In accordance with the requirements of the General Pretreatment Regulations (40 CFR Part 403) established by the U.S. EPA, the City developed an extensive industrial waste control program to:

- safeguard the public's health;
- protect the wastewater systems and its employees; and
- prevent deterioration of the receiving waters and lands.

The final report outlining the details of the industrial waste program identified 4,700 sources or potential sources of nondomestic wastewater, of which about 220 are EPA-designated categorical industries. A program of this magnitude requires a significant committment in terms of personnel, equipment, office space, and supplies. Annual operating costs are expected to exceed $2.5 million by fiscal year 1989.

As part of an initial sampling effort, 35 nonconventional organic and inorganic pollutants were identified in the influent to the two POTWs. Based on these data, the following industrial discharge criteria were recommended or reiterated in the industrial waste control program final report:

Parameter	Limitations (mg/l, except pH)
pH	6.0
FOG	100
CN^-	0.2
Cd	0.18
Cr (Total)	5
Cu	1.9
Pb	0.7
Hg	0.01
Ni	2.5
Zn	2.6
Explosivity	10% LEL

The report further specifies the need for continual monitoring of influent and effluent toxicity through the use of Microtox and bioassay methods at both treatment plants.

One of the more interesting aspects of the Baltimore program is the computer coding of the sewer collection system. By knowing the constituents of each industry's discharge, the flow rate and their location in the coded sewer system, a contaminant discovered at either POTW can theoretically be traced back to its potential source or sources. While such a backtracking program is of little use for isolated discharges, it could prove beneficial in locating chronic dischargers of specific compounds.

BACK RIVER WASTEWATER TREATMENT PLANT
Baltimore, Maryland

The Back River facility is hydraulically and organically overloaded, resulting in effluent BOD and SS consistently in excess of the 45 mg/l interim limits. The plant is currently undergoing a major renovation to replace the 30 acres of trickling filter rock media with complete-mix activated sludge, along with significant alteration and expansion of most process units. The renovation work is in preparation for new NPDES Permit limits of 10/10 and 2 mg/l (NH_3), which will require the addition of powdered activated carbon as an aid for nitrification. Industrial flows to Back River total approximately 27 mgd, and are dominated by metals and solvents in the discharge.

The primary source of metals in the system is from the 12 metal plating operations identified by the industrial waste survey. The problem with the metals content in the wastewater is that it restricts the ultimate disposal options for the digested and dewatered sludge. When local limits were calculated based on unrestricted distribution of the sludge, the limits were occasionally one-fourth of the achievable levels. Consequently, the City of Baltimore opted for the less stringent 10,000 gpd electroplater standards for the noncategorical industries. A compost facility now under construction is expected to process 150 wet tons of the 450 tons produced each day, beginning in March 1987. The metals content continues to remain a concern for this disposal option.

The benefits of pretreatment for metals removal have been demonstrated at Back River. An incinerator had been discharging 2 tons of fly ash per hour into the collection system, which was high in metal content and was responsible for 90 percent of the cadmium in the POTW influent. Other wastewater containing metals were from steel and automobile manufacturing. In each case, pretreatment facilities have come on-line during the past year, with a measureable drop in influent and sludge concentrations. The situation has improved to the point where the City is reevaluating limits and granting exemptions to some industries on a case-by-case basis.

The second major area of concern at the Back River plant stems from the large, batch discharges of solvents, petroleum hydrocarbons and other toxic organics. A 2:00 am discharge of ethylbenzene, xylene and toluene resulted in the evacuation of the largest pump station and other buildings in town. The problem was traced to a paint and chemicals manufacturer, who has since improved its in-house solvent recovery system. A similar evacuation resulted from a 4,000 gallon discharge of xylene by a waste hauler, which was traced to a specific location in the collection system. Tetrachloroethylene has been discovered and traced to dry cleaning operations. While such discharges have not resulted in process inhibition at the plant, the health and safety issues and potential for explosion are of serious concern to the City.

BACK RIVER WASTEWATER TREATMENT PLANT
BALTIMORE, MARYLAND

Design Flow: 180 mgd
Secondary Treatment: Trickling Filters and Activated Sludge

INFLUENT WASTEWATER

	Typical (Upset)
Ave. Flow, mgd	210 (235)
% Industrial	15
BOD_5, mg/l	230
SS, mg/l	190

SIGNIFICANT INDUSTRIES

Industry	Flowrate (mgd)	Problem Pollutants
Metal Plating (12)	0.18	Metals
Auto Mfr.	1.5	Cr, Cu, Ni, Zn
Plant and Chemical	N/A	Ethyl benzene, toluene, xylene
Incinerator	N/A	Cd, Hg
Waste Haulers	N/A	Solvents, petroleum hydrocarbons

PLANT LOADING

Primary Clarifiers	Typical (Upset)
Overflow Rate, gal/sf/day	1,300 (1,500)
Detention Time, hours	1.6
Effluent BOD_5, mg/l	180
Effluent SS, mg/l	135

Aeration Basins	Typical (Upset)
Ave. Flow, mgd	60
F/M, lbs BOD_5/lbs MLSS/day	0.5
MCRT, days	6.1
MLSS, mg/l	2,000
Detention Time, hours	3.5
Return Flow, %	30-35
D.O. Level, mg/l	2-3

Secondary Clarifiers (A.S./T.F)	Typical (Upset)
Overflow Rate, gal/sf/day	750/950
Detention Time, hours	2.5/2.1
SVI, ml/gm	95

Trickling Filters	Typical (Upset)
Ave. Flow, mgd	150 (170)
Hydraulic Loadings, ga/sf/d	120 (136)
Organic Loading, lbs BOD/1000 cf/d	20
Return Flow, %	10

PLANT PERFORMANCE

	Permit Limit	Typical (Upset)
BOD_5, mg/l	45	70 (85)
SS, mg/l	45	55 (70)

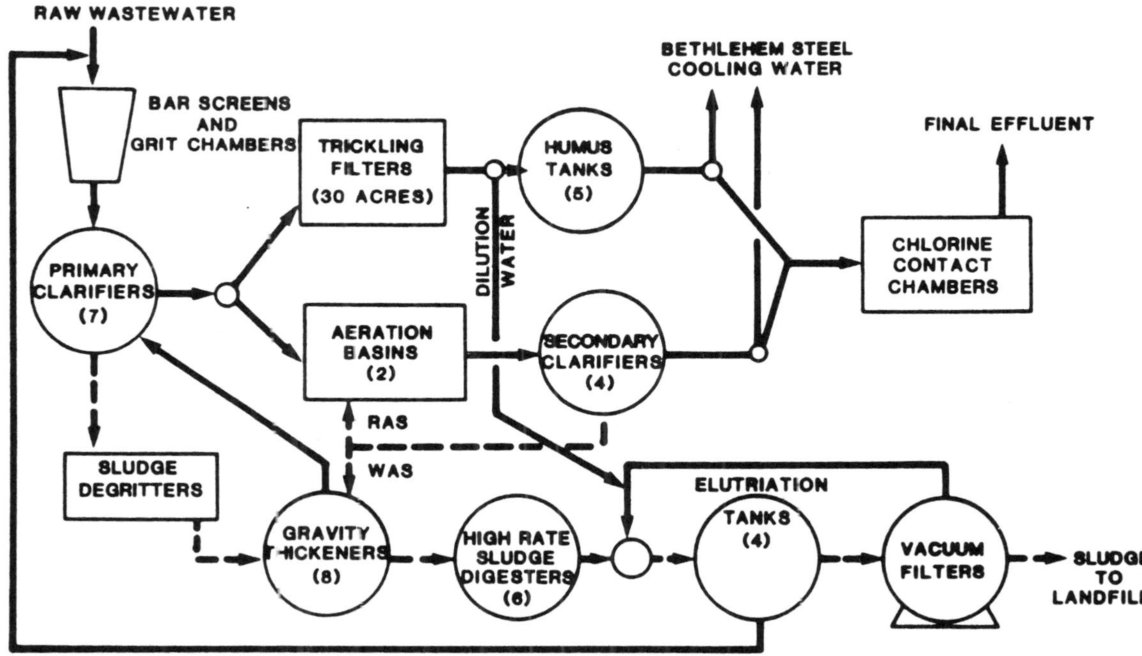

PATAPSCO WASTEWATER TREATMENT PLANT
Baltimore, Maryland

A 1981 EPA-sponsored project on biomonitoring of direct discharges rated the Patapsco plant as having the most toxic effluent of those surveyed. Ironically, the second most toxic discharge came from an agricultural chemicals manufacturer who, in 1983, ceased direct discharging and now sends their pretreated wastewater to Patapsco. The high level of toxicity has resulted in the collection of much bioassay, acute toxicity and respirometer data over the past four years. Despite a high level of measured toxicity in the influent, the plant currently meets its discharge limit for BOD and SS, indicating the ability of activated sludge to acclimate to consistent levels of many organic compounds. Acute toxicity data using a Beckman Microtox unit have been collected since November 1980. Some of the results of these analyses are shown on Figure C-3. The data are on an inverse scale, with the lowest values indicating highest toxicity and approximately 45 percent corresponding to no toxic effect.

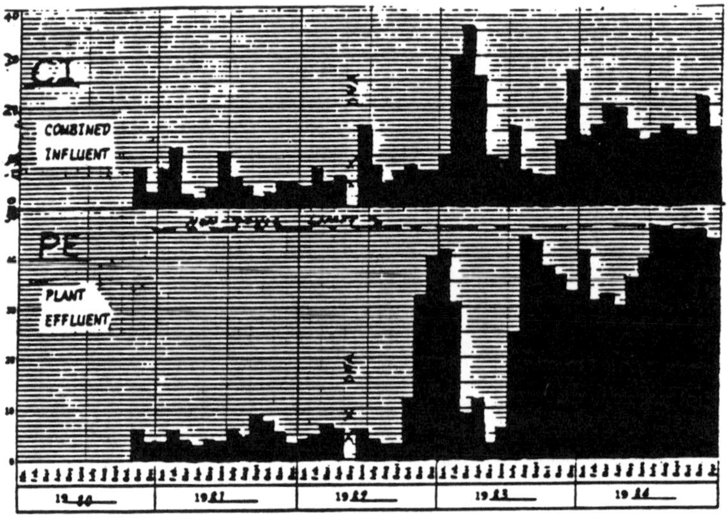

FIGURE C-3
MONTHLY ACUTE TOXICITY
(Courtesy G.H. Slattery, City of Baltimore)

Figure C-3 illustrates the toxic nature of the plant influent and effluent until September 1982, at which time the secondary treatment system went on-line. The acclimation of the activated sludge improved effluent toxicity from 5 percent to 40 percent by December, where it remained until secondary shutdown in February 1983. The effluent again became toxic until the secondaries were returned to service on June 15, providing clear evidence of the detoxification capability of acclimated activated sludge.

The attached data sheet indicates that Patapsco's noncompliance has resulted from the discharge of excess phosphorus and an effluent pH below 6.5. The phosphorus problem is being dealt with by installing A/O technology in the oxygenation basins as a means of biological phosphorus removal. The low pH is inherent in oxygen activated sludge systems, typically producing an effluent in excess of 250 mg/l of CO_2 and a pH of 6.2. The problem can be corrected with either chemical adjustment or post-aeration of the wastewater.

Although compliance with the NPDES Permit has been achieved for BOD and SS at Patapsco, the plant flow is well below the 70 mgd design capacity. Toxic inhibition is still present despite the improvement since 1983 (see Figure C-3). Evidence of this inhibition is provided by the operating F/M of 0.3, which is significantly less than the design value of 0.5. As a means of further improving the situation, the State of Maryland included the following in the consent order issued to the City in 1984:

- install on-line toxicity monitoring of the plant influent
- develop a toxics emergency response plan
- enlarge the scope of the City sewer ordinance to include specifics on toxicity and flammability for industrial effluents.

PATAPSCO WASTEWATER TREATMENT PLANT
BALTIMORE, MARYLAND

Design Flow:	70 mgd
Secondary Treatment:	Activated Sludge (Pure Oxygen)

INFLUENT WASTEWATER

	Typical (Upset)
Ave. Flow, mgd	42
% Industrial	30
BOD_5, mg/l	265 (320)
SS, mg/l	325 (470)
TOX, %	15

SIGNIFICANT INDUSTRIES

Industry	Flowrate (mgd)	Problem Pollutants
Chemicals	1.0	Insecticides, Volatiles, phenols, metals
Metal Finishing	0.13	pH, solvents, metals

PLANT LOADING

Primary Clarifiers	Typical (Upset)
Overflow Rate, gal/sf/day	1,150
Detention Time, hours	1.5
Effluent BOD_5, mg/l	190
Effluent SS, mg/l	80

Aeration Basins	Typical (Upset)
F/M, lbs BOD_5/lbs MLSS/day	0.3
MCRT, days	10-15
MLSS, mg/l	5,000
Detention Time, hours	2
Return Flow, %	30
D.O. Level, mg/l	2-4

Secondary Clarifiers	Typical (Upset)
Overflow Rate, gal/sf/day	450
Detention Time, hours	6.3
SVI, ml/gm	50-75

PLANT PERFORMANCE

	Permit Limit	Typical (Upset)
BOD_5, mg/l	30	18 (40)
SS, mg/l	30	25 (40)
Total-P, mg/l	2.0	3.5
pH	6.5-8.5	6-6.5
TOX, %		40

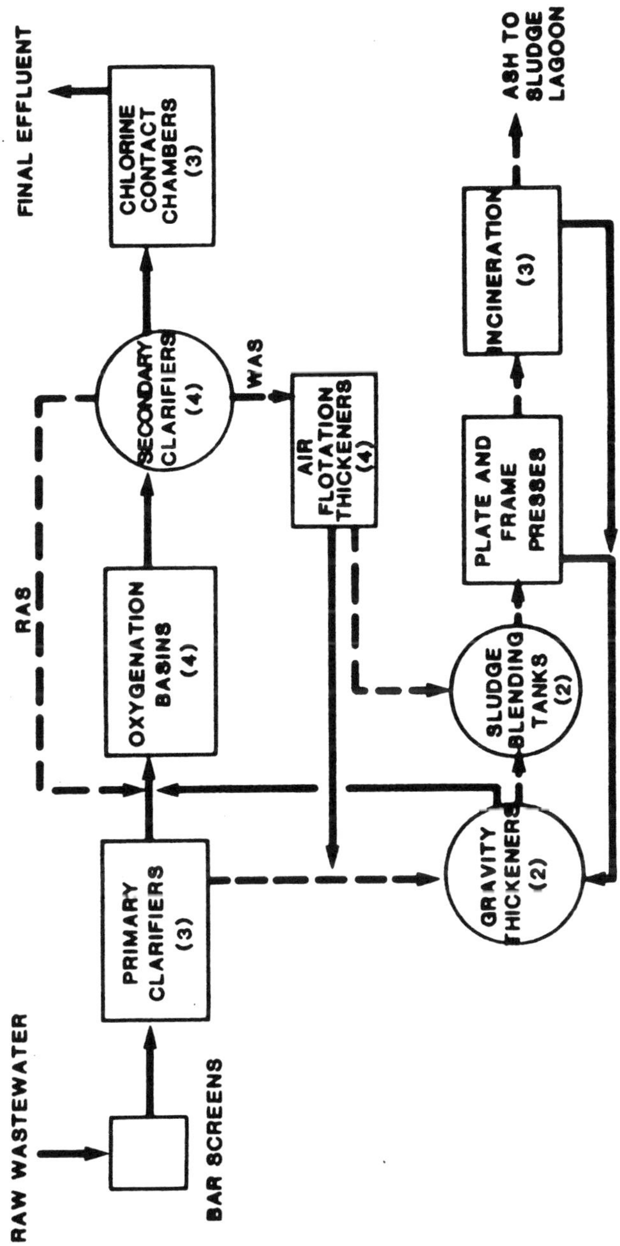
RAW WASTEWATER
BAR SCREENS
PRIMARY CLARIFIERS (3)
OXYGENATION BASINS (4)
RAS
SECONDARY CLARIFIERS (4)
WAS
CHLORINE CONTACT CHAMBERS (3)
FINAL EFFLUENT
AIR FLOTATION THICKENERS (4)
GRAVITY THICKENERS (2)
SLUDGE BLENDING TANKS (2)
PLATE AND FRAME PRESSES
INCINERATION (3)
ASH TO SLUDGE LAGOON

CITY OF RAEFORD WASTEWATER TREATMENT PLANT
Raeford, North Carolina

The single most significant discharger to the Raeford Wastewater Treatment Plant is a turkey processor who contributes 30 percent of the flow volume. Until two years ago, the industry was discharging high concentrations of oil and grease (1000 to 1200 mg/l) and large quantities of feathers to the POTW. The problem was so prevalent that flotation thickeners were used in lieu of primary clarifiers in the original plant design during the 1950's. The problem had become unmanageable from a plant operations perspective, hence the municipality required the industry to install flotation on-site, thereby reducing the FOG level to under 100 mg/l.

Raeford's pretreatment program defines a set of surchargeable and prohibitive limits for five parameters:

Parameter	Surchargeable (mg/l)	Prohibitive (mg/l)
BOD	400	800
COD	1000	1600
TSS	350	600
TKN	40	80
FOG	--	100

If an industry's wastewater exceeds the limits defined in the first column during their twice-per-month sampling, the sewer use fees are computed using mass-based unit costs in addition to the flow-based rates. Should an industry discharge wastewater concentrations in excess of the prohibitive values, a notice of violation is issued and the industry is given 30 days to correct the problem prior to the initiation of a five consecutive day sampling program. Consistent noncompliance with these limits can result in a shut-off of services, but such a drastic step has not been necessary for any of the three industries to date.

The turkey processor is routinely surcharged for their discharge, and at the time of the site visit, were in violation of the prohibitive BOD limit. Their most recent sampling analysis indicated the following concentrations (in mg/l):

BOD = 1000
COD = 1400
TSS = 500
TKN = 70

The two other industries in town (a textile mill and a cosmetics manufacturer) each contribute high BOD (600-800) and COD (1200-1400) wastewater to the plant. Since Raeford is only at 2/3 hydraulic capacity, these organic loads do not adversely affect the extended aeration process.

The cosmetics manufacturer has discharged low pH (1.5) wastewater to the plant in the past, which can be toxic to the biological population in the activated sludge. Effluent BOD has climbed as high as 90 mg/l on occasion. When operations personnel became aware of the problem either by industry notification or increased D.O. in the aeration basins, two procedures have been implemented to mitigate the impact:

- Recycle portions of the aerobic digester to the plant influent, which serves to dilute the low pH wastewater and return healthy organisms to the aeration basins.

- Add lime to the aeration basins to elevate the mixed liquor pH above 6.0.

The other major stumbling block to consistent compliance at Raeford had been the high infiltration/inflow in the collection system. During the summer of 1985, influent flows reached 4 mgd, resulting in a substantial washout of the biological populations. Performing a television survey revealed that one of the two main trunk lines to the plant (an 18" pipeline) had collapsed. By simply closing off the collapsed line (the parallel 24" pipeline was adequate), the I/I flow increment was reduced to 100,000 gpd.

RAEFORD WASTEWATER TREATMENT PLANT
RAEFORD, NORTH CAROLINA

Design Flow:	**3.0 mgd**	Location:	South-Central North Carolina
Secondary Treatment:	**Activated Sludge (Extended Aeration)**	Population Served:	4500

INFLUENT WASTEWATER

	Typical (Upset)
Ave. Flow, mgd	2.0
% Industrial	60
BOD_5, mg/l	300 (750)
SS, mg/l	200 (500)

SIGNIFICANT INDUSTRIES

Industry	Flowrate (1000 gpd)	Problem Pollutants
Poultry	650	COD, BOD, TSS, TKN
Textile	500	COD, BOD
Cosmetics	35	pH, COD, BOD

PLANT LOADING

Primary Flotation	Typical (Upset)
Overflow Rate, gal/sf/day	1050
Detention Time, hours	1.5

Aeration Basins	Typical (Upset)
F/M, lbs BOD_5/lbs MLSS/day	0.1-0.15
MCRT, days	5-10
MLSS, mg/l	3000
Detention Time, hours	9.2
D.O. Level, mg/l	3-5

Secondary Clarifiers	Typical (Upset)
Overflow Rate, gal/sf/day	230
Detention Time, hours	9
SVI, ml/gm	150-250

PLANT PERFORMANCE

	Permit Limit	Typical (Upset)
BOD_5, mg/l	30	25 (50)
SS, mg/l	30	25 (50)

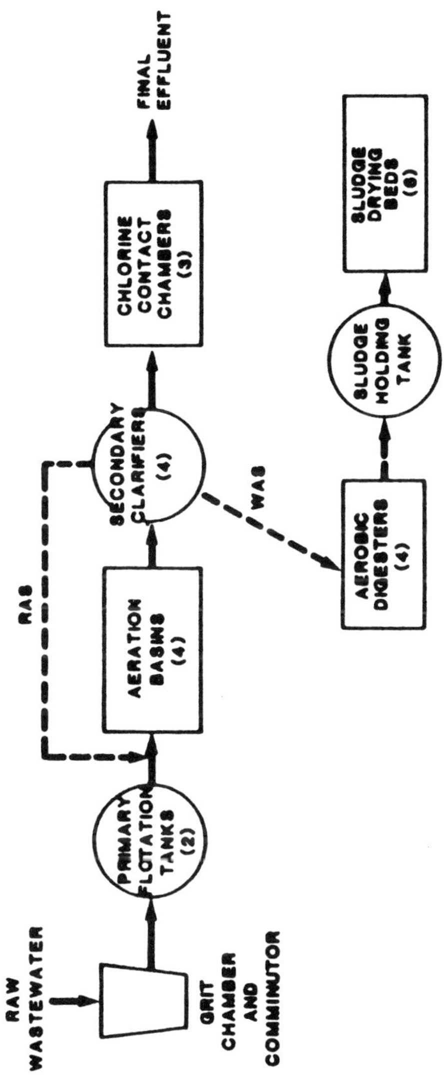
RAW WASTEWATER
GRIT CHAMBER AND COMMINUTOR
PRIMARY FLOTATION TANKS (2)
RAS
AERATION BASINS (4)
SECONDARY CLARIFIERS (4)
WAS
CHLORINE CONTACT CHAMBERS (3)
FINAL EFFLUENT
AEROBIC DIGESTERS (4)
SLUDGE HOLDING TANK
SLUDGE DRYING BEDS (6)

NEUSE RIVER WASTEWATER TREATMENT PLANT
Raleigh, North Carolina

In 1976, the 30 mgd Neuse River Wastewater Treatment Plant (NRWTP) went on-line to replace the overloaded 16 mgd Walnut Creek plant. The City of Raleigh has historically been a community that embraced industry. In the early 1960's, influent BODs exceeded 300 mg/l at Walnut Creek, with the effluent ranging from 35 to 55 mg/l. Industries were encouraged to conserve and recycle wastes, resulting in a 250 mg/l BOD by the mid-1960's. The City's first Sewer Use Ordinance was enacted in 1972, with continual modification to comply with changes in the Federal regulations. The net effect is a current influent BOD consistently below 200 mg/l, despite an industrial flow volume representing 25 percent of the plant flow.

The only significant industrial discharge to the Walnut Creek plant was a large electroplater whose occasional plating bath dumps were not prohibited by a sewer use ordinance during the 1950's. Digester upsets (decreased gas production) and high sludge metals content were traced to this particular industry. Since dried sludge was being made available to the community for landscaping purposes at the time, concern for the metals levels prompted adoption of a proposed ordinance which directed the industry to construct a physical-chemical pretreatment facility.

Two other metals-related industries have been responsible for high sludge metals since the construction of the NRWTP. In the current facility, wet sludge is land applied to farmland adjacent to the POTW, hence metal content is critical. In each case (an electroplater and a printed circuit board manufacturer), the industries were discharging levels of Cr, Ni, Zn, Pb and Cu sometimes in excess of 1,000 mg/l, with highly variable effluent pH, and were uncooperative in dealing with the City of Raleigh. Fining the former industry $ 1,000, and threatening the latter with same, provided sufficient incentive to install pretreatment.

In the early 1980's a producer of amino acids for pharmaceuticals was attracted to Raleigh and given the false impression they would be able to discharge slug loads totaling 1,000 lbs of NH_3 to the POTW each day. Fortunately, an activated sludge system had been constructed for their facility for BOD reduction, which possessed sufficient capacity to nitrify their wastewater to an ammonia concentration of 50 mg/l. On one occasion, the NH_3 levels became toxic to the pretreatment activated sludge, resulting in a gradual loss of nitrification at the POTW. Continual monitoring of alkalinity and NH_3 allowed the City to preserve their own nitrifier population while at the same time re-seeding the industry's activated sludge with a viable nitrifier population for a speedy recovery. The rapid response prevented the monthly effluent NH_3 levels from exceeding the permit limit, despite high daily concentrations following the incident.

A dairy product manufacturer who cleans the stainless steel tanker trucks on-site had previously discharged these wastes directly to the sewer. Average BODs of 10,000 mg/l, with occasional values in the 30,000 to 40,000 mg/l range were typical. Working with the North Carolina State University, a vacuum recovery system was developed and a market identified for the collected whey waste. The effluent BOD now averages 2,000 mg/l, still resulting in a high surcharge payment. The City of Raleigh has waived the prohibitive BOD limit of 1,500 mg/l in this case because space limitations on the industry's property prevents them from installing additional pretreatment.

An unusual case at the NRWTP was the discovery of high zinc levels (1,000 mg/l) in the discharge from an office building with no manufacturing component. Through discussions with maintenance personnel, the City of Raleigh discovered that the contaminated discharges corresponded to floor stripping activities in the building. It turns out that a Zn-based floor wax had been used, and stripping an entire office building over the course of a week discharged enough Zn to the POTW to significantly raise the level in their sludge.

The Raleigh plant is currently under construction to increase the hydraulic capacity from 30 to 40 mgd, with an additional expansion to 60 mgd planned for the near future (the schematic shown on the next page is for the 40 mgd facility). The rapid growth of this community will continue to bring with it a variety of challenging new industrial wastewaters with, in some cases, unpredictable impacts on the POTW. The Raleigh case study illustrates the need for continuous survey and monitoring even after the implementation of a successful industrial waste program in any dynamic population center.

NEUSE RIVER WASTEWATER TREATMENT PLANT
RALEIGH, NORTH CAROLINA

Design Flow:	**40 mgd**	Location:	Central North Carolina
Secondary Treatment:	**Activated Sludge (Extended Aeration)**	Population Served:	195,000

INFLUENT WASTEWATER

	Typical (Upset)
Ave. Flow, mgd	25
% Industrial	25
BOD_5, mg/l	165 (350)
SS, mg/l	170 (500)

SIGNIFICANT INDUSTRIES

Industry	Flowrate (1000 gpd)	Problem Pollutants
Electroplaters, Metal Finishers (5)	750	Cd, Cr, Cu, Ni, Pb, Zn, Cn^-, Fe, pH
Pharmaceutical	400	NH_3
Dairy	110	BOD
Snack Foods	100	BOD

PLANT LOADING

Primary Clarifiers	Typical (Upset)
Overflow Rate, gal/sf/day	650
Detention Time, hours	3.0

Secondary Clarifiers	Typical (Upset)
Overflow Rate, gal/sf/day	680
Detention Time, Hours	3.2
SVI, ml/gm	150-200
Effluent BOD_5, mg/l	5
Effluent SS, mg/l	15
Effluent NH_3, mg/l	3

Aeration Basins	Typical (Upset)
F/M, lbs BOD_5/lbs MLSS/day	.08-1.0
MCRT, days	12-20
MLSS, mg/l	2500
Detention Time, hours	15
Return Flow, %	50
D.O. Level, mg/l	2

Multi-Media Filters	Typical (Upset)
Hydraulic loading, gpm/sf	2.5

PLANT PERFORMANCE

	Permit Limit		Typical (Upset)
	Summer	Winter	
BOD_5, mg/l	6	12	3 (15)
SS, mg/l	30	30	4
NH_3, mg/l	3	6	1.5 (8)

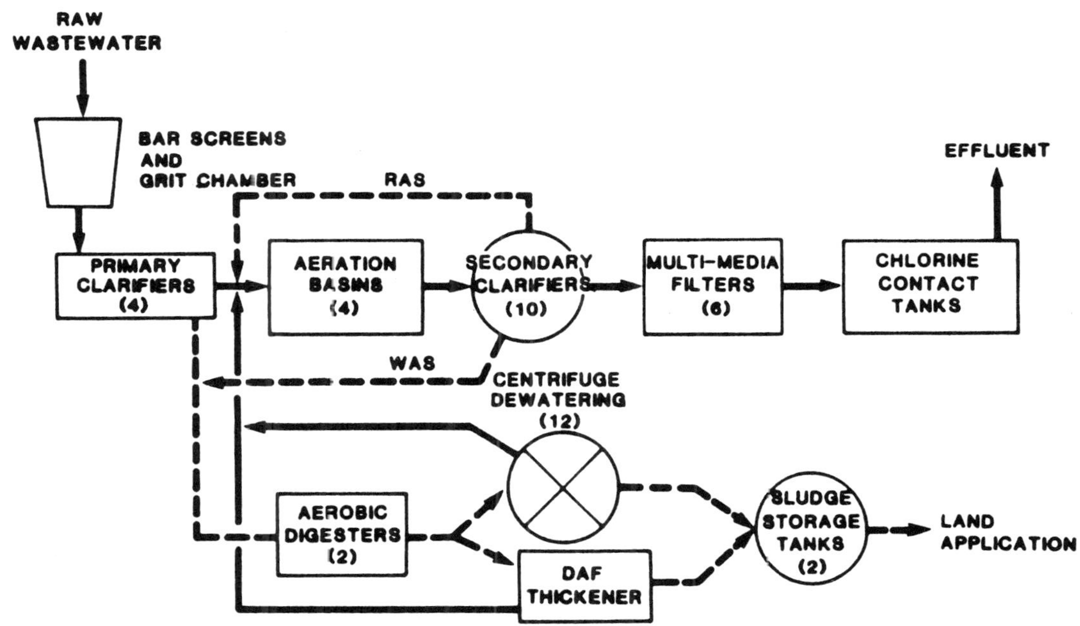

HORSE CREEK POLLUTION CONTROL FACILITY
North Augusta, South Carolina

The Horse Creek Pollution Control Facility (HCPCF) is a regional plant, operated by the Aiken County Public Service Authority (ACPSA), treating a predominantly industrial wastewater. Ninety five percent of the industrial wasteload is contributed by several large textile mills and is characterized by high COD, BOD, alkalinity and pH. Combined domestic/industrial influent wastewater pH fluctuations of up to 2.5 units per day and alkalinity fluctuations of up to 600 mg/l per day caused inhibition of the biomass, poorly settling sludge and caused effluent suspended solids permit violations. Since implementing a pretreatment program and issuing industrial wastewater discharge permits, the treatability of the industrial waste has improved, the result being that HCPCF has been free of NPDES permit violations for over eight months.

Local textile processes include grading operations, finishing processes utilizing dyes, and specialized textile chemical manufacturing. The textile wastewater is highly caustic with alkalinity as high as 2400 mg/l, and pH exceeding 12.5. Prior to pretreatment the combined industrial/domestic influent to the HCPCF had the following characteristics:

pH	>11
BOD	360 mg/l
COD	910 mg/l
Alkalinity	1100 mg/l
TSS	210 mg/l

Other distinguishing characteristics of the influent wastewater included the extremely light nature of the suspended solids and a dark blue/black color, typical of textile wastewater from washing and dying operations.

Prior to the summer of 1985, the textile industries employed a limited type of pretreatment and flow equalization. This limited pretreatment and flow equalization resulted in the previously mentioned plant influent pH fluctuations of 2 to 2.5 units and alkalinity fluctuations of up to 600 mg/l in a given day. These fluctuations caused some inhibition of the biomass, but because the hydraulic detention time in the aeration basins was in excess of 3.5 days, effluent BOD was within the permit limit of 33 mg/l. These pH and alkalinity fluctuations had their most detrimental effect on biomass settling characteristics and solids carryover in the secondary clarifier often resulted, lasting for 24-36 hours. During these episodes, filamentous organisms were occasionally observed in the biomass. The solids carryover problem worsened in the winter months when wastewater temperatures were lower, but chlorination of the return activated sludge, the influent to the secondary clarifier and the contents of the aeration basin was somewhat successful at improving settleability. Despite this, the HCPCF still experienced effluent suspended solids violations in 15 of the 19 months prior to September, 1985.

The State of South Carolina mandated that the ACPSA implement and enforce a pretreatment program in the spring of 1984. The ACPSA responded by developing such a program and issuing draft industrial wastewater discharge permits. Final State approval came in May, 1985. As presently written, the industrial wastewater discharge permits are not restrictive, allowing BOD, COD and alkalinity levels as high as 600 mg/l, 1300 mg/l and 1500 mg/l, respectively. However, the permits have caused the textile industries to make small, but meaningful alterations to their wastewater discharge practices, resulting in average plant influent pH levels dropping from 11-12 to 10 and alkalinity from 1100 mg/l to 700 mg/l. More importantly, maximum daily influent pH fluctuations have been reduced to 0.5 units or less. Figure C-4 shows the magnitude of pH fluctuations both before and after the implementation of pretreatment. Simple modifications at textile facilities to process operations and waste pumping schedules were typical of the changes that were necessary to realize the described results. Because of the more stable wastewater discharge, the HCPCF has realized more consistent plant operation and has not violated its NPDES permit in over eight months.

Some of the textile dischargers do not currently meet the pH and alkalinity limits of their industrial wastewater discharge permits and are under a compliance schedule to do so. The facilities are installing pretreatment works for caustic recovery that should significantly lower pH and alkalinity levels. The HCPCF is also presently studying the addition of floating mixing units to augment the turbine surface aerators in the aeration basins. To date, evidence indicates that a more consistent secondary clarifier solids feed is achieved which improves the quality of the secondary effluent.

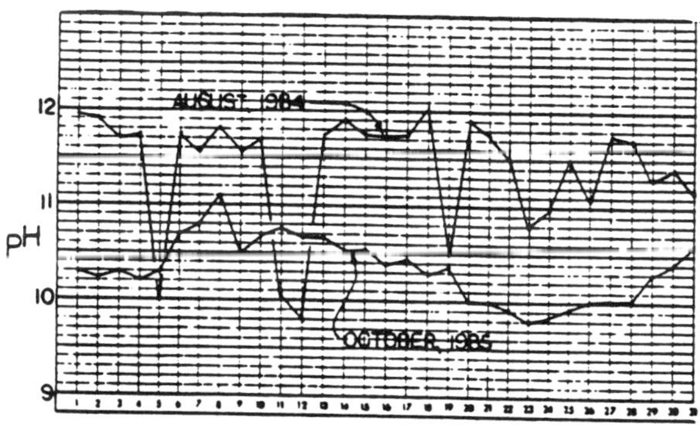

FIGURE C-4
HCPCF INFLUENT pH

HORSE CREEK POLLUTION CONTROL FACILITY
Aiken County, South Carolina

Design Flow:	**20 mgd**	Location:	**West-central South Carolina**
Secondary Treatment:	**Extended Aeration Activated Sludge**	Population Served:	**70,000**

INFLUENT WASTEWATER

	Typical (Upset)
Ave. Flow, mgd	10.4
% Industrial	80
BOD_5, mg/l	360
SS, mg/l	210
COD, mg/l	910
Alkalinity, mg/l	1100 (1600)
pH	10-11 (12.5)

SIGNIFICANT INDUSTRIES

Industry	Flowrate (1000 gpd)	Problem Pollutants
Textile	8,400	COD, Alkalinity, pH
Textile chemicals	300	COD, pH

PLANT LOADING

Primary Clarifiers	Typical (Upset)
Overflow Rate, gal/sf/day	300
Detention Time, hours	4.4

Secondary Clarifiers	Typical (Upset)
Overflow Rate, gal/sf/day	195
Detention Time, hours	9.1

Aeration Basins	Typical (Upset)
F/M, lbs BOD_5/lbs MLSS/day	0.05-0.10
MCRT, days	50-90
MLSS, mg/l	3800-4500
Detention Time, hours	92
Return Flow, %	40-60
D.O. Level, mg/l	1-3 (4)

PLANT PERFORMANCE

	Permit Limit	Typical (Upset)
BOD_5, mg/l	33	15
SS, mg/l	57	40 (85)
COD, mg/l	-	175
pH	9	9 (10)

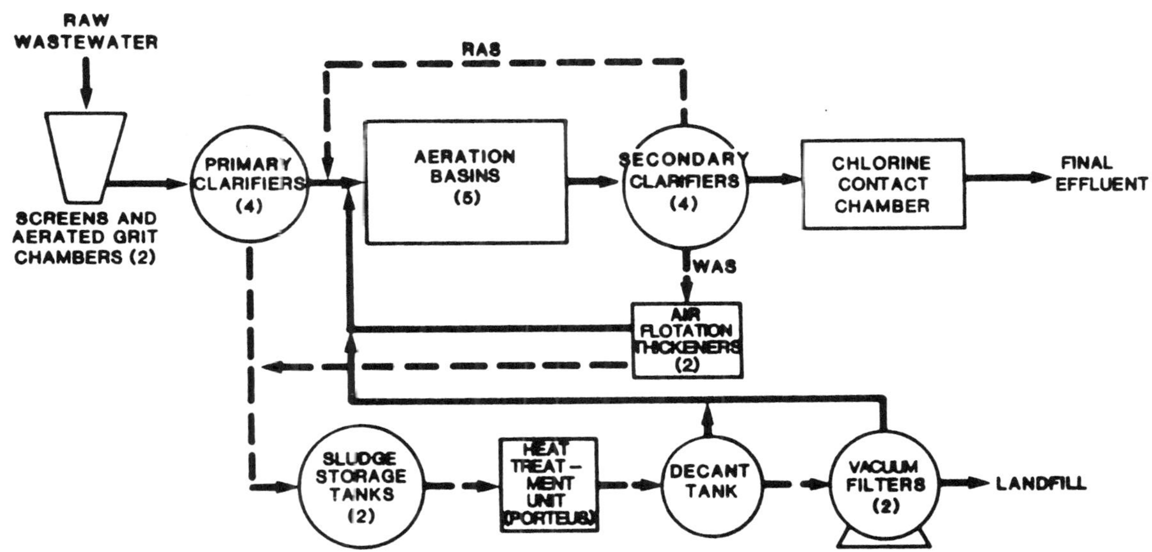

NORTH SHORE SANITARY DISTRICT GURNEE PLANT
Gurnee, Illinois

The Gurnee Plant of the North Shore Sanitary District (NSSDGP) receives an average daily wastewater flow of 12.4 mgd from a variety of sources. Those sources include a major naval installation, domestic sewage discharges, secondary effluent from the District's North Chicago Sewage Treatment Plant, and other industries which contribute 17 percent of the total flow.

Since startup in 1976, the NSSDGP has experienced periodic failures at achieving nitrification in the two-stage activated sludge system. The failures to achieve nitrification to the ammonia levels of the District's NPDES effluent limits have also, at times, been accompanied by general process upsets which have resulted in effluent SS and BOD_5 violations. One of the major industrial contributors to the Gurnee Plant, a pharmaceutical manufacturer discharging an average flow of 750,000 gpd, has similarly experienced upsets of its own activated sludge pretreatment system which have resulted in violations of the District's local sewer use ordinance. It was initially believed that the observed interferences at the NSSDGP were the result of the discharge of filamentous organisms and other solids by the manufacturer. The initiation of in-plant solids control methods (which significantly lessened the quantity of solids entering the industrial wastewater pretreatment system) and pretreatment system upgrades did not, however, eliminate interferences at the NSSDGP.

In 1980, District personnel began to suspect that the presence of a nitrification inhibiting antibiotic, erythromycin, in the pharmaceutical wastewater was the main cause of the process upsets at the NSSDGP. By 1983, test and control bench-scale activated sludge reactors were placed in operation and the effects of the pharmaceutical wastewater and erythromycin on the NSSDGP were investigated. A bioassay test for the presence of erythromycin and other nitrification inhibitors was also developed, along with a Direct Insertion Probe/Mass Spectrometric technique for confirmation. The results of the bench-scale testing indicated that the presence of soluble and/or solid constituents of the pretreated pharmaceutical wastewater inhibited nitrification and, at high levels, could completely suppress nitrification. Additionally, it was found that although erythromycin inhibited nitrification, acclimation to low concentrations of erythromycin could occur in the absence of extreme concentration fluctuations.

During January of 1984, an observed average industrial pretreatment effluent erythromycin concentration of 53 mg/l with mass loading fluctuations of greater than two orders of magnitude completely inhibited nitrification in the Gurnee Plant. The resulting BOD_5 and SS concentrations were as high as 26 mg/l and 67 mg/l, respectively. Lower concentrations of erythromycin in the absence of such strong concentration fluctuations did not interfere with the performance of the Gurnee Plant during August of 1984, with average effluent BOD_5 and SS concentrations of 11 mg/l and 8 mg/l, respectively, and effluent ammonia concentrations ranging from 0.4 mg/l to 1.5 mg/l as N. Experience at the Gurnee Plant and with the bench-scale test systems has also indicated that a lag period of two to three mean cell residence times is required before the effects of erythromycin on the activated sludge process become apparent. Erythromycin also was found to disrupt the settling of the first-stage carbonaceous organisms.

Measures undertaken by District personnel to lessen the effect of the pharmaceutical discharge on plant performance have included:

- The addition of inorganic coagulants to aid primary clarifier performance;
- the addition of polymer to the first-stage activated sludge system,
- daily bioassays of industrial wastewaters for the presence of inhibiting substances; and
- the development of an ordinance governing the discharge of erythromycin to the NSSDGP.

Since passage of the ordinance in November, 1985, in which the discharge limits of erythromycin were established, the NSSDGP has substantially been in compliance with its NPDES permit and ammonia levels of 0.25 mg/l to 1 mg/l as N have been consistently achieved.

NORTH SHORE SANITARY DISTRICT GURNEE PLANT
GURNEE, ILLIONOIS

Design Flow:	**13.8 mgd**	Location:	**Northeastern Illinois**
Secondary Treatment:	**Activated Sludge (Two-Stage, Modified-Contact)**	Population Served:	**65,000**

INFLUENT WASTEWATER

	Typical (Upset)
Ave. Flow, mgd	12.4
% Industrial	37
BOD_5, mg/l	140
SS, mg/l	180
NH_3, mg/l	15

SIGNIFICANT INDUSTRIES

Industry	Flowrate (1000 gpd)	Problem Pollutants
Pharmaceutical	750	Antibiotics, SS
Electroplating	100	Cu, CN
Chemical	170	Organics
Nonferrous Metals	90	W
Military Installation	3,500	pH

PLANT LOADING

Primary Clarifiers	Typical (Upset)
Overflow Rate, gal/sf/day	695
Detention Time, hours	2.7
Effluent BOD_5, mg/l	100
Effluent SS, mg/l	100

First Stage Clarifiers	Typical (Upset)
Overflow Rate, gal/sf/day	780
Detention Time, hours	2.5

Second Stage Clarifiers	Typical (upset)
Overflow Rate, gal/sf/day	645
Detention Time, hours	3.1

First Stage Aeration Basins	Typical (Upset)
F/M, lbs BOD_5/lbs MLVSS/day	0.95
MCRT, days	7
MLSS, mg/l	3000
Detention Time, hours	4.2
Return Flow, %	25
D.O. Level, mg/l	2.5

Second Stage Aeration Basins	Typical (Upset)
F/M, lbs NH_3-N/lbs MLVSS/day	0.07
MCRT, days	13
MLSS, mg/l	3500
Detention Time, hours	5.8
Return Flow, %	50
D.O. Levels, mg/l	2.5

PLANT PERFORMANCE

	Permit Limit	Typical	(Upset)
BOD_5, mg/l	10	5	(17)
SS, mg/l	12	5	(23)
NH_3, mg/l (summer)	1.5	0.5	(15)

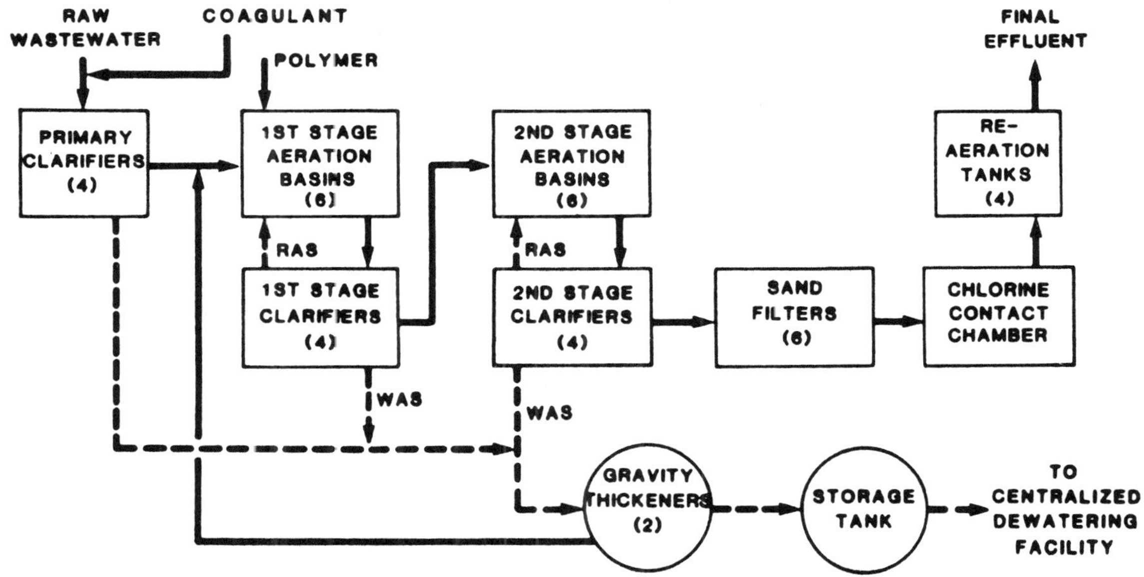

SANITARY DISTRICT OF ROCKFORD SEWAGE TREATMENT PLANT
Rockford, Illinois

The Sanitary District of Rockford operates the Rockford Sewage Treatment Plant (SDRSTP) which serves a population of 240,000 and more than 400 industries. Forty-five metal finishers, two dairies, three food processing plants, several large machine tool manufacturers, twenty-five permitted batch waste haulers, a contract waste treatment facility, and several paint manufacturing plants are among the major sources of industrial wastewater. Industrial wastewater contributes 45 percent of the daily average treatment plant flow of 35 mgd. Over the years, the District has experienced sludge disposal problems and isolated excursions of their NPDES permit discharge limits that were related to the industrial discharges to the POTW.

Upon passage of the Resource Conservation and Recovery Act in 1980, the District's thickened sludge was found to be classified as hazardous because of the cadmium content. The local industrial discharge limits, which had been in existence for several years, were therefore tightened for cadmium from 2.0 mg/l down to 0.9 mg/l and sewer use surcharges were applied. The result was that from initial sludge cadmium concentrations of 800 mg/kg in 1980, a level of 50 mg/kg was achieved by 1984. The addition of excess amounts of lime prior to vacuum filtration was practiced as an interim method of rendering the vacuum filter cake nonhazardous (EP-Toxicity method) and therefore acceptable for land filling. This was done to allow contributing industries time to install pretreatment systems and come into compliance with the lower local discharge limits. Presently, the District is investigating the feasibility of land application for ultimate disposal of sludge.

Isolated incidents of batch discharges of concentrated manufacturing process solutions to the POTW have resulted in process upsets and effluent discharge violations. A batch discharge of a nickel plating solution to the POTW in 1981 resulted in a treatment system upset and effluent BOD and SS concentrations of 38 mg/l and 34 mg/l, respectively. Upon notification of the incident by the industry, the POTW personnel attempted to isolate the contaminated incoming wastewater and confine the nickel slug within the primary clarifiers. However, most of the nickel had passed through the primary treatment units by the time of notification and the process recovered after only a few days. Prior to the nickel spill incident the POTW was subject to a shock load of cyanide in 1977. Upon the arrival of an unknown amount of cyanide at the POTW, it was found to be difficult to maintain a chlorine residual after the disinfection of the secondary effluent. Subsequent analyses revealed effluent cyanide concentrations as high as 1.06 mg/l which were in violation of the NPDES permitted limits of 0.2 mg/l CN^-. Extensive industrial site visits, sampling and interviews were undertaken, but the source of the spill was not verified. The most intensive shock level of cyanide may have occurred in 1970 when the raw wastewater cyanide concentration increased from less than 0.5 mg/l to 34 mg/l in a period of one hour. Although such shock load interferences as experienced by the SDRSTP are isolated events, they have not been uncommon at the SDRSTP. For example, in 1979 there were 36 NPDES permit violations of the concentrations or mass loadings of cyanide, chromium and zinc.

The District's pretreatment program has been characterized by the establishment of effective local limits for the discharge of metals and toxic substances, an extensive industrial monitoring program, the development of spill notification procedures, and cooperation between the District and the local industries. Over the past ten to fifteen years, 34 industrial pretreatment systems have been installed and the metal finishing industry has reduced its discharge by more than one-half to about 3 mgd. The overall result has been a reduction of the number of toxic, noncompatible-pollutant NPDES violations from the historical numbers of excursions, such as 36 in 1979, to zero in 1984.

ROCKFORD SANITARY DISTRICT SEWAGE TREATMENT PLANT
ROCKFORD, ILLINOIS

Design Flow:	**60 mgd**	Location:	**Northern Illinois**
Secondary Treatment:	**Activated Sludge (Conventional)**	Population Served:	**240,000**

INFLUENT WASTEWATER

	Typical (Upset)
Ave. Flow, mgd	35
% Industrial	45
BOD_5, mg/l	164
SS, mg/l	245

SIGNIFICANT INDUSTRIES

Industry	Flowrate (1000 gpd)	Problem Pollutants
Metal Finishing	725	Zn, Cu, Cd, Ni, Cr
Metal Plating	587	CN, Cu, Zn
Dairy	470	BOD_5, SS

PLANT LOADING

Primary Clarifiers	Typical (Upset)
Overflow Rate, gal/sf/day	563
Detention Time, hours	3.4
Effluent BOD_5, mg/l	110
Effluent SS, mg/l	77

Aeration Basins	Typical (Upset)
F/M, lbs BOD_5/lbs MLSS/day	0.2
MCRT, days	14
MLSS, mg/l	2500
Detention Time, hours	4.8
Return Flow, %	33
D.O. Level, mg/l	1.5

Secondary Clarifiers	Typical (Upset)
Overflow Rate, gal/sf/day	486
Detention Time, hours	3.6
SVI, ml/gm	125

PLANT PERFORMANCE

	Permit Limit	Typical (Upset)
BOD_5, mg/l	20	20
SS, mg/l	25	13

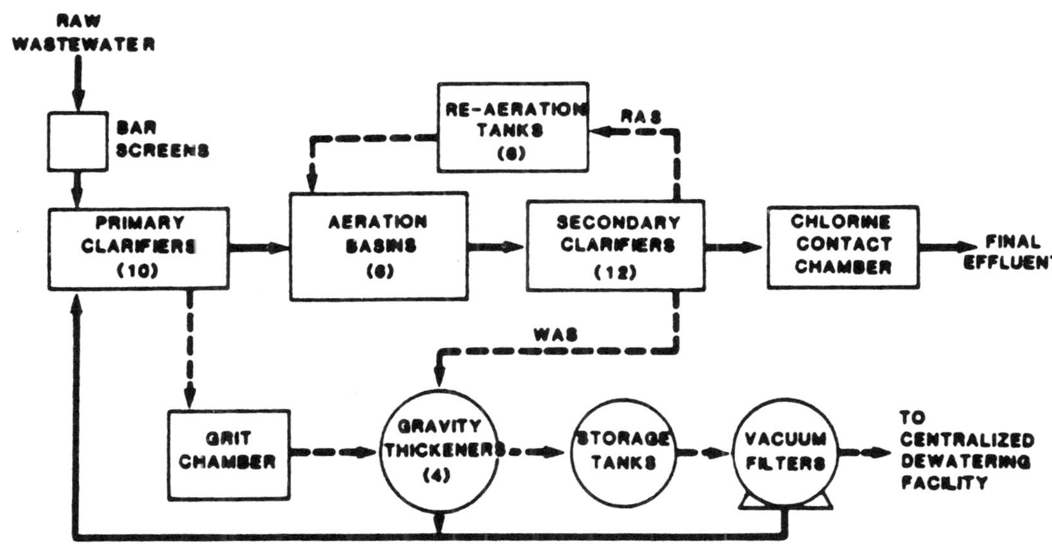

LAKE MILLS WASTEWATER TREATMENT PLANT
Lake Mills, Iowa

The Lake Mills Wastewater Treatment Plant (LMWTP) serves a community of 2,200 persons in North-Central Iowa. Approximately 37 percent of the average flow of 0.35 mgd is industrial in nature and arises from the two major industries within the community. A printed circuit board manufacturer discharges an average flow of 80,000 gpd which contains copper, lead, chromium, nickel and zinc. The other major manufacturing concern discharges 50,000 gpd to the POTW. Presently, there are no problem pollutants associated with this second discharge.

The circuit board manufacturer experienced growth during the past decade which resulted in increased discharges of copper, lead and other metals to the POTW. In 1980, the copper and lead levels in the anaerobically digested sludge were observed to be 4,300 mg/kg and 1,100 mg/kg of dried sludge, respectively. The State then intervened and halted the LMWTP's prior disposal practice of spreading the sludge on agricultural land. Because of a lack of a disposal option, the sludge solids were allowed to accumulate within the single-stage digester of the treatment works with digester supernatant recirculation to the head of the facility. A one-time disposal of 90,000 gal of sludge to a landfill was allowed by the State in 1982 after which solids were again held within the treatment works. Prior to the receipt of a high-rate land application permit in the fall of 1984, the sludge held in the digester contained as much as 16,000 mg/kg of copper.

A program of monitoring the circuit board manufacturer's discharge was initiated in May of 1984 at which time the average copper and lead concentrations being discharged to the POTW were 2 to 4 mg/l and 0.4 to 1.2 mg/l, respectfully. The municipality and the manufacturer entered into a pretreatment agreement which resulted in the installation of an ion exchange and precipitation metals removal system. The pretreatment program has resulted in lower metals loading to the POTW. Monthly average copper and lead concentrations of 2.64 mg/l and 0.3 mg/l have been observed in the pretreated industrial discharge. Current municipal regulations require pretreatment discharge limits for metals which are consistent with those of 40 CFR 433 and fines are imposed for each incidence of noncompliance.

As a result of the reduced metals loading to the POTW and of the ability to dispose of digested sludge solids on a regular basis, the metals content of the digested sludge has been reduced to the vicinity of 5100 mg/kg as was observed in December of 1985.

During the periods for which sludge disposal was not practiced on a regular basis, the overall plant performance was found to deteriorate. From average effluent BOD_5 concentrations of 29 mg/l and 24 mg/l in 1982 and 1983, respectively, a yearly average effluent BOD_5 of 40 mg/l was observed for 1984. The effluent BOD_5 for the last quarter of 1985 averaged 22 mg/l. Although the results are not conclusive, it is believed by some that the poor overall plant performance of recent years was the result of high organic and solids loadings (because of digestor supernatant recycling) to the trickling filter and possible metal toxicity.

LAKE MILLS WASTEWATER TREATMENT PLANT
LAKE MILLS, IOWA

Design Flow:	**0.36 (1.1 hydraulic)**	Location:	**North Iowa**
Secondary Treatment:	Trickling Filter **(High Rate; Rock Media)**	Population Served:	**2,200**

INFLUENT WASTEWATER

	Typical (Upset)
Ave. Flow, mgd	0.35
% Industrial	37
BOD_5, mg/l	460
SS, mg/l	50

SIGNIFICANT INDUSTRIES

Industry	**Flowrate (1000 gpd)**	**Problem Pollutants**
Electroplating	80	Cu, Pb, Cr, Ni, Zn
Automotive Products	50	None

PLANT LOADING

Primary Clarifiers	**Typical (Upset)**
Overflow Rate, gal/sf/day	680
Detention Time, hours	3.7

Trickling Filter	**Typical (Upset)**
Organic Loading, lb BOD_5/1000 cf/day	47
Recirculation, %	123
Covered; Rock Media	

Secondary Clarifiers	**Typical (Upset)**
Overflow Rate, gal/sf/day	500
Detention Time, hours	2.3

Polishing Pond	**Typical (Upset)**
Detention Time, days	54
Effluent BOD_5, mg/l	22 (40)
Effluent BOD_5, mg/l	38 (64)

PLANT PERFORMANCE

	Permit Limit	Typical (Upset)
BOD_5, mg/l	30	22 (40)
SS, mg/l	30	38 (64)
NH_3 mg/l (Summer)	10	3 (4)

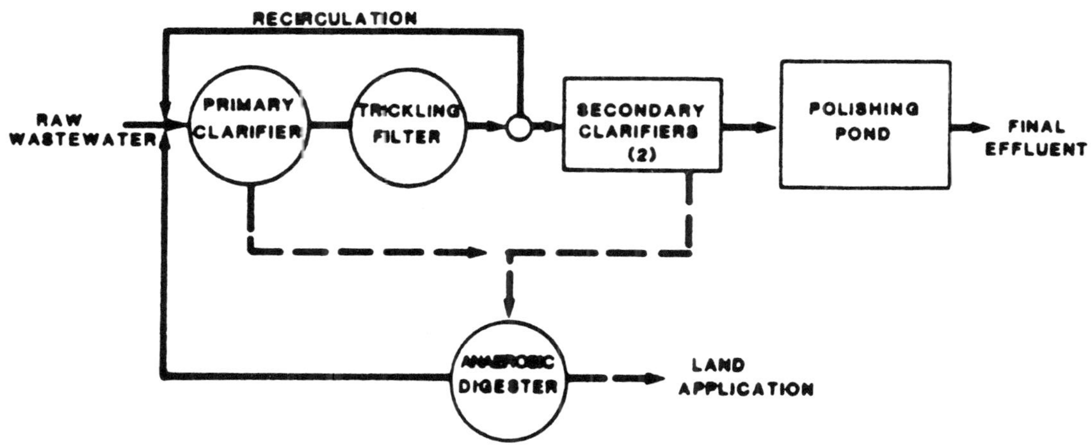

MARSHALLTOWN WATER POLLUTION CONTROL PLANT
Marshalltown, Iowa

The Marshalltown Water Pollution Control Plant (MWPCP) experienced brief periods of effluent BOD limitation violations prior to 1982. The violations were the combined result of high hydraulic loadings to the plant because of infiltration and inflow and excessive BOD_5 loadings which exceeded the capacity of the treatment facilities. Whereas the MWPCP was designed for average daily and peak hydraulic flows of 5.5 and 8.0 mgd, respectively, extreme wet weather flows as great as 20 mgd were experienced. Flows in excess of 8 mgd were found to result in an excessive loss of microorganisms from the activated sludge system and it was necessary to provide only primary treatment for the total flow and bypass secondary treatment for the excess wastewater flows. At the same time that high hydraulic loadings were experienced, BOD_5 loads averaging 14,000 lb BOD_5/day were contributed by a meat packing facility. The average BOD_5 of the industrial loadings represented 65 percent of the average wastewater strength and maximum industrial contributions of 33,000 lb BOD_5/day were observed. Additionally, excessive discharges of grease (up to 15,000 lb/day) had also occurred. Prior to 1982, the industrial wastewater in question was not receiving pretreatment. The effluent limitation violations arose during periods of high hydraulic and high organic loadings at which times plant effluent (secondary effluent plus bypassed primary effluent) BOD_5 concentrations of up to 170 mg/l occurred.

A significant upgrade of the MWPCP was completed in 1982. Included in the upgrade were additional treatment units for increased hydraulic capacity, the installation of a jet aeration system which substantially increased the organic loading capacity, and new sludge handling facilities. The typical data presented in the following table characterize the MWPCP as it now exists. The reported upset parameter values represent the process conditions prior to the 1982 plant upgrade and during periods of high hydraulic and organic loadings.

In conjunction with expansion of the capacity of the MWPCP, the meat packing industrial concern instituted a waste minimization/pretreatment program, upon State intervention, which has reduced the average organic load of the industrial wastewater to 8000 lb BOD_5/day with a solids loading of 4000 lb/day. Livestock holding pen runoff is now subject to primary sedimentation. A blood collection and drying system was installed. The remaining wastewaters are strained and subjected to dissolved air flotation for grease removal. Although other wastewater pretreatment alternatives were proposed, the selected system was economically advantageous because of the potential for protein and grease recovery.

A total of 37 industries were identified in a city-wide industrial survey. Of the 37 industries, 20 were classified as sources of industrial wastewater and of these, six are monitored regularly. One categorical electroplater is subject to a compliance schedule. The total industrial wastewater flow averages 1.2 mgd. The present industrial pretreatment program consists of sampling and analysis of the industrial discharges, as conducted by MWPCP personnel, and close co-operation between the municipality and the various industries. There have been no recent events of noncompliance on behalf of the MWPCP because of industrial waste discharges. There have, however, been several instances of potential interferences to MWPCP operation because of industrial waste discharges. For example, elevated but noninterfering concentrations of lead in the treatment plant influent were traced to the batch dumping of a lead-acetate solution used in the manufacture of latex paint. The lead was voluntarily eliminated from the wastewater initially by reuse and more recently by process substitution.

MARSHALLTOWN WATER POLLUTION CONTROL PLANT
MARSHALLTOWN, IOWA

Design Flow:	**7.5 mgd (11.0 hydraulic)**	Location:	**Central Iowa**
Secondary Treatment:	**Activated Sludge (Conventional)**	Population Served:	**27,000**

INFLUENT WASTEWATER

	Typical (Upset)
Ave. Flow, mgd	6.5 (> 8)
% Industrial	18
BOD_5, mg/l	400 (900)
SS, mg/l	370

SIGNIFICANT INDUSTRIES

Industry	Flowrate (1000 gpd)	Problem Pollutants
Meat Packing	800	BOD_5, grease
Electroplating	30	Cu, Ni, Cr
Metal Finishing	18	Zn
Paint Manufacturing	1	Hg

PLANT LOADING

Primary Clarifiers	Typical (Upset)
Overflow Rate, gal/sf/day	1000 (> 1450)
Detention Time, hours	1.25 (< 0.86)
Effluent BOD_5, mg/l	320 (700)
Effluent SS, mg/l	185

Secondary Clarifiers	Typical (Upset)
Overflow Rate, gal/sf/day	450 (650)
Detention Time, hours	2.2 (1.5)

Aeration Basins	Typical (Upset)
F/M, lbs BOD_5/lbs MLSS/day	0.4
MCRT, days	13
MLSS, mg/l	2900
Detention Time, hours	7 (4.8)
D.O. Level, mg/l	1.0

PLANT PERFORMANCE

	Permit Limit	Typical (Upset)
BOD_5, mg/l	25	<25 (170)
SS, mg/l	25	<25

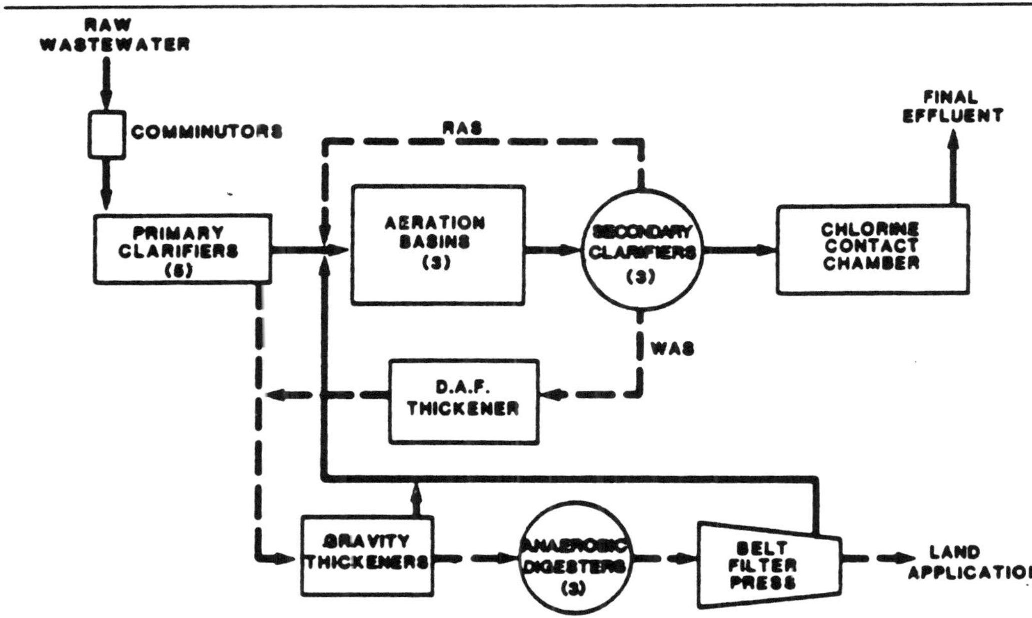

SIOUX CITY WASTE TREATMENT PLANT (SCWTP)
Sioux City, Iowa

The Sioux City Waste Treatment Plant (SCWTP) treats a combined industrial and municipal wastewater average flow of 13.5 mgd and discharges to the Missouri River. More than 140 industries were identified by an industrial survey as potential sources of wastewater. Of these, four are categorical metal finishing or electroplating industries and, as of recently, eleven industries contributed significantly to the suspended solids, BOD and oil and grease discharged to the SCWTP. Although the total volumetric load of the industrial wastewater is typically less than 10 percent of the total flow, the industrial organic loads to the plant account for greater than 50 percent of the observed loads.

The SCWTP has experienced two separate instances in which industrial discharges have interfered with normal plant operations. Isolated slug loads of zinc were experienced by the SCWTP in March and again in April of 1984. Levels as high as 16 mg/l Zn were observed in the treatment plant influent and both slug-load incidences resulted in an upset of the activated sludge process and violations of the NPDES discharge limits. Effluent BOD_5 concentrations exceeded 60 mg/l and effluent suspended solids concentrations in excess of 200 mg/l were observed. The investigation of the first slug load of zinc was somewhat hampered by the lack of in-house capabilities for metals analysis and the first indication of a contamination problem was the process upset itself. Upon confirmation of the nature of the interference, a temporary system for the continuous addition of lime to the primary clarifiers, which would result in the precipitation of subsequent slug loads of zinc, was installed and operated until such time that frequent and periodic monitoring and analysis of the influent for metals could be performed at the SCWTP.

The source of the metal discharge was identified from the City's industrial use survey and from samples of wastewater and solids collected at specific locations in the wastewater collection system. In addition to the process upsets, sludge held in storage lagoons at the facilities became contaminated with zinc and plans to dispose of several years accumulation of sludge by spreading on agricultural land were modified upon receipt of special permitting from the State.

In 1985, a pharmaceutical extractor came on line discharging batches of high strength waste without pretreatment. The strength of the waste ranged from 10,000 to 100,000 mg BOD_5/l and the waste contained high levels of salt and sulfite. The average BOD_5 of the waste was 35,000 mg/l and the batch dumps represented 45 percent of the total organic load to the SCWTP. The activated sludge process was severely overloaded and intermittent depressions of the D.O. level occurred. It was possible to operate the activated sludge process to accommodate the severe organic loads, but the process would again be upset during the weekends when the pharmaceutical extractor was not discharging waste and the organic loads were reduced. Throughout 1985, the SCWTP experienced severe violations of their NPDES BOD_5 and suspended solids discharge limits. Frequent violations of the pharmaceutical extractor's discharge permit occurred with respect to the organic strength and daily mass loading of the waste. The industrial user was placed on a compliance schedule and continued violations of the discharge permit necessitated actions that would result in flow equalization and reductions in the levels of methyl mercaptan, sulfite and sulfide. Presently, all batch waste dumps are transported by bulk to the SCWTP where they are metered, by SCWTP personnel, into the plant influent under controlled conditions.

The upset conditions presented in the following table represent conditions related to the discharge of the pharmaceutical wastewater. The reported upset conditions represent averages for several months of 1985 whereas the typical conditions were based on data for 1984 which spanned nine months and included those months in which the slug loads of zinc were experienced.

SIOUX CITY WASTE TREATMENT PLANT
SIOUX CITY, IOWA

Design Flow:	**30 mgd**	Location:	**Northwest Iowa**
Secondary Treatment:	**Activated Sludge (Conventional)**	Population Served:	**135,000**

INFLUENT WASTEWATER

	Typical (Upset)
Ave. Flow, mgd	13.5
% Industrial	7
BOD_5, mg/l	380 (612)
SS, mg/l	550 (630)

SIGNIFICANT INDUSTRIES

Industry	Flowrate (1000 gpd)	Problem Pollutants
Meat Processing	1,000	BOD_5, oil and grease, SS
Pharmaceutical	70	BOD_5, methyl mecaptan, sulfite
Metal Finishing	20	Zn, Cr, Ni

PLANT LOADING

Primary Clarifiers	Typical (Upset)
Overflow Rate, gal/sf/day	577
Detention Time, hours	2.9
Effluent BOD_5, mg/l	220 (370)
Effluent SS, mg/l	240 (235)

Secondary Clarifiers	Typical (Upset)
Overflow Rate, gal/sf/day	722
Detention Time, hours	3
SVI, ml/gm	150

Aeration Basins	Typical (Upset)
F/M, lbs BOD_5/lbs MLSS/day	0.2 (0.3)
MCRT, days	10
MLSS, mg/l	2500
Detention Time, hours	15
Return Flow, %	40
D.O. Level, mg/l	2.5

PLANT PERFORMANCE

	Permit Limit	Typical (Upset)
BOD_5, mg/l	30	34 (37)
SS, mg/l	30	33 (45)

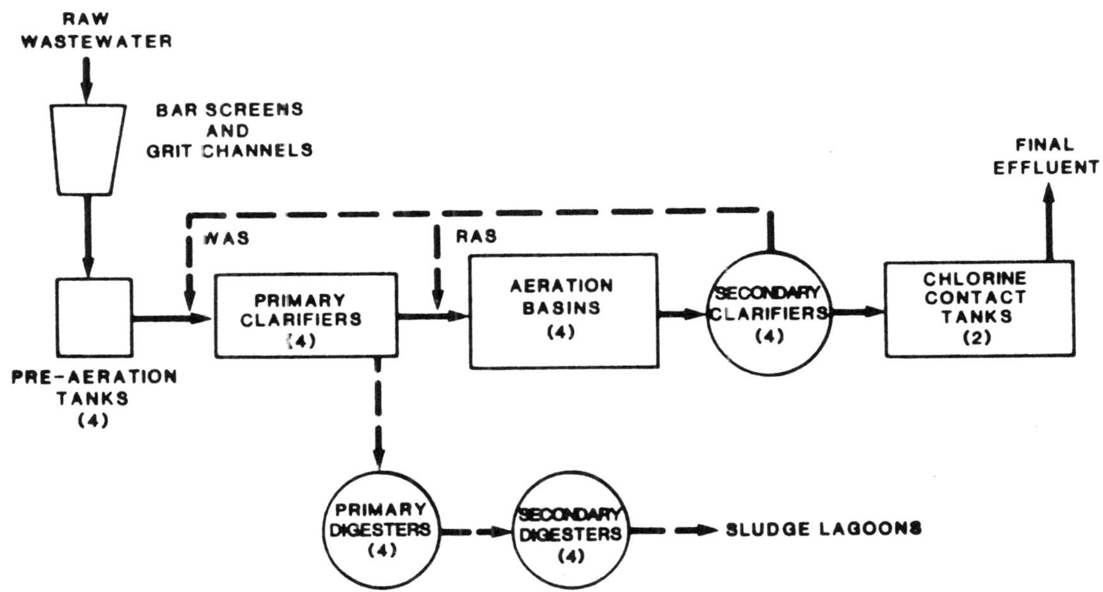

NEWARK WASTEWATER TREATMENT PLANT (NWTP)
Newark, Ohio

The NWTP had been in substantial non-compliance of their 1981 NPDES Permit from the beginning of 1983 until the middle of 1984. This consistent violation had resulted primarily from increased waste loads on the POTW from industrial sources. Between 1979 and 1984, the percentage of industrial wastewater increased from 12 to 22 percent by volume, with influent BOD increasing from 220 to 330 mg/l, while suspended solids increased from 200 to 350 mg/l. To complicate the non-compliance problem, four separate ammonia discharge episodes occurred from August to October, 1983 which resulted in the killing of 80,000 fish in the Licking River. The fish kill precipitated the submission of Verified Complaints to the Ohio EPA on August 6, 1984 by the Black Hand Gorge Preservation Association, against the City of Newark and the NWTP. Following an investigation, the Ohio EPA issued Director's Final Findings and Orders, specifying a compliance schedule and interim discharge limits until a planned facility upgrade is completed by July 1988.

There are two significant industrial contributors to the NWTP who were also issued Director's Final Findings and Orders in May, 1985. A fiberglas insulation manufacturer had been discharging high concentrations of phenol (2-5 mg/l) and NH_3 (up to 500 mg/l), with occasional spills of formaldehyde into the collection system. The activated sludge bacteria were acclimated to the phenol in the wastewater, but were susceptible to shock loadings of the NH_3 and formaldehyde. Fortunately, the industry was responsive to the problems of the NWTP, and instituted a corrective program to:

- conserve and recycle plant flows, which have reduced their discharge by 60 percent (from 1.22 to 0.45 mgd) over the past two years;

- construct an aerated equalization basin to air-strip phenol and distribute diurnal fluctuations; and

- construct a pretreatment facility for their landfill leachate.

The POTW is still subject to occasionally high NH_3 loads from the industry, which is currently the only identifiable cause of interference problems in the plant. The municipality and industry continue to work cooperatively to resolve this problem through the implementation of a spill prevention and control program. Additionally, the renovated POTW will use some of the existing clarifier tankage for off-line storage in the event of future spill episodes.

A second major industry is a dairy which came on-line in 1976. Initially, the dairy stored their whey waste in a silo and typically bled it into the sewer system. The discharge was high in both BOD and suspended solids (2,000 mg/l), and would occasionally be batch discharged to the POTW, resulting in a shock loading to the activated sludge. The industry has since installed a reverse osmosis treatment system for the whey waste which has reduced the solids and organic loading to the plant.

The only categorical industry that currently discharges to NWTP is an electroplater who constructed a metals removal system in conformance with federal pretreatment regulations. In the past, dewatered sludge had been applied to corn fields adjacent to the plant property. However, when heavy metals were detected in seven of ten monitoring wells, Newark began hauling liquid sludge off-site. The planned facility upgrade will include installation of belt filter presses, so that the existing sludge (with acceptable levels of heavy metals) can once again be dewatered and more economically hauled off-site to farm land.

The replacement of coarse bubble aerators with fine bubble equipment in mid-1984 significantly improved BOD removals and the NWTP compliance record. Nitrification, which did not occur previously, now takes place in the last two aeration basins. The only incident of non-compliance with the interim permit in 1985 resulted from an NH_3 discharge from the fiberglass manufacturer. In this case, even though the average monthly BOD measured 29 mg/l, the carbonaceous component was less than 10 mg/l. The final permit will have a more stringent NH_3 requirement and will also designate CBOD as a permitted parameter.

NEWARK WASTEWATER TREATMENT PLANT
NEWARK, OHIO

Design Flow:	**8.0 (12.0 Hydraulic) mgd**	Location:	**Central Ohio**
Secondary Treatment:	**Activated Sludge (Conventional)**	Population Served:	**41,000**

INFLUENT WASTEWATER

	Typical (Upset)
Ave. Flow, mgd	7.5
% Industrial	15
BOD_5, mg/l	305 (450)
SS, mg/l	360 (550)
NH_3, mg/l	35 (60)

SIGNIFICANT INDUSTRIES

Industry	Flowrate (1000 gpd)	Problem Pollutants
Fiberglass	450	Phenol, NH_3, Formaldehyde
Dairy	223	BOD, Phosphorus, SS
Electroplater	97	Cr, Cd, Pb, Ni, Zn, Cyanide

PLANT LOADING

Primary Clarifiers	Typical (Upset)
Overflow Rate, gal/sf/day	560
Detention Time, hours	3.2
Effluent BOD_5, mg/l	194 (280)
Effluent SS, mg/l	147 (218)

Secondary Clarifiers	Typical (Upset)
Overflow Rate, gal/sf/day	500
Detention Time, hours	3.7
SVI, ml/gm	150 (350)

Aeration Basins	Typical (Upset)
F/M, lbs BOD_5/lbs MLSS/day	0.25 (0.4)
MCRT, days	5-6
MLSS, mg/l	2,000
Detention Time, hours	6.3
Return Flow, %	50
D.O. Level, mg/l	2.0

PLANT PERFORMANCE

	Permit Limit	Typical (Upset)
BOD_5 mg/l	20	15 (60)
SS, mg/l	40	15 (95)
NH_3, mg/l (Summer)	25	15 (30)

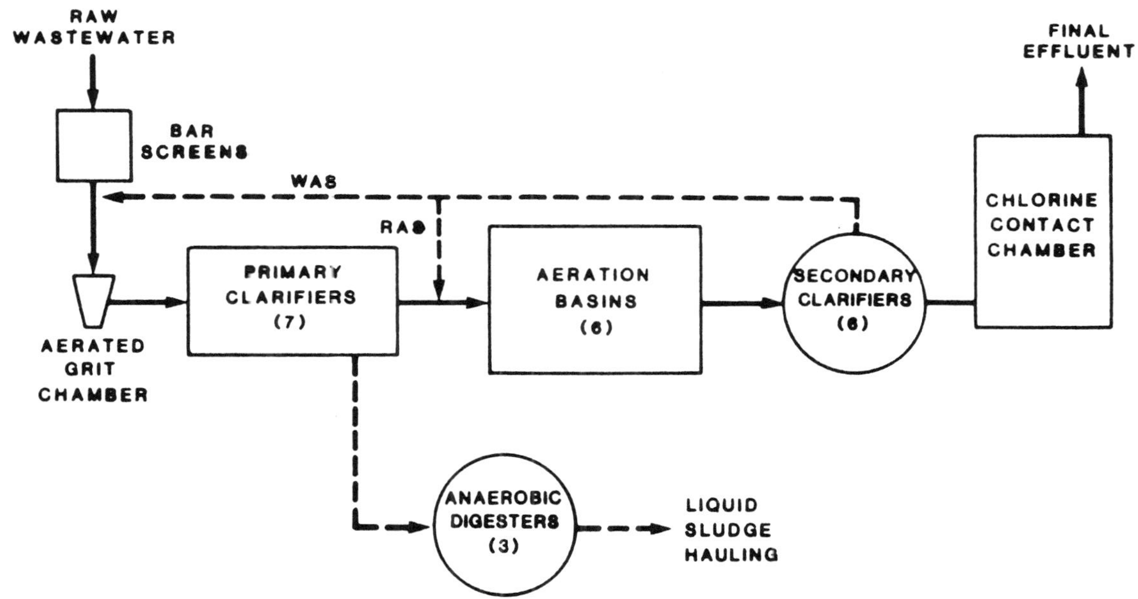

91st AVENUE WASTEWATER TREATMENT PLANT
Phoenix, Arizona

The 91st Avenue Wastewater Treatment Plant (NAWTP) provides secondary treatment for a major portion of the wastewater flow from the greater Phoenix area. The most significant industrial contributors to the NAWTP are electroplaters and metal finishers--their principal pollutants being cadmium, copper, chromium and cyanide. Because of acclimation, the effect of these metals has not been measureably detrimental to the NAWTP's biological system, although occasionally Cu and Cd pass through the plant to the effluent in concentrations violating permit limits of 0.05 mg/l and 0.01 mg/l, respectively. Most of the metals entering the NAWTP partition to the sludge, which prevented land disposal as an option in the past. An industrial pretreatment program, developed over the last four years and approved in July, 1985, has markedly decreased the amount of metals entering the NAWTP and consequently the pass through and sludge disposal problems have been nearly eliminated.

Prior to industrial pretreatment, influent copper and cadmium concentrations at the NAWTP were approximately 0.25-0.32 mg/l and 0.03 mg/l, respectively. Six to eight percent of the influent wastewater was industrial, nearly all of which originated at metal finishing and plating operations. Typical copper discharge concentrations for some circuit board manufacturers were as high as 40-60 mg/l.

Heavy metals removal from the wastestream was generally greater than 75-80 percent; copper and cadmium concentrations in the digested sludge were measured at 2,020 mg/l and 44 mg/l, respectively in 1983. The concentrations precluded disposal of the sludge on agricultural lands. Fortunately for the City of Phoenix, at about the time these metal concentrations were discovered, a precious metals processor became interested in utilizing the sludge and for five years incinerated all the sludge produced by the NAWTP, recovered the metal content and disposed of the ash to a landfill. Despite the high metals partitioning to the sludge, pass through of copper and cadmium in excess of permitted effluent concentrations was not uncommon. In response, an industrial pretreatment program was developed in 1982 to decrease the influent metal concentrations to the NAWTP. Industries were involved by the City in the program development, and the City of Phoenix offered technical knowledge, short of design, to the industries trying to meet the reduced metal discharge limits. Prior to the implementation of a pretreatment program, most industries had no pretreatment other than flow equalization, and many installed pretreatment works in order to meet the new copper and cadmium discharge limits of 4.5 mg/l and 0.1 mg/l, respectively. A pretreatment and metal recovery system at one large circuit board manufacturer cost in excess of $ 2.5 million.

As a result of the pretreatment program, typical treatment plant influent copper and cadmium concentrations have been cut to 0.15 mg/l and 0.012 mg/l, respectively, and treatment plant Cu and Cd effluent limits are generally not exceeded. With the reduction in influent wastewater metals concentrations a corresponding reduction in the sludge metal concentrations occurred and it was no longer profitable for the precious metal recovery firm to continue processing and disposing of the NAWTP sludge. However, the metal concentrations were reduced to a level where the sludge was acceptable as a soil conditioner, and as a result another company has begun marketing the dried sludge commercially. Industries are presently self-monitoring their discharge, with the City sampling most industries at least eight times per year as well. The monitoring frequency is increased when an industry is in non-compliance with its permit. Figure C-5 shows the change in influent metals levels at the 91st Avenue Plant between 1982 and 1985.

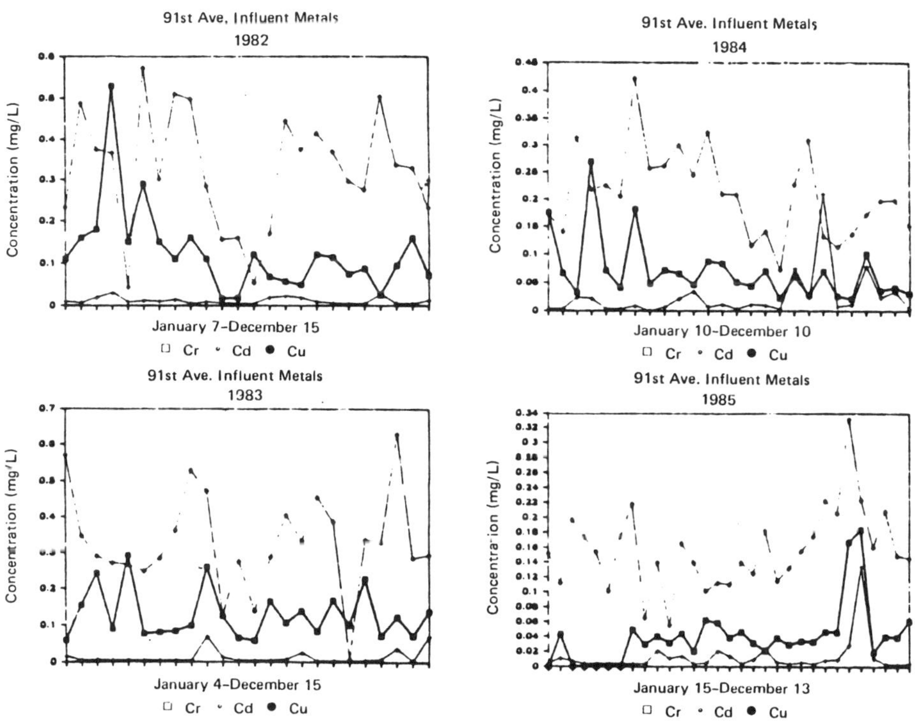

FIGURE C-5
91st AVENUE INFLUENT METALS

91st AVENUE WASTEWATER TREATMENT PLANT
PHOENIX, ARIZONA

Design Flow:	**120 mgd**	Location:	**South Central Arizona**
Secondary Treatment:	**Activated Sludge (Complete Mix)**	Population Served:	**1,100,000**

INFLUENT WASTEWATER

	Typical
Ave. Flow, mgd	140
% Industrial	7
BOD_5, mg/l	220
SS, mg/l	150
Cd, mg/l	0.012 (0.03)[a]
Cu, mg/l	0.15 (0.3)[a]

SIGNIFICANT INDUSTRIES

Industry	Flowrate (mgd)	Problem Pollutants
Electroplaters and Metal Finishers	10	Cd, Cr, Cu, CN^-

PLANT LOADING

Primary Clarifiers	Typical
Overflow Rate, gal/sf/day	1,050
Detention Time, hours	1.8
Effluent BOD_5, mg/l	180
Effluent SS, mg/l	80

Secondary Clarifiers	Typical
Overflow Rate, gal/sf/day	700
Detention Time, hours	2.2

Aeration Basins	Typical
MCRT, days	1-2
MLSS, mg/l	500-800
Detention Time, hours	5
Return Flow, %	35
D.O. Level, mg/l	2.0

PLANT PERFORMANCE

	Permit Limit	Typical
BOD_5, mg/l	30	16
SS, mg/l	30	12
Cd, mg/l	0.01	0.005 (0.03)[a]
Cu, mg/l	0.05	0.03 (0.10)[a]

a = Before Pretreatment

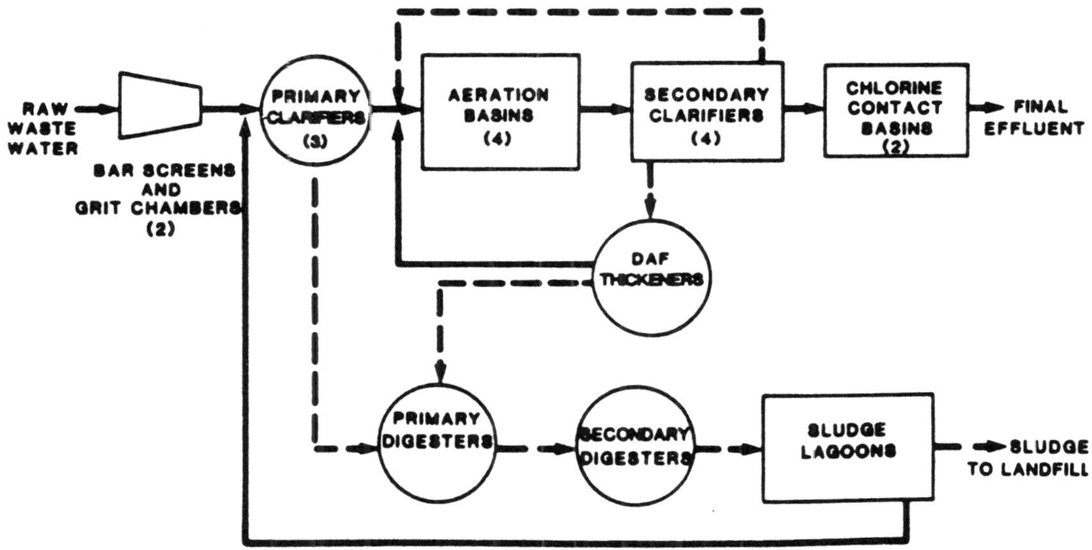

TOLLESON WASTEWATER TREATMENT PLANT
Tolleson, Arizona

The Tolleson Wastewater Treatment Plant (TWTP) is a two stage trickling filter plant that treats a predominantly domestic wastewater from Phoenix, Arizona suburbs. The successful operation of the TWTP is dependent on the one significant industrial contributor to the treatment plant, a meatpacker who processes 1,000 to 1,400 head of beef per day. The treatment plant typically discharges effluent with BOD_5 and SS levels both below 10 mg/l, but has been upset on occasion to the point of effluent permit non-compliance when it receives slug loads of blood and grease from the meatpacker with BOD_5 and SS levels of up to 2,200 mg/l and 1,375 mg/l, respectively. Upset frequency and severity have been reduced in recent years through improved industrial waste monitoring and treatment process monitoring, respectively.

The influent to the TWTP could be typified as medium to high-strength municipal wastewater with average BOD_5 and SS levels being 275 mg/l and 225 mg/l, respectively. Approximately 25 to 30 percent of the organic and solids loading is contributed by the meatpacker on an average basis at levels of 1,100-1,600 mg/l BOD_5 and 700-1,400 mg/l SS, for wastewater flows of 0.8-1.0 mgd. In general, the domestic/industrial waste stream can be treated to well within 30/30 discharge limits, but in the past the meatpacker would upset the treatment process by slug discharging blood or some other high strength organic slaughter by-product. Prior to 1982, these upset conditions would last for several days and result in weekly and monthly effluent suspended solids of 30-40 mg/l, in violation of permit limits.

Treatment upsets have diminished in frequency and intensity since 1982 for two reasons:

- A legal contract with the meatpacker limits flow to 0.8 mgd, BOD_5 to 10,675 lbs per day (1,600 mg/l) and SS to 6,670 lbs per day (1,000 mg/l), and provides for fines or disconnection if these limits are exceeded, and

- Improved treatment plant process monitoring has enabled operators to better detect, and thus act on, a potentially upsetting condition.

The contract with the meat packer attempts to prevent waste blood from being stored for more than about eight hours at a time before discharging to the sewer. Prior practice resulted in blood being held back for up to a week at a time before being discharged all at once.

Primarily through trial and error, the operators of the TWTP have established several operating parameters that help them in detecting upset conditions in the plant. The depth of sludge in the primary clarifiers is monitored closely; a high or rapidly increasing sludge depth is indicative of upset conditions and is caused by the high solids content of the meatpacking waste. The mixed liquor in the solids contact basin following the second trickling filter is monitored closely as well, with levels above 500 mg/l signaling possible problems. Mixed liquor concentrations of 1,500 mg/l generally result in effluent suspended solids of greater than 30 mg/l. To remedy an upset condition, primary sludge pumping rates are manually increased above their normal levels to reduce the solids inventory and prevent escape in the effluent.

As a result of all industrial wastewater interference prevention work, the TWTP has gone from experiencing periodic effluent permit vilolations to experiencing infrequent upsets, seldom resulting in NPDES Permit violations.

TOLLESON WASTEWATER TREATMENT PLANT
TOLLESON, ARIZONA

Design Flow;	8.3 mgd	Location:	South Central Arizona
Secondary Treatment:	2 Stage Trickling Filter with Solids Contact	Population Served:	65,000

INFLUENT WASTEWATER

	Typical (Upset)
Ave. Flow, mgd	7.4
% Industrial	14
BOD_5, mg/l	275 (340)
SS, mg/l	225 (280)

SIGNIFICANT INDUSTRIES

Industry	Flowrate (1000 gpd)	Problem Pollutants
Meat Packer	1000	BOD, SS

PLANT LOADING

Primary Clarifiers	Typical (Upset)
Overflow Rate, gal/sf/day	860
Detention Time, hours	1.9
Effluent BOD_5, mg/l	160
Effluent SS, mg/l	95

Intermediate Clarifiers	Typical (Upset)
Overflow Rate, gal/sf/day	735
Detention Time, hours	2.4
Effluent BOD_5, mg/l	30
Effluent SS, mg/l	30

Secondary Clarifiers	Typical (Upset)
Overflow Rate, gal/sf/day	480
Detention Time, hours	7.4

First Stage Trickling Filter	Typical
Hydraulic Loading, gal/sf/day	1,000
Organic Loading, lbs BOD_5/1000 cf/day	45
Recirculation, %	100

Second Stage Trickling Filter	Typical
Hydraulic Loading, gal/sf/day	500
Recirculation, %	100

PLANT PERFORMANCE

	Permit Limit	Typical (Upset)
BOD_5, mg/l	30	9 (25)
SS, mg/l	30	9 (35)

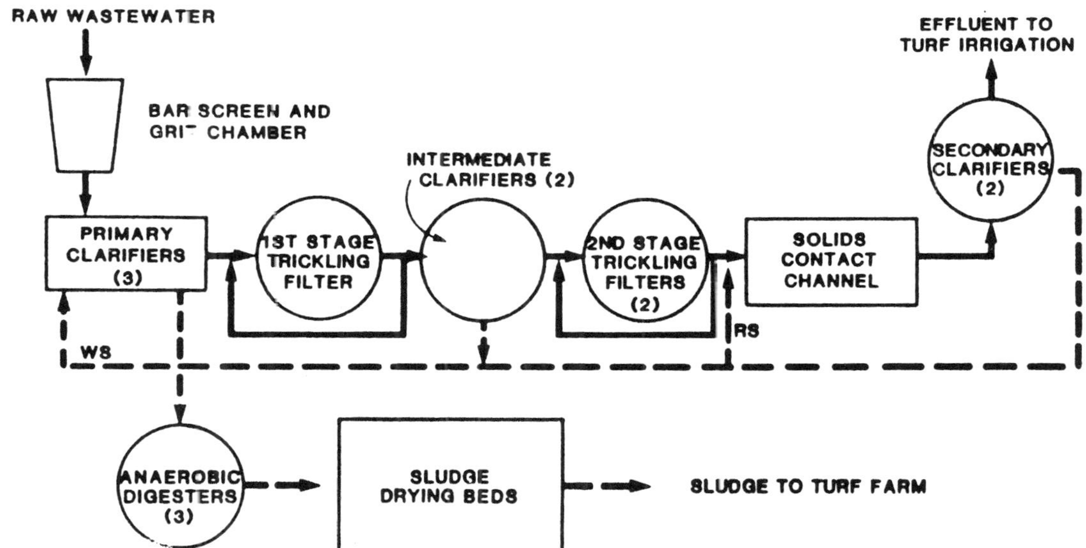

VICTOR VALLEY REGIONAL WATER RECLAMATION PLANT (VVRWRP)
Victorville, California

The VVRWRP has, since 1981, experienced periodic activated sludge upsets accompanied by chronic aeration basin and anaerobic digester foaming problems, believed to be caused by solvent-type chemicals. During this same period, COD levels up to three times the permitted effluent limit of 15 mg/l have also been discharged. Initial efforts to discern the cause of the upsets and foaming, and document the source of the pollutants were limited to visual and olfactory investigation of the treatment facilities and sewer interceptors. Recently, more thorough attempts to document the upsets (wastewater sampling and laboratory analyses) have resulted in positive identification of the upsetting pollutants and source, and have established the framework for the correction of the problems.

Start-up of the VVRWRP, treating primarily domestic wastewater, was completed in June, 1981 with the connection of an Air Force Base (AFB) sewer interceptor. The AFB contributes both domestic and industrial wastewater with vehicle and plane washing, jet fueling and paint stripping facilities producing the largest industrial flows. The VVRWRP began experiencing effluent COD permit violations, aeration basin foaming and occasional biological upsets shortly after the connection of the AFB sewer interceptor. The foaming and upset problems continued into 1985 without significant efforts made to document the cause or source of the problems. Chemical addition and variation of the food to microorganism ratio and the mixed liquor suspended solids were unsuccessful at mitigating the foaming problems. Periodically, strong solvent or oil odors were detected at the treatment facility and in the influent wastewater, coinciding with two-fold effluent BOD and COD increases. Attempts to trace the odor of the pollutants to the source generally implicated the AFB, but no further action to substantiate the AFB as the pollutant source was immediately initiated.

Decisive steps were taken to document and correct the problems following a February, 1985 "spill" of pollutants with a strong solvent smell into the treatment plant. Wastewater samples were immediately taken at the plant influent, the AFB interceptor and the sewer above the AFB connection. Laboratory analyses showed significant concentrations of a number of pollutants in both the plant influent and in the AFB wastewater. Other similar events were sampled and analyzed from July through September, 1985. The ranges of concentrations detected for six compounds during the July-September upset sampling are shown below:

Compound	VVRWRP Influent (ug/l)	AFB Effluent (ug/l)
Chloroform	15-23	10-55
Methylene Chloride	11-43	11-1600
Toluene	11-29	43-400
m, p-Xylene	11-19	56-320
Phenol	11-12	11-230
bis (2-Ethylhexyl) phthalate	39-210	17-830

During upset conditions, effluent COD values doubled to approximately 35 mg/l and turbidity levels exceeded 2 NTU, also in violation of effluent limits. Methylene chloride concentrations as high as 68 mg/l were measured in the February analysis of the AFB effluent.

The documented upsets resulted in discussions between VVRWRP and the AFB officials, with the district requesting that base practices causing the discharge of inhibitory levels of contaminants be stopped. The AFB pretreatment currently consists only of poorly operated oil-water separation units.

Despite the VVRWRP-AFB dialog, the treatment plant continues to experience foaming problems and violate effluent COD limits, presumably because of the AFB discharges. A formal "Cease and Desist" order was issued to the AFB in September, 1985. As of this writing, a wastewater sampling and analysis program is being completed by the AFB as the first step of a negotiated agreement to correct the sewer discharge/treatment plant interference problem.

VICTOR VALLEY REGIONAL WATER RECLAMATION PLANT
VICTORVILLE, CALIFORNIA

Design Flow:	**4.8 mgd**	Location:	**Southcentral California**
Secondary Treatment:	**Activated Sludge (Modified Step Feed)**	Population Served:	**40,000**

INFLUENT WASTEWATER

	Typical (Upset)
Ave. Flow, mgd	3.8
% Industrial	15
BOD_5, mg/l	200 (300)
COD, mg/l	350 (600)
SS, mg/l	200 (440)

SIGNIFICANT INDUSTRIES

Industry	Flowrate (1000 gpd)	Problem Pollutants
Military Base	850	Methylene chloride; Toluene; m, p-Xylene
Cement Mfr.	110	cement dust
Restaurants	-	fats and grease

PLANT LOADING

Primary Clarifiers	Typical (Upset)
Overflow Rate, gal/sf/day	972
Detention Time, hours	1.71

Aeration Basins	Typical (Upset)
F/M, lbs BOD_5/lbs MLSS/day	0.3-0.4
MCRT, days	5-10
MLSS, mg/l	2000
Detention Time, hours	7.3
Return Flow, %	50-100
D.O. Level, mg/l	0.5-1.0

Secondary Clarifiers	Typical (Upset)
Overflow Rate, gal/sf/day	533
Detention Time, hours	4.71

Pressure Filters	Typical
Filtration Rate, gal/sf/min	2.54

PLANT PERFORMANCE

	Permit Limit	Typical (Upset)
BOD_5, mg/l	10	3-4 (6-7)
COD, mg/l	15	20 (35)
SS, mg/l	10	1-3 (1-3)
Turbidity, NTU	2	1 (2-5)

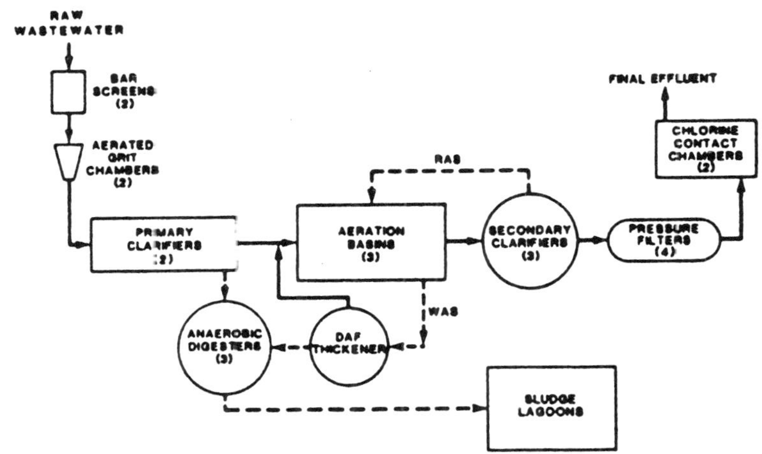

DUCK CREEK SEWAGE TREATMENT PLANT, PAW PAW SEWAGE TREATMENT PLANT
Denison, Texas

The Duck Creek Sewage Treatment Plant (DCSTP) is one of four small treatment facilities owned, operated and maintained by the City of Denison, Texas. The plant is located on the far north end of town, adjacent to the Red River. The Paw Paw Sewage Treatment Plant (PPSTP), located on the east end of Denison, is one the older sewage treatment facilities operated by the City.

The DCSTP and PPSTP had consistently failed to meet NPDES discharge standards for BOD and suspended solids prior to 1978. A state court order required Denison to monitor the industrial waste discharges from the four largest industries in town on a 5 day per week basis. Two of these industries were deleted from the court order when they initiated their own pretreatment program and constructed pretreatment facilities. In 1985, Denison was issued an EPA Administrative Order to implement a pretreatment program. A revised City Sewer Ordinance was approved by the City Council on January 6, 1986.

The four largest industries in Denison are all food processors; of these, the two that do not pretreat are the major source of industrial interferences at the PPSTP. These two industries are a food oil refinery, and an oily-type food processor (margarine, salad oil, etc.). These two facilities have a common discharge pump station and have the capability to flow by gravity to the DCSTP.

The influent to the DCSTP has a BOD_5 concentration of 400-500 mg/l and a TSS concentration ranging from 50-300 mg/l. Concentrations of fats, oils and grease (FOG) cause problems at the DCSTP, particularly when one of those industries releases a batch dump of their waste. Apparently, such an incident had occurred on Monday, February 3, 1986, and the residual effects of this batch dump were noted at the plant during the JMM site visit on Tuesday, February 4, 1986. Major effects included clogging of the bar screen, and scum on the secondary clarifier. There would also have been a thick grease layer on the oxidation ditch, but a previous day's rain (4 inches in 6 hours) had caused the headworks to overflow, and a substantial quantity of grease was noted on the ground adjacent to the oxidation ditch. Due to the overloaded condition and lack of parallel units at this plant, there are no process control alternatives for responding to these batch discharges other than bypassing the bar screen and running the influent comminutor. The DCSTP is in compliance with the NPDES discharge permit about 65 percent of the time. When it is in an upset condition, effluent BOD and TSS concentrations exceed 140 mg/l and 200 mg/l, respectively, on the average.

The most significant industrial flows are processed through the PPSTP. This plant receives upwards of 400,000 gpd of industrial waste with a BOD ranging from 1,200 to 2,000 mg/l, TSS range of 400 to 650 mg/l and FOG of 300 to 400 mg/l. The FOG is noted to be extremely high. In addition, the flow contributed by these industries is only a City estimate, based on their flow measurements when the flow from these industries has been diverted to the DCSTP due to pump station problems. The two industries in question claim total combined discharge of 140,000 gpd, based on pump station wet well size and pump cycling. There is no flow meter on this pump station, nor have water consumption records been used to gage discharge flow due to uncertainties regarding in-plant consumptive use values.

At the time of the administrative order, the City was asked to guarantee NPDES discharge permit compliance for the PPSTP. To accomplish this, a chemical addition system was added to feed cationic polymer and liquid alum into the PPSTP influent flow. The result of this program is the reduction of effluent BOD and TSS from averages around 35 mg/l each, to less than 20 mg/l each. The cost of this chemical addition program is approximately $ 8,000 per month, essentially due to the cost of alum ($137/ton, 8 gpd feed rate) and polymer ($1.85/lb, 6.1 gpd feed rate). Due to mechanical problems at this plant, one of two trickling filters is out of service, and has been so for over two months, awaiting arrival of replacement parts from the manufacturer.

Other problems noted at PPSTP are the grease accumluation, and foam due to detergents. The grease can accumulate on the bar screen, primary clarifier, or chlorine contact tanks, and can plug the trickling filter distributor ports. Several ports were noted to be plugged during the site visit, and the wastewater superintendent noted that if the ports are cleaned, they generally plug up again within less than one hour's time. The greatest accumulation of grease was noted on the chlorine contact tanks' water surface. The City has received permission from the EPA to periodically pump these tanks down to remove accumulated grease. Foaming was noted at the downstream side of the mechanical grit chamber, and the influent splitter box to the chlorine contact tanks. The oil and grease do not appear to significantly impact effluent quality.

In mid-1985, the City hired a consultant to examine their wastewater treatment system and recommend any necessary modifications. The consultant recommended a new 2.5 mgd Trickling Filter/Activated Sludge plant to be constructed by late 1987, replacing both the Paw Paw and Duck Creek Sewage Treatment Plants.

PAW PAW SEWAGE TREATMENT PLANT
DENISON, TEXAS

Design Flow:	**2.5 mgd**	**Location:**	North Texas
Secondary Treatment:	**Trickling Filters**	**Population Served:**	5,000

INFLUENT WASTEWATER

	Typical (Upset)
Ave. Flow, mgd	1
% Industrial	50
BOD_5, mg/l	480
SS, mg/l	440

SIGNIFICANT INDUSTRIES

Industry	Flowrate (1000 gpd)	Problem Pollutants
Food Processor	24	BOD, TSS, FOG
Food Oil Processor	200	BOD, TSS, FOG

PLANT LOADING

Primary Clarifier	Typical (Upset)
Overflow Rate, gal/sf/day	230
Detention Time, hours	7
Effluent BOD_5, mg/l	100
Effluent SS, mg/l	80

Trickling Filters	Typical (Upset)
Loading lbs BOD_5/1,000 cu ft	5 (15)
Return Flow, %	400

Secondary Clarifiers	Typical (Upset)
Overflow Rate, gal/sf/day	350
Detention Time, hours	4.5

PLANT PERFORMANCE

	Permit Limit	Typical (Upset)
BOD_5, mg/l	20	18 (34)
SS, mg/l	20	12 (39)

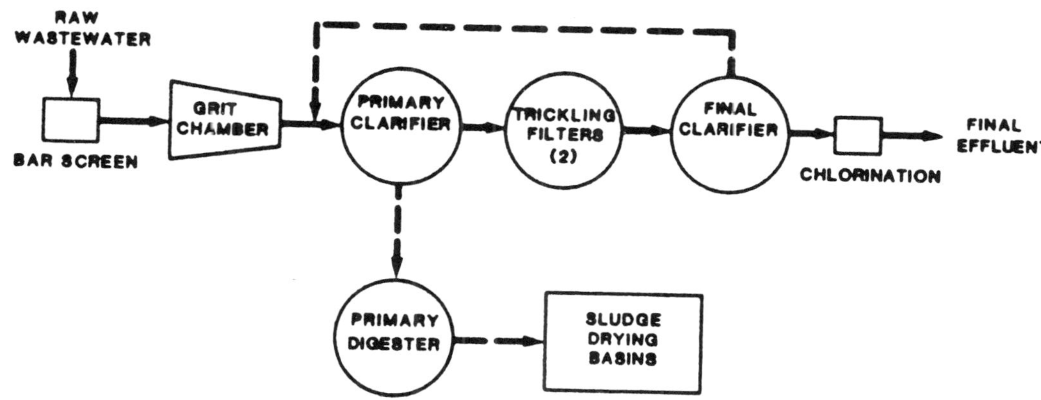

DUCK CREEK SEWAGE TREATMENT PLANT
DENISON, TEXAS

Design Flow:	**1.0 mgd**	Location:	North Texas
Secondary Treatment:	**Activated Sludge (Oxidation Ditch Ext. Aeration)**	Population Served:	7,000

INFLUENT WASTEWATER

	Typical (Upset)
Ave. Flow, mgd	1.2
% Industrial	50
BOD_5, mg/l	320
SS, mg/l	330

SIGNIFICANT INDUSTRIES

Industry	Flowrate (1000 gpd)	Problem Pollutants
Food Processor	39	BOD, O & G
Food Processor	4 (up to 30)	BOD, TSS, O & G
Wood Preserver	30	Creosote

PLANT LOADING

Primary Clarifiers

No Primary Clarifiers

Aeration Basins	Typical (Upset)
F/M, lbs BOD_5/lbs MLSS/day	0.13
MCRT, days	7.7
MLSS, mg/l	1,800
Detention Time, hours	26
Return Flow, %	50
D.O. Level, mg/l	1.5 - 2.0

Secondary Clarifiers	Typical (Upset)
Overflow Rate, gal/sf/day	6.0
Detention Time, hours	2.9

PLANT PERFORMANCE

	Permit Limit	Typical (Upset)
BOD_5, mg/l	20	35 (140)
SS, mg/l	20	45 (200)

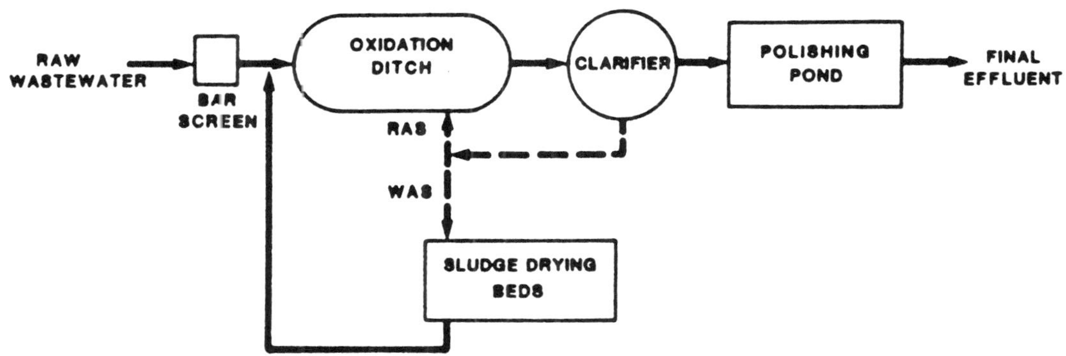

PARIS WASTEWATER TREATMENT PLANT
Paris, Texas

The Paris Wastewater Treatment Plant (PWTP) was constructed in 1972 to serve the municipal and industrial needs of this north Texas community. The City fathers had adopted a policy of bringing industries into the City by making generous allowances in the areas of municipal taxes and utilities. There were no industrial waste discharge requirements for industries in Paris until 1983, when the EPA issued an administrative order for the City to improve discharges from the PWTP such that they would comply with their NPDES permit. The City Utilities Department staff then set up a comprehensive 90 day industrial discharge monitoring program to determine which industrial discharges were responsible for the treatment plant overloading. The plant was designed to treat a maximum BOD loading of 8,000 lbs/day, but was receiving an average of 10,000 lbs/day with peaks of over 15,000 lbs/day. The sampling program revealed that greater than 53 percent of the influent BOD loading was attributable to four large industries.

The Utilities Department developed an industrial sewer use ordinance which was put into effect by the City Council in late 1983. This ordinance is strictly enforced by the Utilities Department. In the first year, over $ 350,000 in surcharge fees were collected. This has subsequently dropped to about $ 190,000 per year. The Utilities Department required the four largest industries to install permanent recording flow meters and refrigerated composite samplers. A second set of industries was required to install flow meters, flow monitors, and manholes for the City to take samples. A third group of industries was required to install a Parshall Flume, with the City making periodic flow measurements and taking samples. Finally, the smallest industrial dischargers were required to install an effluent manhole from which the City could withdraw samples. The result of the ordinance and strict enforcement of the surcharge program is that the majority of the large industries have all installed their own pretreatment systems. Two have their own NPDES permits, and one of these operates a 6 mgd overland flow treatment system.

Of the major industries in Paris, most are food processors or paper products manufacturers. One of the four largest, however, is a categorical (metal finishing) industry. This discharger still contributes an average of 40-50 mg/l of ammonia, 30 mg/l of copper and 17 mg/l of zinc into the PWTP. The City has worked with this industry to develop a timetable whereby it will reduce its ammonia discharge to 30 mg/l and its heavy metals discharge to EPA categorical standards by June 1986. The presence of heavy metals in the sludge limits sludge disposal to non-agricultural lands.

The PWTP now receives an average of 6,000 lbs/day of BOD_5, which it can easily handle with its existing facilities. Occasional slug loads of up to 27,000 lbs/day have been received at the PWTP. These are typically handled by increasing the MCRT and MLSS concentrations. The most recent episode of a slug of high strength waste did not have any adverse effects on the PWTP.

The City is starting construction of an upgrade to the PWTP which is scheduled to be completed by December, 1986. The major feature of the upgrading will be the addition of an 80 foot diameter by 15 foot deep plastic media roughing filter ahead of the activated sludge system. This is being paid for by the largest industry in town, and will increase the design BOD_5 loading for the PWTP to 15,000 lbs/day.

One other major feature of the City's ordinance is their refusal to take wastes which are high in oil and grease content. The PWTP has no primary clarifiers and can therefore not easily remove oily wastes from the flow stream. All local restaurants are required to have a grease trap to remove grease from their waste flows before discharge to the City sewer system.

At the core of the City's successful pretreatment program is their willingness to enforce the City ordinance, and their laboratory. The City has its own water/wastewater laboratory which produces duplicate analyses of all samples. The City provides effluent analyses of industrial discharges to industries at no charge to that industry. The City data are consistently accurate, and match quarterly EPA sampling data.

PARIS WASTEWATER TREATMENT PLANT
PARIS, TEXAS

Design Flow:	6 mgd	**Location:**	North Texas
Secondary Treatment:	Activated Sludge (Oxidation Ditch Extended Aeration)	**Population Served:**	26,000

INFLUENT WASTEWATER

	Typical (Upset)
Ave. Flow, mgd	4
% Industrial	15
BOD_5, mg/l	200 (1,000)
SS, mg/l	200

SIGNIFICANT INDUSTRIES

Industry	Flowrate (1000 gpd)	Problem Pollutants
Food Processor	210	BOD_5, TSS
Metal Finisher	100	NH_3, Cu, Zn

PLANT LOADING

Primary Clarifiers

No Primary Clarifiers

Secondary Clarifiers

	Typical (Upset)
Overflow Rate, gal/sf/day	800
Detention Time, hours	2.5
SVI, ml/gm	120 (160)

Aeration Basins	Typical (Upset)
F/M, lbs BOD_5/lbs MLSS/day	0.09 (0.33)
MCRT, days	16 (20)
MLSS, mg/l	2,500 (3,200)
Detention Time, hours	22.5
Return Flow, %	50
D.O. Level, mg/l	0.5

PLANT PERFORMANCE

	Permit Limit	Typical (Upset)
BOD_5, mg/l	20	10 (30)
SS, mg/l	20	10 (30)

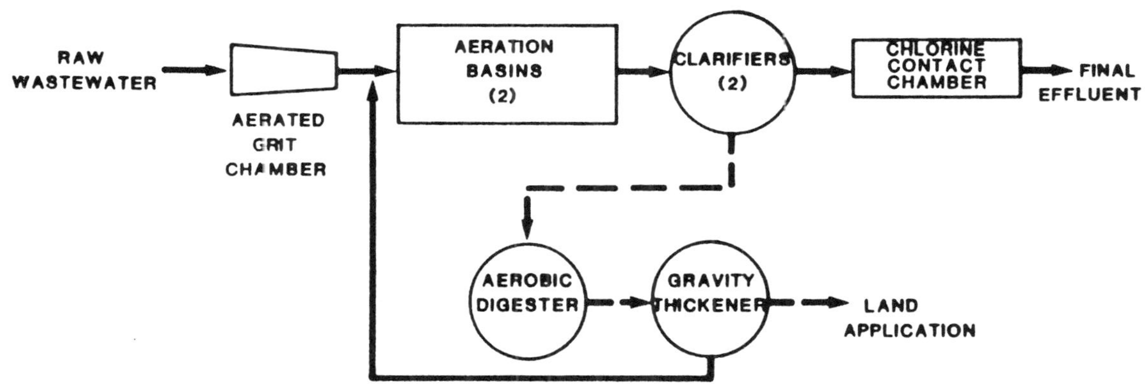

NEWBERG WASTEWATER TREATMENT PLANT
Newberg, Oregon

The Newberg Wastewater Treatment Plant (NWTP) has experienced periodic episodes of non-compliance with their NPDES Permit for approximately ten years due to fluctuating BOD loadings, and biological upsets caused by excessive copper discharges. In general, copper discharges have not limited sludge disposal options. With implementation of a pretreatment program and tighter industrial discharger limits, interference incidences have become more infrequent since mid-1984. NPDES discharge permit compliance should increase in the next two years as some industrial waste permit limits are tightened and a completely new treatment facility with greater hydraulic capacity is brought on line.

The two main industrial contributors to the NWTP are a circuit board manufacturer (cbm) and a fruit processor (fp), both of which operate year-round. The cbm has been discharging wastewater to the NWTP since 1974, but the flowrate was increased in 1978 when it was discovered by the Oregon Department of Environmental Quality (DEQ) that a wastestream with copper concentrations as high as 50-80 ppm was being directly discharged to a local stream. Subsequent biological failure of the NWTP showed copper levels as high as 100 ppm in the primary clarifier sludge. The incident required reseeding of the biological population and 45 days to completely recover. Because of regular upsets, the City began sampling and testing for pH and copper in 1981 at the first manhole downstream of the cbm facility. The City experienced great difficulty in working with the cbm to reduce copper levels and periodic discharge problems continued until May, 1984 when copper discharges caused a complete activated sludge and anaerobic digester failure. With pressure on the City by the DEQ and an updated sewer ordinance with more "teeth", the City aggressively pursued compliance by the cbm. A "show cause" hearing resulted and rather than address the pretreatment issue, the cbm chose to cease production and lay off 50-60 people.

The cbm reopened later that summer with the new pretreatment equipment required to comply with its industrial waste discharge permit. Since installing pretreatment, the cbm has been in constant compliance. Wastewater monitoring of the cbm continues.

The second major industrial discharger which has caused the NWTP to violate its NPDES Permit is a processor of pie cherries. The fp requires on the average, one third of the NWTP BOD treatment capacity which normally does not present a problem. However, waste strength variability can result in the fp contributing 2900 pounds of BOD in one day to the plant which is designed to handle 3200 pounds/day. With the addition of the domestic BOD load, the biological process is overwhelmed. Typically, this occurs in the summer at low plant flowrates and can cause the effluent BOD to rise to 50 mg/l. The fp has been responsive to the City's pretreatment program and tighter industrial waste discharge limits, which have attempted to solve the problem. Stricter industrial discharge limits may be imposed by the City when the fp's permit is renewed in the near future. A good working relationship between the City and the fp has made corrective actions easier to implement than in the case of the cbm.

In dealing with all industries in town, the City has attempted to work cooperatively to implement pretreatment and issue industrial waste discharge permits. The City has paid up to one half of all consulting fees associated with industries implementing pretreatment, and has paid for most laboratory analysis of wastewater samples.

Newberg experiences substantial I/I which on occasion contributes to non-compliance problems by hydraulically overloading the treatment process. The I/I problem is being addressed by the City.

NEWBERG WASTEWATER TREATMENT PLANT
NEWBERG, OREGON

Design Flow:	2.0 mgd	Location:	Northwestern Oregon
Secondary Treatment:	Complete Mix Activated Sludge	Population Served:	11,000

INFLUENT WASTEWATER

	Typical (Upset) Summer	Typical (Upset) Winter
Ave. Flow, mgd	1.26	2.0
% Industrial	10	
BOD_5, mg/l	230 (380)	130 (230)
SS, mg/l	170	90
Cu, mg/l	< 1 (10-50)	

SIGNIFICANT INDUSTRIES

Industry	Flowrate (1000 gpd)	Problem Pollutants
Circuit Board Manuf.	65	Cu, pH
Fruit processing	58	BOD

PLANT LOADING

Primary Clarifiers	Typical (Upset)
Overflow Rate, gal/sf/day	1,070
Detention Time, hours	3.0

Secondary Clarifiers	Typical (Upset)
Overflow Rate, gal/sf/day	530 and 470
Detention Time, hours	3.4 and 5

Aeration Basins	Typical (Upset)
MCRT, days	4-8
MLSS, mg/l	1200-2000
Detention Time, hours	4.7
Return Flow, %	50-70
D.O. Level, mg/l	1.5-2.5

PLANT PERFORMANCE

	Permit Limit Summer	Permit Limit Winter	Typical (Upset)
BOD_5, mg/l	20	30	25 (50-100)
SS, mg/l	20	30	15

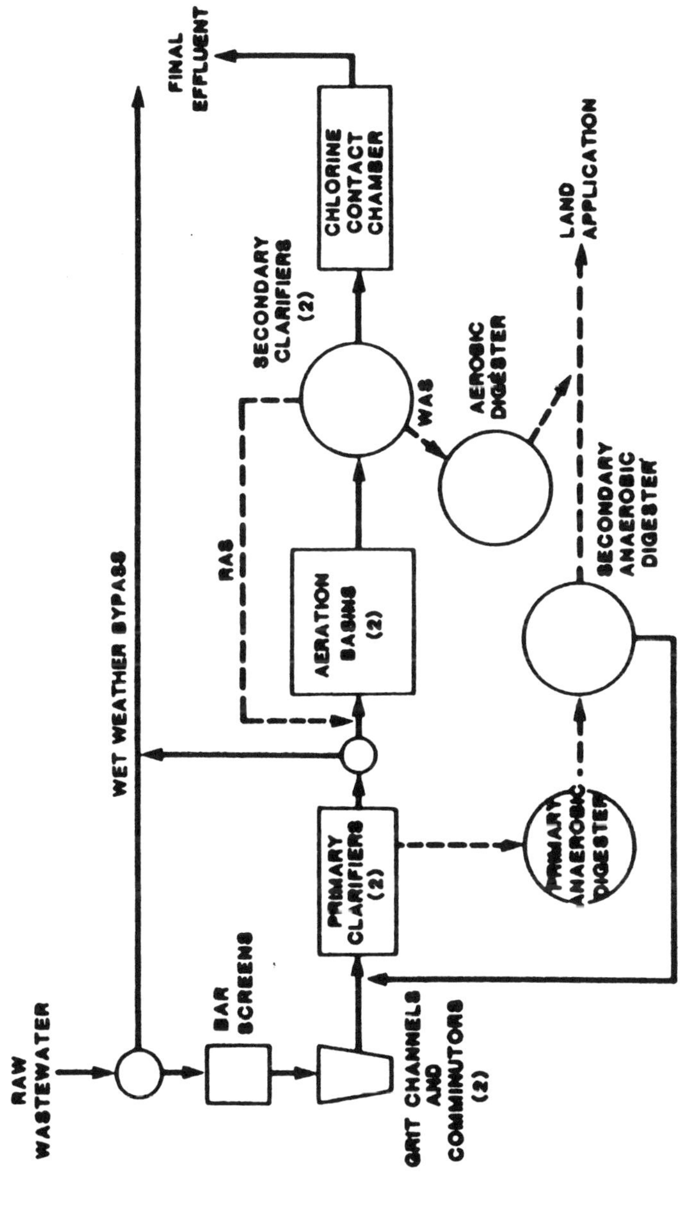
RAW WASTEWATER
BAR SCREENS
GRIT CHANNELS AND COMMINUTORS (2)
WET WEATHER BYPASS
PRIMARY CLARIFIERS (2)
PRIMARY ANAEROBIC DIGESTER
RAS
AERATION BASINS (2)
SECONDARY ANAEROBIC DIGESTER
SECONDARY CLARIFIERS (2)
WAS
AEROBIC DIGESTER
CHLORINE CONTACT CHAMBER
LAND APPLICATION
FINAL EFFLUENT

METRO-WEST POINT TREATMENT PLANT
Seattle, Washington

The Municipality of Metropolitan Seattle (METRO) has had an operational industrial pretreatment program since 1969. With minor modifications, the program was EPA-approved in 1981 as one of the first in the nation. Successful reductions in influent wastewater and primary sludge heavy metal concentrations during the last five years can, to a great extent, be attributed to implementation and enforcement of pretreatment standards. As an outcome of this, self-monitoring by industrial dischargers augmented with year-round spot monitoring by Metro's Industrial Waste Section has reduced the incidences of toxic upsets in the anaerobic digesters of the West Point Treatment Plant and in the activated sludge process of the neighboring Renton Treatment Plant.

The Metro-West Point Treatment Plant provides primary treatment and sludge digestion for an average daily wastewater flow of 132 mgd, 4.7 percent originating from industrial sources. Approximately 70 metal finishing/electroplating industries discharge to the sewer system in addition to a variety of other categorical and non-categorical industries. Records of periodic digester upsets go back as early as 1967, but their occurrences have become less frequent since 1980, coinciding with substantial overall reductions in heavy metal concentrations. Past upsets directly linked to toxic metals (generally chromium) caused increased volatile acid concentrations, increased carbon dioxide content of the gas produced, reduced gas production, and in a few cases caused complete failure of the digesters. An October, 1980 chromium spill to the West Point facility caused a typical upset and resulted in the plant influent chromium concentration jumping 10 fold to greater than 2 mg/l. Primary sludge concentrations of chromium reached 710 mg/l, resulting in a 30 mg/l increase in digester concentrations above their normal 16-17 mg/l level. Land application of the sludge was not altered, as presently there are no established allowable metals application rates for silvercultural use.

Figure C-6 below typifies the reduction in metals realized during the 1981-1985 time period. Plant influent chromium levels dropped approximately 55 percent while the digested sludge concentrations were reduced by more than 40 percent. The magnitude of these decreases are typical of other heavy metals as well, averaging 44 percent for chromium, cadmium, copper, lead, nickel and zinc combined (see the accompanying data sheet). The primary reason for the reduction of cadmium and chromium concentrations is improved industrial pretreatment. In addition to pretreatment, a less corrosive city water supply has also resulted in lower background metal concentrations for the other metals, especially for copper. The city recently began chemically conditioning its water in an attempt to extend conduit life.

Success of the Metro Industrial Pretreatment Program can be attributed to a number of important factors including:

- development of stringent local limits for industrial discharges;

- year-round industrial waste sampling programs supported financially by industry; and
- follow-up procedures to industrial waste spills, taking enforcement action and levying fines when necessary.

Metro has recently implemented the following steps to improve their pretreatment program:

- information exchange with industries through the use of quarterly newsletters and personal communication, and
- increasing public awareness of industrial discharge violators by publishing the names of violating companies in local papers along with a statement of Metro's enforcement policy.

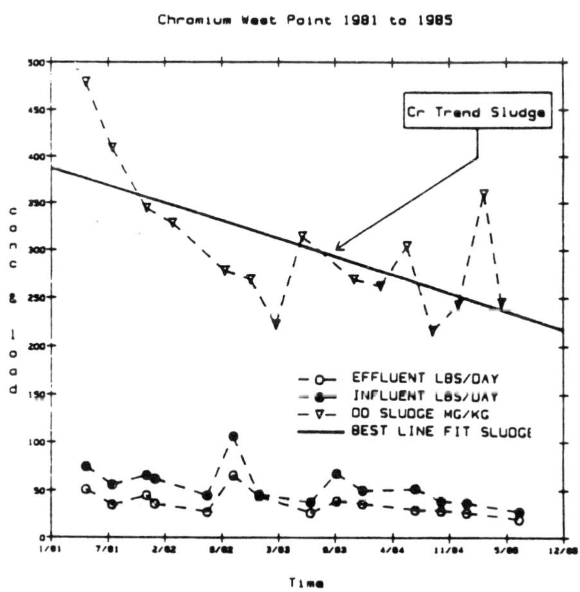

FIGURE C-6
WEST POINT CHROMIUM CONCENTRATIONS

WEST POINT TREATMENT PLANT
SEATTLE, WASHINGTON

Design Flow:	125 mgd	Location:	West-Central Washington
Primary Treatment		Population Served:	500,000

INFLUENT WASTEWATER

	Typical (Upset)
Ave. Flow, mgd	132
% Industrial	5
BOD_5, mg/l	160
SS, mg/l	260
Cr, Mg/l	0.05 (2.0)

SIGNIFICANT INDUSTRIES

Industry	Flowrate (mgd)	Problem Pollutants
Metal finishing and electroplating	1.1	Cd, Cr, Cu, Ni, Zr

PLANT LOADING

Primary Clarifiers	Typical (Upset)
Overflow Rate, gal/sf/day	1080
Detention Time, hours	1.58
Effluent BOD_5, mg/l	75-110
Effluent SS, mg/l	60-90

Digested Sludge Metal Concentrations

	1981 Level	1985 Level
Cadmium, mg/kg	45	28
Chromium, mg/kg	480	250
Copper, mg/kg	1300	700
Nickel, mg/kg	160	120
Lead, mg/kg	800	400

PLANT PERFORMANCE

	Permit Limit		Typical	
	Summer	Winter	Summer	Winter
BOD_5, mg/l	135	85	110	75
SS, mg/l	125	65	90	60
Cr, mg/L	0.07	0.07	<0.05	(0.15)

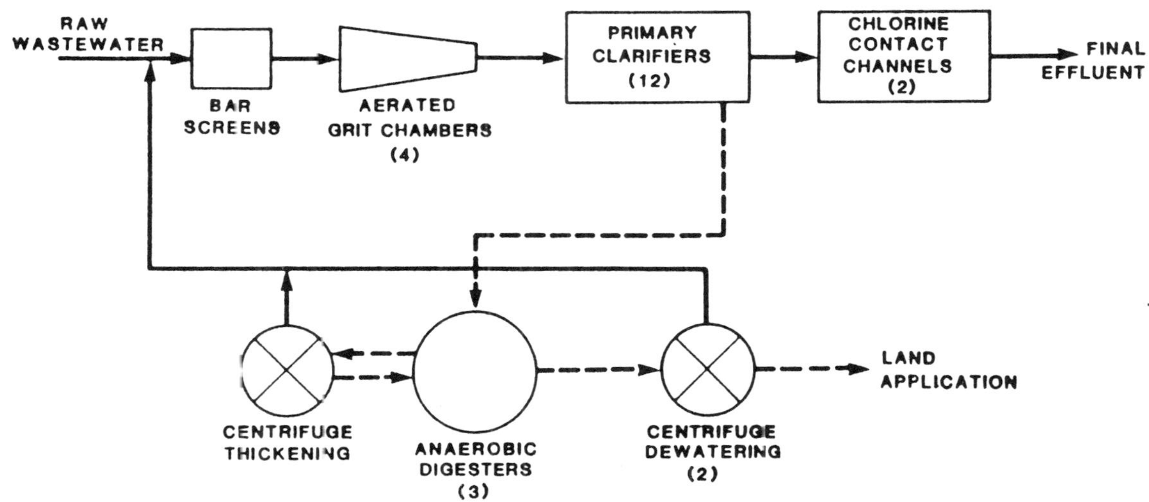

POST OAK WASTEWATER TREATMENT PLANT
SHERMAN, TEXAS

Design Flow:	8 mgd	**Location:**	North Texas
Secondary Treatment:	Two-stage; Trickling Filters/ Activated Sludge	**Population Served:**	30,000

INFLUENT WASTEWATER

	Typical (Upset)
Ave. Flow, mgd	6.5
% Industrial	60
BOD_5, mg/l	200
SS, mg/l	180

SIGNIFICANT INDUSTRIES

Industry	Flowrate (1000 gpd)	Problem Pollutants
Food Processor	610	BOD, TSS, FOG
First Aid Products Mfr.	440	BOD, FOG, pH
Coffee Processor	260	BOD, TSS, FOG, Color
Electronics Mfr.	1,020	

PLANT LOADING

Primary Clarifiers	Typical (Upset)
Overflow Rate, gal/sf/day	500
Detention Time, hours	3.6

Trickling Filters	Typical (Upset)
BOD_5 loading, lbs/1,000 cu ft/day	33
Recycle, %	100

Secondary Clarifiers	Typical (Upset)
Overflow Rate, gal/sf/day	415
Detention Time, hours	4.3

Aeration Basins	
F/M, lbs BOD_5/lbs MLSS/day	0.27 (0.30)
MCRT, days	5-6
MLSS, mg/l	1,800 (2,000)
Detention Time, hours	7.5
Return Flow, %	50 (60)
D.O. Level, mg/l	1

PLANT PERFORMANCE

	Permit Limit	Typical (Upset)
BOD_5, mg/l	20	8 (36)
SS, mg/l	20	12 (37)

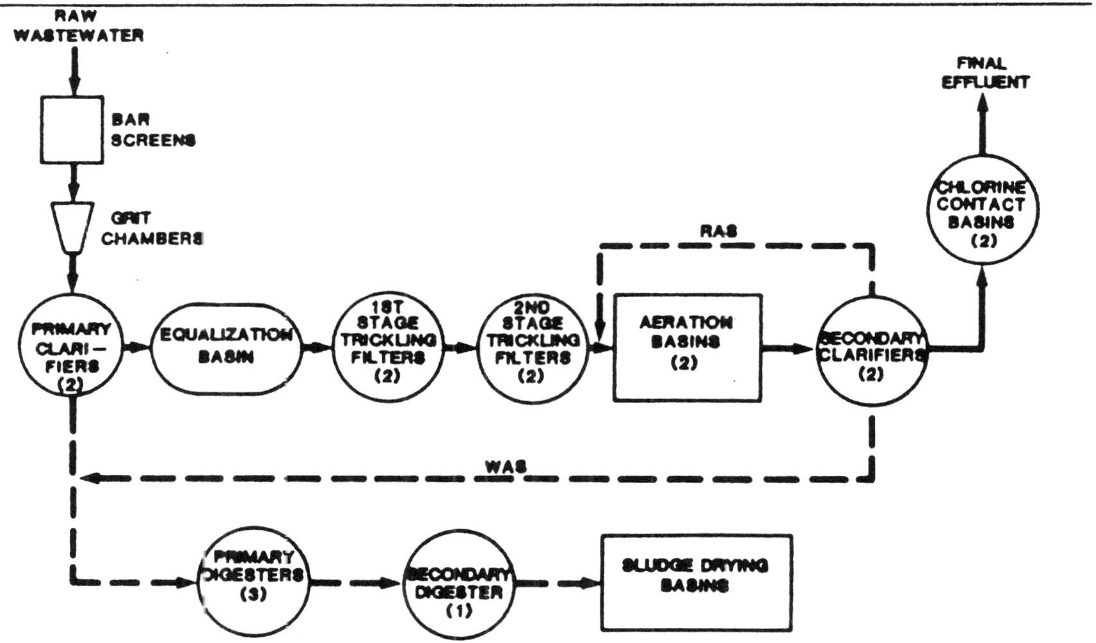

POST OAK WASTEWATER TREATMENT PLANT
Sherman, Texas

The Post Oak Wastewater Treatment Plant (POWTP) treats all municipal and industrial wastewater generated in the City of Sherman, Texas. In 1982, the City was issued an EPA Administrative Order to institute a pretreatment program. With the assistance of the Director of Utilities for the City of Paris, a sewer use ordinance was developed and passed by the City Council in 1983.

In addition to the implementation of the sewer use ordinance, the POWTP was upgraded in 1983 with the addition of an activated sludge system following the existing trickling filters. As soon as this system was placed on-line, the plant effluent concentrations of BOD and TSS dropped from over 30 for each, to less than 13 for each. Prior to the addition of the activated sludge system, the POWTP was never able to respond to industrial discharges due to a lack of control on the trickling filter recycle pumps. The POWTP now consistently meets its discharge permit requirements.

The implementation of the sewer use ordinance led to the installation of pretreatment plants at all five of the City's major industries. One of these, a coffee processor, utilizes an overland flow system during dry weather, and discharges to the POWTP during wet weather. Wet weather flows to the POTW typically peak in excess of 17 mgd.

Another industry of interest is an edible oil - food processor. This industry has installed a pH adjustment/heat treatment system to remove fats, oils and grease from its discharge, but still discharges in excess of 2,000 mg/l BOD, 600 mg/l TSS and 600 mg/l FOG to the POWTP. This user pays approximately $ 250,000 per year in surcharge fees. Influent FOG to the POWTP averages 40-50 mg/l due to dilution, and does not cause any significant process problems.

The one major categorical industry is a chromium plater who discharges in excess of 1 mgd to the POWTP. This user has installed a chromium reduction, pH adjustment, metal hydroxide precipitation pretreatment facility to reduce its surcharge liability. Dewatered metal hydroxide sludge is trucked to Houston for appropriate disposal.

Conversion Factors

This document is written primarily using English (customary) units. The following multipliers can be used to convert from English to the applicable SI units for parameters used in this report.

Customary Unit	Multiplier		SI Unit
foot (ft)	x 0.3048	=	meter (m)
gallon (gal)	x 3.785	=	liter (l)
gal/cf/day	x 0.1337	=	$m^3/m^3/d$
gal/sf/day	x 0.04074	=	$m^3/m^2/d$
gal/min (gpm)	x 5.450	=	m^3/d
lb/cf/day	x 16.02	=	$kg/m^3/d$
lb/sf/day	x 4.883	=	$kg/m^2/d$
mgd	x 3785	=	m^3/d

Other Noyes Publications

CONTROLLING VOLATILE ORGANIC COMPOUND EMISSIONS FROM INDUSTRIAL WASTEWATER

by

Jeffrey Elliott **Sheryl Watkins**
Radian Corporation

Pollution Technology Review No. 197

This book presents a description of the sources of organic containing wastewater, VOC emission estimation procedures for treatment and collection system units, and available VOC emission control strategies. In addition, secondary impacts and the control costs associated with steam stripping are presented.

Technical information is provided to industry and governmental organizations for: (1) Estimating emissions of VOCs from the collection and treatment of industrial wastewaters; and (2) Best Available Control Technology (BACT), and Lowest Achievable Emission Rate (LAER) determinations for controlling emissions of VOCs from industrial wastewaters.

Although this book applies to four industries in particular: organic chemicals, plastics and synthetic fibers; pesticides; pharmaceuticals; and hazardous waste, treatment, storage and disposal—the technical information is applicable to a broad range of other industries.

Facilities within these industries have the potential to generate wastewaters containing high concentrations of organic compounds. These wastewaters typically pass through a series of collection and primary treatment units, before treatment is applied to reduce the concentration of organic compounds prior to discharge. Many of these collection and treatment units are open to the atmosphere and allow organic-containing wastewaters to contact ambient air. Atmospheric exposure of these organic-containing wastewaters results in significant volatilization of VOCs from the wastewater.

These emissions can be reduced by applying one of three control strategies. The most effective strategy is to apply waste minimization techniques to reduce the organic concentration of the wastewaters, or to produce a more manageable wastewater stream through waste segregation or recycling. Even with waste minimization, some waste streams will be generated. Emissions from these wastewater streams can be reduced by applying treatment at the point of generation. Numerous controls are suitable in specific cases, but the most universally applicable treatment technology for controlling emissions from wastewater generated by these industries is steam stripping. A third control strategy that may be appropriate for some situations is to enclose the wastewater collection system and cover all treatment units up through removal or destruction of the organic compounds.

A condensed table of contents listing **chapter titles and selected subtitles** is given below.

1. **INDUSTRY DESCRIPTIONS**
2. **VOC EMISSIONS DURING WASTEWATER COLLECTION AND TREATMENT**
   Sources or Organic-Containing Wastewater
   Sources of Air Emissions
   Example Waste Stream Collection and Treatment System
3. **VOC EMISSION CONTROL EQUIPMENT**
   Waste Minimization
   Organic Compound Removal
   VOC Emission Control
   Add-On Controls
4. **STEAM STRIPPING**
   Impacts on VOC Emissions
   Secondary Air Impacts
   Cross-Media Impacts
   Energy Impacts
5. **COST ANALYSIS OF STEAM STRIPPER SYSTEM**

APPENDIX A—Emission Estimates
APPENDIX B—Estimation of Air Emission Factors from Airflow in Wastewater Collection Systems
APPENDIX C—Physical/Chemical Properties
APPENDIX D—Steam Stripper Performance Data
APPENDIX E—Determination of the Volatile Organic Concentration in Waste Samples
APPENDIX F—Facility Analysis
   Emissions Estimates
   Steam Stripper VOC Removal
   Test Method Concentration

ISBN 0-8155-1261-9 (1990) **6" x 9"** **362 pages**

Other Noyes Publications

FINE PORE AERATION FOR WASTEWATER TREATMENT

by

W.C. Boyle, G.T. Daigger, J.A. Heidman, G.L. Huibregste, J.J. Marx, J.J. McKeown, B. Newbry, T.C. Rooney, L.A. Rossman, G. Speirs, H.D. Stensel

Pollution Technology Review No. 191

This book provides designers, end users, and regulators, information on the nature of fine pore aeration devices, their performance, and related operation and maintenance (O&M) requirements to promote the intelligent application of fine pore aeration technology.

The supply of oxygen for aeration is the single largest energy consumer at activated sludge wastewater treatment plants, representing 50-90% of total plant energy requirements. Replacement of less efficient aeration systems with fine pore aeration devices can save up to 50% of aeration energy costs and has resulted in typical simple payback periods of 2-6 years. As a result of these very impressive cost savings, more than 1,300 municipal and industrial wastewater treatment facilities in the United States and Canada use fine pore aeration.

Typically, fine pore diffusers will produce a headloss due to surface tension in clean water of greater than about 5 cm (2 in) water gauge. For the purposes of this book, fine pore diffusers are defined as including the following devices:

- Porous ceramic plates, discs, domes, and tubes
- Rigid porous plastic plates, discs, and tubes
- Nonrigid porous plastic tubes
- Perforated membrane tubes and discs

The major objective of this book is to provide the technical community with the most current information available on fine pore aeration systems. It presents what are considered the best current practices for selecting, designing, operating, maintaining, and controlling fine pore aeration systems.

Example calculations are presented thoughout to illustrate concepts, and case histories are presented to give the user an appreciation for the highly site-specific nature of this technology.

The partial table of contents below lists **chapter headings and selected subtitles.**

1. INTRODUCTION AND OVERVIEW

2. FINE PORE DIFFUSER CHARACTERISTICS
Types of Fine Pore Media
Types of Fine Pore Diffusers
Diffuser Layout
Characteristics of Fine Pore Media
Clean Water Performance

3. PROCESS WATER PERFORMANCE
Factors Affecting Performance
Diffuser Fouling
Process Water Data Base

4. OPERATION AND MAINTENANCE
Operation
Maintenance

5. SYSTEM DESIGN AND INSTALLATION
Process and O&M Considerations
Process Oxygen and Mixing Requirements
Air Diffusion System
Air Supply System
Summary of Aeration System Design Procedure

6. AERATION CONTROL
Benefits of Aeration Control
Control Strategy Developments
Control System Components
Aeration Control Example
Experiences with Automated Aeration Control
Summary

7. ECONOMIC ANALYSIS
Cost Components
Economic Analysis Procedures
Sample Desktop Economic Analysis
Lotus Spreadsheet
Compendium of Empirical Cost Data

8. CASE HISTORIES
Performance Evaluation by Off-Gas Testing (8)
Performance Evaluation by Means Other Than Off-Gas Testing (5)
Sources of Information

APPENDICES

ISBN 0-8155-1252-X (1990) **7" x 10"** **296 pages**

Other Noyes Publications

CONTAMINATION OF GROUND WATER
Prevention, Assessment, Restoration

by

Michael Barcelona, Allen Wehrmann
Illinois State Water Survey

Joseph F. Keely
U.S.E.P.A.

Wayne A. Pettyjohn
Oklahoma State University

Pollution Technology Review No. 184

The need exists for a resource document that brings together available technical information on ground-water management in a form convenient for ground-water personnel at all levels of government, as well as in the private sector. The information contained in this handbook of ground-water contamination meets that need. The book covers measures for (1) prevention of contamination, (2) assessment of extent of contamination, and (3) restoration of ground-water quality.

Because contamination of ground water has occurred in every state and is being detected with increasing frequency, regulatory agencies and courts have been developing guidelines, laws and rules to protect this resource. Ground-water quality laws deal with both the prevention of ground-water contamination and assigning responsibility for ground-water protection or cleanup and legal liability for damages where contamination has occurred. Provisions aimed at prevention of contamination regulate the conduct of activities which could have the effect of polluting ground-water or posing risks to human health. Other statutory provisions call for government or private party response to incidents of contamination; they also may assign penalties or other legal liability to polluters. The operation of these provisions is generally triggered by the release of certain harmful substances, identified by statute or government regulation, into the environment.

This book will provide research information to decision makers, field managers, and the scientific community which will help satisfy the immediate need for technology transfer applicable to ground-water contamination control and prevention.

A condensed contents listing **part titles, chapter titles, and selected subtitles** is given below.

I. FRAMEWORK FOR PROTECTING GROUND-WATER RESOURCES

1. **GROUND-WATER CONTAMINATION**
   Extent of Ground-Water Contamination
   Sources of Ground-Water Contamination
   Movement of Contaminants in Ground Water
2. **GROUND-WATER QUALITY INVESTIGATIONS**
   Types of Ground-Water Quality Investigations
   Conducting the Investigation
3. **GROUND-WATER RESTORATION**
   Subsurface Effects of Contaminant Mobility
   Physical Containment Techniques
   Hydrodynamic Controls
   Withdrawal and Treatment
   In-Situ Treatment Techniques
   Treatment Trains

II. SCIENTIFIC AND TECHNICAL BACKGROUND FOR ASSESSING AND PROTECTING THE QUALITY OF GROUND-WATER RESOURCES

4. **BASIC HYDROGEOLOGY**
5. **MONITORING WELL DESIGN AND CONSTRUCTION**
   Monitoring Well Design Components
   Monitoring Well Drilling Methods
6. **GROUND-WATER SAMPLING**
   Establishing a Sampling Point
   Elements of the Sampling Protocol
7. **GROUND-WATER TRACERS**
   Public Health Considerations
   Direction of Water Movement
   Travel Time
   Sorption of Tracers and Related Phenomena
   Practical Aspects
   Types of Tracers
8. **USE OF MODELS IN MANAGING GROUND-WATER PROTECTION PROGRAMS**
   Utility of Models
   Assumptions, Limitations, and Quality Control
   Applications in Practical Settings
9. **BASIC GEOLOGY**
   Geologic Maps and Cross-Sections
   Relationship Between Geology, Climate, and Ground-Water Quality

APPENDIX—SOURCES OF INFORMATION ABOUT GROUND-WATER CONTAMINATION INVESTIGATIONS

ISBN 0-8155-1243-0 (1990) **7" x 10"** **213 pages**